Stochastic Calculus and Recurrence Relations

Stochastic Calculus and Recurrence Relations

Edited by **Charlin Chester**

Willford Press
New York

Published by Willford Press,
118-35 Queens Blvd., Suite 400,
Forest Hills, NY 11375, USA
www.willfordpress.com

Stochastic Calculus and Recurrence Relations
Edited by Charlin Chester

International Standard Book Number: 978-1-68285-196-8 (Hardback)

Printed in China.

Contents

Preface

The main aim of this book is to educate learners and enhance their research focus by presenting diverse topics covering this vast field. This is an advanced book which compiles significant studies by distinguished experts. This book addresses successive solutions to the challenges arising in the area of application, along with it; the book provides scope for future developments.

Stochastic calculus has contributed majorly towards problem solving in fields such as engineering, quantitative finance, etc. This book attempts to understand the multiple branches that fall under this discipline and how such concepts have practical applications. It comprises of research from across the globe to keep the readers updated with the usage of stochastic equations in upcoming fields. This book is highly recommended for graduate and post graduate students as well as research scholars pursuing mathematics, finance, and allied fields of study.

It was a great honour to edit this book, though there were challenges, as it involved a lot of communication and networking between me and the editorial team. However, the end result was this all-inclusive book covering diverse themes in the field.

Finally, it is important to acknowledge the efforts of the contributors for their excellent chapters, through which a wide variety of issues have been addressed. I would also like to thank my colleagues for their valuable feedback during the making of this book.

Editor

Random attractors for the stochastic damped Klein-Gordon-Schrödinger system

Xin Zhao[1] and Yin Li[2,3*]

*Correspondence: liyin2009521@163.com
[2]School of Mathematics and Statistics, Shaoguan University, Shaoguan, 512005, P.R. China
[3]Department of Mathematics, Sun Yat-sen (Zhongshan) University, Guangzhou, 510275, P.R. China
Full list of author information is available at the end of the article

Abstract

This paper is concerned with random attractors for the stochastic Klein-Gordon-Schrödinger system. By using the tail-estimates method, we prove the asymptotic compactness of random dynamical systems and obtain the existence of random attractor on unbounded domain in $\mathbb{R}^n$.

1 Introduction

In recent decades, much attention has been paid to the existence of random attractor for stochastic partial differential equations. To the authors' knowledge, Crauel and Flandoli [1] and Flandoli and Schmalfuß [2] first introduced a corresponding generalization of the attractor (random attractor) to the stochastic partial differential equations. After that, the study of random attractors has gained considerable attention, see [3–6] for a comprehensive survey. There are also many other papers concerning the existence of random attractor for some SDEs on bounded domains, see [7–9] for stochastic damped sine-Gordon equation, [10] for Ladyzhenskaya model. In 2009, Bates *et al.* [11] introduced a new method to study the existence of random attractors for stochastic reaction-diffusion equations in unbounded domains, and they proved the asymptotic compactness of the solutions and then obtained random attractor in $\mathbb{L}^2(\mathbb{R}^n)$ by approaching $\mathbb{R}^n$ with a sequence of bounded domains Q_k, and combining the tail estimates in spatial variables with the compactness of Sobolev embedding in the bounded domain Q_k. Wang [12] used this method to study the existence of random attractors for the Benjamin-Bona-Mahony equation.

In this paper, we consider the damped Klein-Gordon-Schrödinger system perturbed by an ϵ-small random term

$$idu + (\Delta u + i\alpha u + uv)\,dt = f\,dt + \epsilon u\,dW_1, \quad x \in \mathbb{R}^n, t > 0,$$
$$dv_t + \left(\nu v_t - \Delta v + \mu v - \beta|u|^2\right)dt = g\,dt + \epsilon\delta\,dW_2, \quad x \in \mathbb{R}^n, t > 0$$

with the initial value conditions

$$u(x,0) = u_0(x), \qquad v(x,0) = v_0(x), \qquad v_t(x,0) = v_1(x), \quad x \in \mathbb{R}^n,$$

where α, ϵ, ν, μ and β are positive constants, $n \le 3, f,g \in \mathbb{L}^2$, $\delta \in \mathbb{H}^1$, W_1 and W_2 are independent two-side real-valued Wiener processes on a probability space which will be specified later.

The coupled Klein-Gordon-Schrödinger (KGS) system is an important model in nonlinear science. It is encountered in several diverse branches of physics, for example, in the description of the interaction between high frequency electron waves and low frequency ion plasma waves in a homogeneous magnetic field, adapted to model the UHH plasma heating scheme. The system also focuses on the vital role of collisions, by considering the non-homogeneous polarization drift for the low frequency coupling [13], as the following nonautonomous KGS equation:

$$
\begin{aligned}
&i(u_t + \alpha u) + \triangle v + uv = f, \\
&v_{tt} - \triangle v + \nu v_t + \mu v - \beta |u|^2 = g,
\end{aligned}
$$

where u denotes a complex scalar nucleon field and v represents a real meson field, the complex-valued function f and the real-valued function g are external sources. By using Galerkin's method, Fukuda and Tsutsumi [14] first studied the coupled KGS system and obtained the existence of global strong solutions. Biler [15] obtained the existence of weak global attractors for the KGS system in a bounded domain. Wang and Lange [16] pointed out that the weak global attractor is actually a strong one. Recently, Guo and Li [17] proved the existence of a global attractor in $\mathbb{H}^2(\mathbb{R}^3) \times \mathbb{H}^2(\mathbb{R}^3)$ which attracts bounded sets of space $\mathbb{H}^3(\mathbb{R}^3) \times \mathbb{H}^3(\mathbb{R}^3)$ in the topology of $\mathbb{H}^2(\mathbb{R}^3) \times \mathbb{H}^2(\mathbb{R}^3)$. Lu and Wang [18] improved their results and obtained global attractors in space $\mathbb{H}^k(\mathbb{R}^3) \times \mathbb{H}^k(\mathbb{R}^3)$ for $k \geq 1$ which attracts all bounded sets of $\mathbb{H}^k(\mathbb{R}^3) \times \mathbb{H}^k(\mathbb{R}^3)$ in norm topology.

However, a system in reality is usually affected by external perturbations which in many cases are of great uncertainty or random influence. These random effects are not only introduced to compensate for the defects in some deterministic models, but also to explain the intrinsic phenomena. In [19], we have investigated the dynamical behavior for stochastic Klein-Gordon-Schrödinger lattice system in one dimension, which can be regarded as an approximation to a stochastic continuous case. Yan *et al.* [20, 21] studied the existence of random attractors for a class of first order stochastic lattice dynamical systems with delay.

Throughout this paper, we denote by $L^2(\mathbb{R}^n)$ both the standard real and complex Hilbert spaces and equip $L^2(\mathbb{R}^n)$ with the inner product and norm as

$$
(u, v) = \int_{\mathbb{R}^n} u(x)\bar{v}(x)\,dx, \qquad \|u\|^2 = (u, u).
$$

Hereafter, we denote by C any positive constants which may change from line to line.

This paper is organized as follows. In Section 2, we recall the basic concepts and some known results related to random dynamical systems and the pullback random attractors. In Section 3, we first derive the uniform estimates of solutions, and then prove the existence of the pullback random attractor for the stochastic KGS system.

2 Preliminaries

First we recall the definitions of the random dynamical systems and random attractors which are taken from [5, 6, 11, 12, 22].

Let $(H, \|\cdot\|_H)$ be a separable Hilbert space with Borel σ-algebra $\mathbb{B}(H)$, and let $(\Omega, \mathbb{F}, \mathbb{P})$ be a probability space.

Definition 2.1 $(\Omega, \mathbb{F}, \mathbb{P}, (\theta_t)_{t\in\mathbb{R}})$ is called a metric dynamical system if $\theta : \mathbb{R} \times \Omega \to \Omega$ is $(\mathbb{B}(\mathbb{R}) \times \mathbb{F}, \mathbb{F})$ measurable, $\theta_0 = id$, $\theta_{t+s} = \theta_t \circ \theta_s$ for all $t, s \in \mathbb{R}$, and $\theta_t\mathbb{P} = \mathbb{P}$ for all $t \in \mathbb{R}$.

Definition 2.2 A stochastic process $\phi(t, \omega)$ is called a random dynamical system (RDS) over $(\Omega, \mathbb{F}, \mathbb{P}, (\theta_t)_{t\in\mathbb{R}})$ if ϕ is $(\mathbb{B}(\mathbb{R}^+) \times \mathbb{F} \times \mathbb{B}(H), \mathbb{B}(H))$-measurable, and for all $\omega \in \Omega$

- the mapping $\phi : \mathbb{R}^+ \times \Omega \times H \to H$ is continuous;
- $\phi(0, \omega, \cdot) = id$ on H;
- $\phi(t + s, \omega, x) = \phi(t, \theta_s\omega, \phi(s, \omega, x))$ for all $t, s \geq 0$ and $x \in H$ (cocycle property).

Definition 2.3 A random bounded set $B(\omega) \subset H$ is called tempered with respect to $(\theta_t)_{t\in\mathbb{R}}$ if for a.e. $\omega \in \Omega$ and all $\epsilon > 0$,

$$\lim_{t\to\infty} e^{-\epsilon t} d\big(B(\theta_{-t}\omega)\big) = 0,$$

where $d(B) = \sup_{x\in B} \|x\|_H$.

Definition 2.4 Let $\mathbb{D}$ be a collection of random subsets of H. Then $\mathbb{D}$ is called inclusion-closed if $D = \{D(\omega)\}_{\omega\in\Omega} \in \mathbb{D}$ and $\tilde{D} = \{\tilde{D}(\omega) \subseteq H : \omega \in \Omega\}$ with $\tilde{D} \subseteq D(\omega)$ for all $\omega \in \Omega$ imply that $\tilde{D} \in \mathbb{D}$.

Definition 2.5 A random set $\mathbb{K}(\omega)$ is called an absorbing set in $\mathbb{D}$ if for all $B \in \mathbb{D}$ and P-a.e. $\omega \in \Omega$, there exists $t_B(\omega) > 0$ such that

$$\phi\big(t, \theta_{-t}\omega, B(\theta_{-t}\omega)\big) \subset \mathbb{K}(\omega), \quad t \geq t_B(\omega).$$

Definition 2.6 A random set $\mathbb{A}(\omega)$ is a random $\mathbb{D}$-attractor for RDS ϕ if

- $\mathbb{A}(\omega)$ is a random compact set, *i.e.*, $\omega \to d(x, \mathbb{A}(\omega))$ is measurable for every $x \in H$ and $\mathbb{A}(\omega)$ is compact for a.e. $\omega \in \Omega$;
- $\mathbb{A}(\omega)$ is strictly invariant, *i.e.*, $\phi(t, \omega, \mathbb{A}(\omega)) = \mathbb{A}(\theta_t\omega)$, $\forall t \geq 0$ and for a.e. $\omega \in \Omega$;
- $\mathbb{A}(\omega)$ attracts all sets in $\mathbb{D}$, *i.e.*, for all $B \in \mathbb{D}$ and a.e. $\omega \in \Omega$ we have

  $$\lim_{t\to\infty} d\big(\phi\big(t, \theta_{-t}\omega, B(\theta_{-t}\omega)\big), \mathbb{A}(\omega)\big) = 0,$$

  where $d(X, Y) = \sup_{x\in X} \inf_{y\in Y} \|x - y\|_H$, $X, Y \subset H$.

  The collection $\mathbb{D}$ is called the domain of attraction of $\mathbb{A}$.

Definition 2.7 Let ϕ be an RDS on a Hilbert space H. ϕ is called $\mathbb{D}$ pullback asymptotically compact in H if for P-a.e. $\omega \in \Omega$, $\{\phi(t_n, \theta_{-t_n}\omega, x_n)\}_{n=1}^{\infty}$ has a convergent subsequence in H whenever $t_n \to \infty$, and $x_n \in B(\theta_{t_n}\omega)$ with $\{B(\omega)\}_{\omega\in\Omega} \in \mathbb{D}$.

From [11, 12, 16] we have the following result.

Proposition 2.8 *Let $\mathbb{D}$ be an inclusion-closed collection of random subsets of H, and let ϕ be a continuous RDS on H over $(\Omega, \mathbb{F}, P, (\theta_t)_{t\mathbb{R}})$. Suppose that $\{\mathbb{K}(\omega)\}_{\omega\in\Omega}$ is a closed absorbing set of ϕ in $\mathbb{D}$ and ϕ is $\mathbb{D}$-pullback asymptotic compact in H. Then ϕ has a unique*

$\mathbb{D}$*-random attractor* $\{\mathbb{A}(\omega)\}_{\omega\in\Omega}$ *which is given by*

$$\mathbb{A}(\omega)=\bigcap_{\kappa\geq 0}\overline{\bigcup_{t\geq\kappa}\phi\big(t,\theta_{-t}\omega,\mathbb{K}(\theta_{-t}\omega)\big)}.$$

3 The existence of random attractors

In this section, we will apply Proposition 2.8 to prove the existence of random attractors for the Klein-Gordon-Schrödinger system with ϵ-small random perturbation. The main tool is the tail-estimates method which is extensively used to prove the existence of random attractor, see [11, 12].

Consider the Klein-Gordon-Schrödinger system with ϵ-small random perturbation

$$\begin{aligned}&i\,du+(\Delta u+i\alpha u+uv)\,dt=f\,dt+\epsilon u\,dW_1, \quad x\in\mathbb{R}^n, t>0,\\ &dv_t+\left(\nu v_t-\Delta v+\mu v-\beta|u|^2\right)dt=g\,dt+\epsilon\delta\,dW_2, \quad x\in\mathbb{R}^n, t>0,\end{aligned} \tag{3.1}$$

with the initial value conditions

$$u(x,0)=u_0(x), \qquad v(x,0)=v_0(x), \qquad v_t(x,0)=v_1(x), \quad x\in\mathbb{R}^n,$$

where α, ϵ, ν, μ and β are positive constants, $n\leq 3$, $f,g\in\mathbb{L}^2$, $\delta\in\mathbb{H}^1$, W_1 and W_2 are independent two-side real-valued Wiener processes on a probability space $(\Omega,\mathbb{F},\mathbb{P})$.

For our purpose, we introduce the probability space

$$\Omega=\left\{\omega=(\omega_1,\omega_2)\in\mathbb{C}\left(\mathbb{R},\mathbb{R}^2\right):\omega(0)=0\right\}$$

endowed with the compact open topology [5]. Then we have $(W_1(t,\omega),W_2(t,\omega))=\omega(t)$, $t\in\mathbb{R}$. Let $\mathbb{P}$ be the corresponding Wiener measure, $\mathbb{F}$ be the $\mathbb{P}$-completion of the Borel σ-algebra on Ω, and $\theta_t\omega(\cdot)=\omega(\cdot+t)-\omega(t)$, $t\in\mathbb{R}$. Then $(\Omega,\mathbb{F},\mathbb{P},(\theta_t)_{t\in\mathbb{R}})$ is a metric dynamical system with the filtration $\mathbb{F}_t:=\bigvee_{s\leq t}\mathbb{F}_s^t$, $t\in\mathbb{R}$, where $\mathbb{F}_s^t=\sigma\{W(t_2)-W(t_1):s\leq t_1\leq t_2\leq t\}$ is the smallest σ-algebra generated by the random variable $W(t_2)-W(t_1)$ for all t_1, t_2 such that $s\leq t_1\leq t_2\leq t$, see [5] for more details.

We introduce an Ornstein-Uhlenbeck process $(\Omega,\mathbb{F},\mathbb{P},\theta_t)$ given by the Wiener process:

$$\begin{aligned}&y_1(\theta_t\omega_1)=-\nu\int_{-\infty}^{0}e^{\nu h}(\theta_t\omega_1)(h)\,dh, \quad t\in\mathbb{R},\\ &y_2(\theta_t\omega_2)=-\lambda\int_{-\infty}^{0}e^{\lambda h}(\theta_t\omega_2)(h)\,dh, \quad t\in\mathbb{R},\end{aligned}$$

where ν and λ are positive. The above integral exists in the sense that for any path ω with a subexponential growth, y_1, y_2 solve the following Itô equations:

$$\begin{aligned}&dy_1+\nu y_1\,dt=dW_1(t), \quad t\in\mathbb{R},\\ &dy_2+\lambda y_2\,dt=dW_2(t), \quad t\in\mathbb{R}.\end{aligned}$$

Furthermore, there exists a θ_t invariant set $\Omega'\subset\Omega$ of full $\mathbb{P}$ measure such that:

(1) the mappings $t\rightarrow y_i(\theta_t\omega_i)$, $i=1,2$, are continuous for each $\omega\in\Omega'$;

(2) the random variables $\|y_i(\omega_i)\|$, $i = 1, 2$, are tempered (for more details, see [11, 12]). Let $z_1(\theta_t\omega) = y_1(\theta_t\omega_1)$ and $z_2(\theta_t\omega) = \delta y_2(\theta_t\omega_2)$. Then we have

$$dz_1 + \nu z_1\,dt = dW_1(t), \quad t \in \mathbb{R},$$

$$dz_2 + \lambda z_2\,dt = \delta\,dW_2(t), \quad t \in \mathbb{R}.$$

Lemma 3.1 ([23]) *There exists a $(\theta_t)_{t\in\mathbb{R}}$-invariant set $\tilde{\Omega} \subset \Omega$ of full measure with sublinear growth*

$$\lim_{t\to\infty} \frac{\|W_i(t)\|}{t} = 0$$

of $\mathbb{P}$-measure one. In addition,

$$\lim_{t\to\infty} \frac{|y_i(\theta_t\omega_i)|}{|t|} = 0 \quad \text{and} \quad \lim_{t\to\infty} \frac{\int_0^t y_i(\theta_s\omega_i)\,ds}{t} = 0,$$

where $\omega \in \tilde{\Omega}$, $i = 1, 2$.

By introducing the transformation $\psi = \frac{dv}{dt} + \rho v$ with ρ a positive constant which satisfies $\rho < \nu$, system (3.1) becomes

$$i\frac{du}{dt} + \Delta u + i\alpha u + uv = f + \epsilon u\frac{dW_1}{dt}, \quad x \in \mathbb{R}^n, \tag{3.2}$$

$$\frac{dv}{dt} = \psi - \rho v, \quad x \in \mathbb{R}^n, \tag{3.3}$$

$$\frac{d\psi}{dt} + (\nu - \rho)\psi + \left[\mu - \rho(\nu - \rho) - \Delta\right]v - \beta|u|^2 = g + \epsilon\delta\frac{dW_2}{dt}, \quad x \in \mathbb{R}^n. \tag{3.4}$$

In order to prove the existence of global solutions of (3.1), we introduce the processes

$$\tilde{u}(t) = z(t)u(t)$$

and

$$\tilde{\psi}(t) = \psi(t) - \epsilon\delta z_2(\theta_t\omega),$$

where $z(t) = e^{i\epsilon z_1(\theta_t\omega)}$ satisfies the stochastic differential equation

$$dz(t) = -\frac{\epsilon^2}{2}z(t)\,dt + i\epsilon z(t)\,dz_1. \tag{3.5}$$

Then system (3.2)-(3.4) can be changed into the following system:

$$\begin{cases} i\frac{d\tilde{u}}{dt} + \Delta\tilde{u} + (i\alpha + \frac{\epsilon^2}{2})\tilde{u} + \tilde{u}v - fz = 0, \\ \frac{dv}{dt} + \rho v = \tilde{\psi} + \epsilon\delta z_2, \\ \frac{d\tilde{\psi}}{dt} + (\nu - \rho)\tilde{\psi} + [\mu - \rho(\nu - \rho) - \Delta]v - \beta|\tilde{u}z^{-1}|^2 + \epsilon(\nu - \rho)\delta z_2 = g, \end{cases} \tag{3.6}$$

with the initial data $\tilde{u}_0 = u_0$, $v_0 = v_0$ and $\tilde{\psi}_0 = \psi_0 - \epsilon\delta z_2(\omega)$.

Lemma 3.2 *Let $f \in \mathbb{L}^2$. Then the solution of the first equation in* (3.6) *satisfies*

$$\|\tilde{u}\|^2 \leq e^{-\alpha t}\|\tilde{u}_0\|^2 + \frac{4}{\alpha^2}\|f\|^2.$$

Proof Taking the imaginary part of the inner product of (3.6) with $\tilde{u}$, we obtain

$$\frac{d}{dt}\|\tilde{u}\|^2 + 2\alpha\|\tilde{u}\|^2 = 2\,\mathrm{Im}\int_{\mathbb{R}^n}(fz,\tilde{u})\,dx. \tag{3.7}$$

Obviously, the right-hand side of (3.7) is bounded by

$$2\,\mathrm{Im}\int_{\mathbb{R}^n}(fz,\tilde{u})\,dx \leq \alpha\|\tilde{u}\|^2 + \frac{4}{\alpha}\|f\|^2\|z\|^2.$$

Thus we have

$$\frac{d}{dt}\|\tilde{u}\|^2 + \alpha\|\tilde{u}\|^2 \leq \frac{4}{\alpha}\|f\|^2\|z\|^2 \leq \frac{4}{\alpha}\|f\|^2.$$

By Gronwall's lemma we get

$$\|\tilde{u}\|^2 \leq e^{-\alpha t}\|\tilde{u}_0\|^2 + \frac{4}{\alpha^2}\|f\|^2.$$

The proof is completed. □

Remark 3.1 By Lemma 3.2, we know that there is $T_1 > 0$ such that $\|\tilde{u}\|$ is bounded for $t > T_1$, *i.e.*, $\|\tilde{u}\| \leq M_1$, $t > T_1$.

Lemma 3.3 *Let $f \in \mathbb{L}^4$. Then, for any $m \geq 0$, the solution of the first equation in* (3.1) *satisfies*

$$\int_{|x|\geq m}|\tilde{u}|^4\,dx \leq e^{-\alpha t}\int_{|x|\geq m}|\tilde{u}_0|^4\,dx + \frac{64}{\alpha^4}\int_{|x|\geq m}|f|^4\,dx$$

for all $\omega \in \Omega$.

Proof The proof is similar to the proof in Lemma 3.2, so we omit it here. □

Here and after, $\mathbb{I}$ denotes the space $\mathbb{L}^2(\mathbb{R}^n) \times \mathbb{H}^1(\mathbb{R}^n) \times \mathbb{L}^2(\mathbb{R}^n)$. By Galerkin's method, it is easy to prove that, for $\mathbb{P}$-a.e. $\omega \in \Omega$ and for all $(\tilde{u}_0, v_0, \tilde{\psi}_0) \in \mathbb{I}$, system (3.6) has a unique solution $(\tilde{u}(\cdot,\omega,\tilde{u}_0), v(\cdot,\omega,v_0), \tilde{\psi}(\cdot,\omega,\tilde{\psi}_0)) \in \mathbb{C}([0,\infty),\mathbb{I})$ with $\tilde{u}(0,\omega,\tilde{u}_0) = \tilde{u}_0$, $v(0,\omega,v_0) = v_0$ and $\tilde{\psi}(0,\omega,\tilde{\psi}_0) = \tilde{\psi}_0$. Furthermore, the solution is continuous with respect to $(\tilde{u}_0, v_0, \tilde{\psi}_0) \in \mathbb{I}$. To indicate the dependence of (u, v, ψ) on the initial data (u_0, v_0, ψ_0), we define a mapping

$$\phi_\epsilon : \mathbb{R}^+ \times \Omega \times \mathbb{I} \to \mathbb{I}$$

by

$$\begin{aligned}\phi_\epsilon\big(t,\omega,(u_0,v_0,\psi_0)\big) &= \big(u(t,\omega,u_0), v(t,\omega,v_0), \psi(t,\omega,\psi_0)\big)\\ &= \big(\tilde{u}(t,\omega,\tilde{u}_0)z^{-1}(t), v(t,\omega,v_0), \tilde{\psi}(t,\omega,\tilde{\psi}_0) + \epsilon\delta z_2(\theta_t\omega)\big)\end{aligned}$$

for all $(t,\omega,(u_0,v_0,\psi_0)) \in \mathbb{R}^+ \times \Omega \times \mathbb{I}$.

It is obvious that ϕ_ϵ satisfies all conditions in Definition 2.2. Therefore, ϕ_ϵ is a continuous random dynamical system associated with (3.1). It is easy to verify that ϕ_ϵ satisfies

$$\phi_\epsilon\big(t,\theta_{-t}\omega,(u_0,v_0,\psi_0)\big) = \big(u(t,\theta_{-t}\omega,u_0),v(t,\theta_{-t}\omega,v_0),\psi(t,\theta_{-t}\omega,\psi_0)\big)$$

for $\mathbb{P}$-a.e. $\omega \in \Omega$ and $t \geq 0$.

The following lemma shows that ϕ_ϵ has a closed bounded random absorbing set in $\mathbb{D}$, which verifies the first condition (I) in Proposition 2.8.

Lemma 3.4 *Let $f,g \in \mathbb{L}^2$, $\delta \in \mathbb{H}^1$. Assume that $\mu - \rho(\nu - \rho) > 0$. Then there exists $\{\mathbb{K}(\omega)\}_{\omega\in\Omega} \in \mathbb{D}$ such that $\{\mathbb{K}(\omega)\}_{\omega\in\Omega}$ is a random absorbing set for ϕ in $\mathbb{D}$, that is, for any $\mathbb{B} = \{\mathbb{B}(\omega)\}_{\omega\in\Omega} \in \mathbb{D}$ and $\mathbb{P}$-a.e. $\omega \in \Omega$, there is $T_{\mathbb{B}}(\omega) > 0$ such that*

$$\phi_\epsilon\big(t,\theta_{-t}\omega,\mathbb{B}(\theta_{-t}\omega)\big) \subseteq \mathbb{K}(\omega) \quad \textit{for all } t \geq T_{\mathbb{B}}(\omega).$$

Proof Taking the imaginary part of the inner product of the first equation of (3.6) with $\tilde{u}$, we have

$$\frac{d}{dt}\|\tilde{u}\|^2 + 2\alpha\|\tilde{u}\|^2 = 2\,\mathrm{Im}(fz,\tilde{u}). \tag{3.8}$$

Taking the inner product of the third equation of (3.6) with $\tilde{\psi}$, we get

$$\begin{aligned}&\frac{d}{dt}\|\tilde{\psi}\|^2 + 2(\nu-\rho)\|\tilde{\psi}\|^2 + 2\big[\mu-\rho(\nu-\rho)\big](v,\tilde{\psi}) - 2(\Delta v,\tilde{\psi})\\ &\quad - 2\beta\big(|\tilde{u}z^{-1}|^2,\tilde{\psi}\big) + 2\epsilon(\nu-\rho)\big(\delta z_2(\theta_t\omega),\tilde{\psi}\big) = 2(g,\tilde{\psi}).\end{aligned} \tag{3.9}$$

Note that

$$2(v,\tilde{\psi}) = 2\left(v,\frac{dv}{dt} + \rho v - \epsilon\delta z_2(\theta_t\omega)\right) = \frac{d}{dt}\|v\|^2 + 2\rho\|v\|^2 - 2\epsilon\big(v,\delta z_2(\theta_t\omega)\big) \tag{3.10}$$

and

$$\begin{aligned}-2(\Delta v,\tilde{\psi}) &= \frac{d}{dt}\|\nabla v\|^2 + 2\rho\|\nabla v\|^2 + 2\epsilon\big(\nabla v, z_2(\theta_t\omega)\nabla\delta\big)\\ &\quad + 2\epsilon\big(\nabla v,\delta\nabla z_2(\theta_t\omega)\big).\end{aligned} \tag{3.11}$$

Summing up (3.8)-(3.11), we have

$$\begin{aligned}&\frac{d}{dt}\big(\|\tilde{u}\|^2 + \big[\mu-\rho(\nu-\rho)\big]\|v\|^2 + \|\nabla v\|^2 + \|\tilde{\psi}\|^2\big) + 2\alpha\|\tilde{u}\|^2\\ &\quad + 2\rho\big[\mu-\rho(\nu-\rho)\big]\|v\|^2 + 2\rho\|\nabla v\|^2 + 2(\nu-\rho)\|\tilde{\psi}\|^2\\ &= 2\,\mathrm{Im}(fz,\tilde{u}) - 2\epsilon\big(\nabla v, z_2(\theta_t\omega)\nabla\delta\big) - 2\epsilon\big(\nabla v,\delta\nabla z_2(\theta_t\omega)\big)\\ &\quad + 2\epsilon\big[\mu-\rho(\nu-\rho)\big]\big(v,\delta z_2(\theta_t\omega)\big) + 2\beta\big(|\tilde{u}z^{-1}|^2,\tilde{\psi}\big)\\ &\quad - 2\epsilon(\nu-\rho)\big(\delta z_2(\theta_t\omega),\tilde{\psi}\big) + 2(g,\tilde{\psi}).\end{aligned} \tag{3.12}$$

Now, we estimate each term on the right-hand side of (3.12). By Young's inequality, we have

$$
\begin{aligned}
&2\left|\operatorname{Im}(fz,\tilde{u})\right| \le \alpha\|\tilde{u}\|^2 + \frac{4}{\alpha}\|f\|^2,\\
&2\left|\epsilon\left(\nabla v, z_2(\theta_t\omega)\nabla\delta\right)\right| \le \frac{\rho}{2}\|\nabla v\|^2 + \frac{8\epsilon^2}{\rho}\|\nabla\delta\|^2\left\|z_2(\theta_t\omega)\right\|^2,\\
&2\left|\epsilon\left(\nabla v, \delta\nabla z_2(\theta_t\omega)\right)\right| \le \frac{\rho}{2}\|\nabla v\|^2 + \frac{8\epsilon^2}{\rho}\|\delta\|^2\left\|\nabla z_2(\theta_t\omega)\right\|^2,\\
&2\left|\epsilon\left[\mu - \rho(\nu-\rho)\right]\left(v, \delta z_2(\theta_t\omega)\right)\right|\\
&\quad \le \rho\left(\mu - \rho(\nu-\rho)\right)\|v\|^2 + \frac{4\epsilon^2}{\rho(\mu - \rho(\nu-\rho))}\|\delta\|^2\left\|z_2(\theta_t\omega)\right\|^2,\\
&2\beta\left|\left(\left|\tilde{u}z^{-1}\right|^2, \tilde{\psi}\right)\right| \le \frac{\nu-\rho}{3}\|\tilde{\psi}\|^2 + \frac{12\beta^2}{\nu-\rho}\|\tilde{u}\|^4,\\
&2\left|\epsilon(\nu-\rho)\left(\delta z_2(\theta_t\omega), \tilde{\psi}\right)\right| \le \frac{\nu-\rho}{3}\|\tilde{\psi}\|^2 + 12\epsilon^2(\nu-\rho)\|\delta\|^2\left\|z_2(\theta_t\omega)\right\|^2,\\
&2\left|(g,\tilde{\psi})\right| \le \frac{\nu-\rho}{3}\|\tilde{\psi}\|^2 + \frac{12}{\nu-\rho}\|g\|^2.
\end{aligned}
$$

By combining the above estimates with (3.12), we obtain

$$
\begin{aligned}
&\frac{d}{dt}\left(\|\tilde{u}\|^2 + \left[\mu - \rho(\rho-\mu)\right]\|v\|^2 + \|\nabla v\|^2 + \|\tilde{\psi}\|^2\right) + \alpha\|\tilde{u}\|^2\\
&\quad + \rho\left[\mu - \rho(\rho-\mu)\right]\|v\|^2 + \rho\|\nabla v\|^2 + (\nu-\rho)\|\tilde{\psi}\|^2\\
&\le \frac{4}{\alpha}\|f\|^2 + \frac{8\epsilon^2}{\rho}\|\delta\|^2\left\|\nabla z_2(\theta_t\omega)\right\|^2 + \frac{12\beta^2}{\nu-\rho}\|\tilde{u}\|^4 + \frac{12}{\nu-\rho}\|g\|^2\\
&\quad + \left[\frac{8\epsilon^2}{\rho}\|\nabla\delta\|^2 + \frac{4\epsilon^2}{\rho(\mu-\rho(\nu-\rho))}\|\delta\|^2 + 12\epsilon^2(\nu-\rho)\|\delta\|^2\right]\left\|z_2(\theta_t\omega)\right\|^2.
\end{aligned}
\tag{3.13}
$$

Furthermore, by Lemma 3.2, for $t \ge 0$

$$
\|\tilde{u}\|^4 \le e^{-2\alpha t}\|\tilde{u}_0\|^4 + \frac{8}{\alpha^2}e^{-\alpha t}\|u_0\|^2\|f\|^2 + \frac{16}{\alpha^4}\|f\|^4. \tag{3.14}
$$

Therefore, by (3.13) and (3.14), we have

$$
\begin{aligned}
&\frac{d}{dt}\left(\|\tilde{u}\|^2 + \left[\mu - \rho(\rho-\mu)\right]\|v\|^2 + \|\nabla v\|^2 + \|\tilde{\psi}\|^2\right) + \alpha\|\tilde{u}\|^2\\
&\quad + \rho\left[\mu - \rho(\rho-\mu)\right]\|v\|^2 + \rho\|\nabla v\|^2 + (\nu-\rho)\|\tilde{\psi}\|^2\\
&\le \frac{4}{\alpha}\|f\|^2 + \frac{8\epsilon^2}{\rho}\|\delta\|^2\left\|\nabla z_2(\theta_t\omega)\right\|^2 + \frac{12\beta^2}{\nu-\rho}\left[e^{-2\alpha t}\|\tilde{u}_0\|^4\right.\\
&\quad \left. + \frac{8}{\alpha^2}e^{-\alpha t}\|u_0\|^2\|f\|^2 + \frac{16}{\alpha^4}\|f\|^4\right] + \frac{12}{\nu-\rho}\|g\|^2\\
&\quad + \left[\frac{8\epsilon^2}{\rho}\|\nabla\delta\|^2 + \frac{4\epsilon^2}{\rho(\mu-\rho(\nu-\rho))}\|\delta\|^2 + 12\epsilon^2(\nu-\rho)\|\delta\|^2\right]\left\|z_2(\theta_t\omega)\right\|^2.
\end{aligned}
\tag{3.15}
$$

Let $C_1 := \min\{\alpha, \rho, (\nu - \rho)\} > 0$. Then, by (3.15), we have

$$\frac{d}{dt}E_1(\tilde{u}, v, \tilde{\psi}) + C_1 E_1(\tilde{u}, v, \tilde{\psi}) \leq F(f, g, z_2(\theta_t\omega), \nabla z_2(\theta_t\omega)), \tag{3.16}$$

where

$$E_1(\tilde{u}, v, \tilde{\psi}) = \|\tilde{u}\|^2 + [\mu - \rho(\rho - \mu)]\|v\|^2 + \|\nabla v\|^2 + \|\tilde{\psi}\|^2,$$

and

$$\begin{aligned} &F(f, g, z_2(\theta_t\omega), \nabla z_2(\theta_t\omega)) \\ &= \frac{4}{\alpha}\|f\|^2 + \frac{8\epsilon^2}{\rho}\|\delta\|^2 \left\|\nabla z_2(\theta_t\omega)\right\|^2 + \frac{12}{\nu - \rho}\|g\|^2 \\ &+ \frac{12\beta^2}{\nu - \rho}\left[e^{-2\alpha t}\|\tilde{u}_0\|^4 + \frac{8}{\alpha^2}e^{-\alpha t}\|u_0\|^2\|f\|^2 + \frac{16}{\alpha^4}\|f\|^4\right] \\ &+ \left[\frac{8\epsilon^2}{\rho}\|\nabla\delta\|^2 + \frac{4\epsilon^2}{\rho(\mu - \rho(\nu - \rho))}\|\delta\|^2 + 12\epsilon^2(\nu - \rho)\|\delta\|^2\right]\left\|z_2(\theta_t\omega)\right\|^2. \end{aligned}$$

Note that $z_1(\theta_t\omega) = y_1(\theta_t\omega_1)$ and $z_2(\theta_t\omega) = \delta y_2(\theta_t\omega_2)$, $\delta \in \mathbb{H}^1(\mathbb{R}^n)$, therefore, by Lemma 3.1, F is bounded by

$$C\sum_{i=1}^{2}\left(\left|y_i(\theta_t\omega_i)\right|^2 + \left|y_i(\theta_t\omega_i)\right|^p\right) + C = P_1(\theta_t\omega) + C. \tag{3.17}$$

By Proposition 4.3.3 in [5], there exists a tempered function $r(\omega) > 0$ such that $r(\theta_t\omega) \leq e^{\frac{C}{2}|t|}r(\omega)$. Therefore, by (3.17), we find that for $\mathbb{P}$-a.e. $\omega \in \Omega$,

$$P_1(\theta_s\omega) \leq Ce^{\frac{C}{2}|s|}r(\omega), \quad \forall s \in \mathbb{R}.$$

By replacing ω by $\theta_{-t}\omega$, we get from (3.16) and (3.17)

$$\begin{aligned} &E_1\left(\tilde{u}(t, \theta_{-t}\omega, \tilde{u}_0(\theta_{-t}\omega)), v(t, \theta_{-t}\omega, v_0(\theta_{-t}\omega)), \tilde{\psi}(t, \theta_{-t}\omega, \tilde{\psi}_0(\theta_{-t}\omega))\right) \\ &\leq e^{-C_1 t}E_1\left(\tilde{u}_0(\theta_{-t}\omega), v_0(\theta_{-t}\omega), \tilde{\psi}_0(\theta_{-t}\omega)\right) + \epsilon\int_0^t e^{C_1(s-t)}P_1(\theta_{s-t}\omega)\,ds + C_2 \\ &\leq e^{-C_1 t}E_1\left(\tilde{u}_0(\theta_{-t}\omega), v_0(\theta_{-t}\omega), \tilde{\psi}_0(\theta_{-t}\omega)\right) + \epsilon\int_{-t}^0 e^{C_1 s}P_1(\theta_s\omega)\,ds + C_2 \\ &\leq e^{-C_1 t}E_1\left(\tilde{u}_0(\theta_{-t}\omega), v_0(\theta_{-t}\omega), \tilde{\psi}_0(\theta_{-t}\omega)\right) + \epsilon C_1\int_{-t}^0 e^{\frac{C_1}{2}s}r(\omega)\,ds + C_2 \\ &\leq e^{-C_1 t}E_1\left(\tilde{u}_0(\theta_{-t}\omega), v_0(\theta_{-t}\omega), \tilde{\psi}_0(\theta_{-t}\omega)\right) + \epsilon C_1 r(\omega) + C_2. \end{aligned} \tag{3.18}$$

Note that

$$\begin{aligned} &\phi(t, \theta_{-t}\omega, (u_0, v_0, \psi_0)) \\ &= \left(\tilde{u}(t, \theta_{-t}\omega, u_0 e^{-i\epsilon z_1(\omega)})z^{-1}, v(t, \theta_{-t}\omega, v_0), \tilde{\psi}(t, \theta_{-t}\omega, \tilde{\psi}_0 + \epsilon\delta z_2(\omega)) + \epsilon\delta z_2(\theta_{-t}\omega)\right). \end{aligned}$$

Therefore, by (3.18) we know that there exists a positive constant C_3 such that, for all $t \geq 0$,

$$\begin{aligned}
&\left\|\phi\left(t,\theta_{-t}\omega,(u_0,v_0,\psi_0)(\theta_{-t}\omega)\right)\right\|_{\mathbb{I}}^2 \\
&\quad = \left\|\tilde{u}\left(t,\theta_{-t}\omega,u_0(\theta_{-t}\omega)e^{-i\epsilon z_1(\theta_{-t}\omega)}\right)z^{-1}\right\|^2 \\
&\qquad + \left\|v\left(t,\theta_{-t}\omega,v_0(\theta_{-t}\omega)\right)\right\|_{\mathbb{H}^1}^2 + \left\|\tilde{\psi}\left(t,\theta_{-t}\omega,\psi_0(\theta_{-t}\omega)+\epsilon\delta z_2(\omega)\right)+\epsilon\delta z_2(\omega)\right\|^2 \\
&\quad \leq \left\|\tilde{u}\left(t,\theta_{-t}\omega,u_0(\theta_{-t}\omega)e^{-i\epsilon z_1(\theta_{-t}\omega)}\right)\right\|^2 + \left\|v\left(t,\theta_{-t}\omega,v_0(\theta_{-t}\omega)\right)\right\|_{\mathbb{H}^1}^2 \\
&\qquad + 2\left\|\tilde{\psi}\left(t,\theta_{-t}\omega,\psi_0(\theta_{-t}\omega)+\epsilon\delta z_2(\theta_{-t}\omega)\right)\right\|^2 + 2\epsilon^2\|\delta\|^2\left\|z_2(\omega)\right\|^2 \\
&\quad \leq C_3e^{-C_1t}E_1\left(u_0(\theta_{-t}\omega),v_0(\theta_{-t}\omega),\psi_0(\theta_{-t}\omega)\right) + C_3e^{-C_1t}\epsilon^2\|\delta\|^2\left\|z_2(\theta_{-t}\omega)\right\|^2 \\
&\qquad + 2\epsilon^2\|\delta\|^2\left\|z_2(\omega)\right\|^2 + \epsilon C_1r(\omega) + C_2. \qquad (3.19)
\end{aligned}$$

By definition of $z_2(\omega)$, it is easy to see that $\|z_2(\omega)\|^2$ is tempered. In addition, $\{\mathbb{B}(\omega)\}_{\omega\in\Omega} \subset \mathbb{D}$ is also tempered by assumption. Therefore, for $(u_0(\theta_{-t}\omega), v_0(\theta_{-t}\omega), \psi_0(\theta_{-t}\omega)) \in \mathbb{B}(\theta_{-t}\omega)$, there is $T_{\mathbb{B}}(\omega) > 0$ such that for all $t \geq T_{\mathbb{B}}(\omega)$,

$$e^{-C_1t}E_1\left(u_0(\theta_{-t}\omega),v_0(\theta_{-t}\omega),\psi_0(\theta_{-t}\omega)\right) + C_3e^{-C_1t}\epsilon^2\|\delta\|^2\left\|z_2(\theta_{-t}\omega)\right\|^2 \leq \epsilon C_4r(\omega) + C_4,$$

which along with (3.19) leads to

$$\left\|\phi\left(t,\theta_{-t}\omega,(u_0,v_0,\psi_0)(\theta_{-t}\omega)\right)\right\|_{\mathbb{I}}^2 \leq C_5 + \epsilon C_5r(\omega) + 2\epsilon^2\|\delta\|^2\left\|z_2(\omega)\right\|^2.$$

Given $\omega \in \Omega$, define

$$\mathbb{K}(\omega) = \left\{(u,v,\psi)\in\mathbb{I} : \|u\|^2 + \|v\|_{\mathbb{H}^1}^2 + \|\psi\|^2 \leq C + \epsilon Cr(\omega) + 2\epsilon^2\|\delta\|^2\left\|z_2(\omega)\right\|^2\right\}.$$

Then $\{\mathbb{K}(\omega)\}_{\omega\in\Omega}$ is an absorbing set for ϕ in $\mathbb{D}$, which completes the proof. □

Lemma 3.5 *Let $f,g \in \mathbb{H}^1$, $\delta \in \mathbb{H}^2$, $\mathbb{B} = \{\mathbb{B}(\omega)\}_{\omega\in\Omega} \in \mathbb{D}$ and $(u_0(\omega), v_0(\omega), \psi_0(\omega)) \in \mathbb{B}(\omega)$. Assume that $\alpha(\nu-\rho) > 12M_1\beta^2$, $\alpha(\mu-\rho(\nu-\rho)) > 8M_1$. Then, for $\mathbb{P}$-a.e. $\omega \in \Omega$, there exists $T'_{\mathbb{B}}(\omega) > 0$ such that for all $t \geq T'_{\mathbb{B}}$,*

$$\left\|u\left(t,\theta_{-t}\omega,u_0(\theta_{-t}\omega)\right)\right\|_{\mathbb{H}^1}^2 + \left\|v\left(t,\theta_{-t}\omega,v_0(\theta_{-t}\omega)\right)\right\|_{\mathbb{H}^2}^2 + \left\|\psi\left(t,\theta_{-t}\omega,\psi_0(\theta_{-t}\omega)\right)\right\|_{\mathbb{H}^1}^2 \leq R(\omega),$$

where $R(\omega)$ is a positive tempered random function.

Proof Taking the real part of the inner product of the first equation in system (3.6) with $-\Delta\tilde{u}$ in $\mathbb{L}^2(\mathbb{R}^n)$, we have that

$$\frac{d}{dt}\|\nabla\tilde{u}\|^2 + 2\alpha\|\nabla\tilde{u}\|^2 + 2(\nabla\tilde{u},\tilde{u}\nabla v) + 2\,\mathrm{Im}\left(\nabla\tilde{u},\nabla(fz)\right) = 0. \qquad (3.20)$$

Taking the inner product of the third equation in system (3.6) with $-\Delta\tilde{\psi}$ in $\mathbb{L}^2(\mathbb{R}^n)$, we find

$$\begin{aligned}
&\frac{d}{dt}\|\nabla\tilde{\psi}\|^2 + 2(\nu-\rho)\|\nabla\tilde{\psi}\|^2 + 2(\Delta\tilde{\psi},\Delta v) - 2\left[\mu-\rho(\nu-\rho)\right](\Delta\tilde{\psi},v) \\
&\qquad + 2\beta\left(\Delta\tilde{\psi},\left|\tilde{u}z^{-1}\right|^2\right) - 2\epsilon(\nu-\rho)\left(\Delta\tilde{\psi},\delta z_2(\theta_{-t}\omega)\right) + 2(\Delta\tilde{\psi},g) = 0. \qquad (3.21)
\end{aligned}$$

By the second equation in system (3.6), we have

$$\begin{aligned}
&2(\Delta\tilde{\psi},\Delta v) - 2[\mu - \rho(\nu-\rho)](\Delta\tilde{\psi},v) \\
&\quad = 2(\Delta v_t + \rho\Delta v - \epsilon\Delta(z_2(\theta_t\omega)\delta),\Delta v) \\
&\qquad - 2[\mu - \rho(\nu-\rho)](\Delta v_t + \rho\Delta v - \epsilon\Delta(z_2(\theta_t\omega)\delta),v) \\
&\quad = \frac{d}{dt}(\|\Delta v\|^2 + (\mu - \rho(\nu-\rho))\|\nabla v\|^2) + 2(\rho\|\Delta v\|^2 + (\mu-\rho(\nu-\rho))\|\nabla v\|^2) \\
&\qquad - 2\epsilon(z_2(\theta_t\omega)\Delta\delta,\Delta v) - 2\epsilon(\delta\Delta z_2(\theta_t\omega),\Delta v) \\
&\qquad - 2\epsilon[\mu-\rho(\nu-\rho)](z_2(\theta_t\omega)\nabla\delta,\nabla v) \\
&\qquad - 2\epsilon[\mu-\rho(\nu-\rho)](\delta\nabla z_2(\theta_t\omega),\nabla v).
\end{aligned} \tag{3.22}$$

Then it follows from (3.20)-(3.22) that

$$\begin{aligned}
&\frac{d}{dt}(\|\nabla\tilde{u}\|^2 + \|\nabla\tilde{\psi}\|^2 + \|\Delta v\|^2 + (\mu-\rho(\nu-\rho))\|\nabla v\|^2) \\
&\qquad + 2\alpha\|\nabla\tilde{u}\|^2 + 2(\nu-\rho)\|\nabla\tilde{\psi}\|^2 + 2\rho\|\Delta v\|^2 + 2(\mu-\rho(\nu-\rho))\|\nabla v\|^2 \\
&\quad = -2\,\mathrm{Im}(\nabla\tilde{u},\nabla(fz)) - 2(\nabla\tilde{u},\tilde{u}\nabla v) - 2\beta(\Delta\tilde{\psi},|\tilde{u}z^{-1}|^2) \\
&\qquad + 2\epsilon(\nu-\rho)(\Delta\tilde{\psi},z_2(\theta_t\omega)\delta) - 2(\Delta\tilde{\psi},g) + 2\epsilon(z_2(\theta_t\omega)\Delta\delta,\Delta v) \\
&\qquad + 2\epsilon(\delta\Delta z_2(\theta_t\omega),\Delta v) + 2\epsilon[\mu-\rho(\nu-\rho)](z_2(\theta_t\omega)\nabla\delta,\nabla v) \\
&\qquad + 2\epsilon[\mu-\rho(\nu-\rho)](\delta\nabla z_2(\theta_t\omega),\nabla v).
\end{aligned} \tag{3.23}$$

Now, we estimate each term on the right-hand side of (3.23). By Young's inequality, we get

$$\begin{aligned}
|-2\,\mathrm{Im}(\nabla\tilde{u},\nabla(fz))| &= 2|\mathrm{Im}(\nabla\tilde{u},z\nabla f) + \mathrm{Im}(\nabla\tilde{u},f\nabla z)| \\
&\le \frac{\alpha}{2}\|\nabla\tilde{u}\|^2 + \frac{8}{\alpha}(\|\nabla f\|^2 + \epsilon^2\|f\|^2\|\nabla z_1\|^2),
\end{aligned}$$

$$|-2(\nabla\tilde{u},\tilde{u}\nabla v)| \le \frac{\alpha}{2}\|\nabla\tilde{u}\|^2 + \frac{8}{\alpha}\|\tilde{u}\|^2\|\nabla v\|^2,$$

$$|-2\beta(\Delta\tilde{\psi},|\tilde{u}z^{-1}|^2)| \le \frac{\nu-\rho}{3}\|\nabla\tilde{\psi}\|^2 + \frac{12\beta^2}{\nu-\rho}\|\tilde{u}\|^2\|\nabla\tilde{u}\|^2,$$

$$\begin{aligned}
&|2\epsilon(\nu-\rho)(\Delta\tilde{\psi},z_2(\theta_t\omega)\delta)| \\
&\quad \le \frac{\nu-\rho}{3}\|\nabla\tilde{\psi}\|^2 + 12\epsilon^2(\nu-\rho)(\|\delta\|^2\|\nabla z_2(\theta_t\omega)\|^2 + \|z_2(\theta_t\omega)\|^2\|\nabla\delta\|^2),
\end{aligned}$$

$$|-2(\Delta\tilde{\psi},g)| \le \frac{\nu-\rho}{3}\|\nabla\tilde{\psi}\|^2 + \frac{12}{\nu-\rho}\|\nabla g\|^2,$$

$$|2\epsilon(z_2(\theta_t\omega)\Delta\delta,\Delta v)| \le \frac{\rho}{2}\|\Delta v\|^2 + \frac{8\epsilon^2}{\rho}\|\Delta\delta\|^2\|z_2(\theta_t\omega)\|^2,$$

$$|2\epsilon(\delta\Delta z_2(\theta_t\omega),\Delta v)| \le \frac{\rho}{2}\|\Delta v\|^2 + \frac{8\epsilon^2}{\rho}\|\delta\|^2\|\Delta z_2(\theta_t\omega)\|^2,$$

$$\begin{aligned}
&|2\epsilon[\mu-\rho(\nu-\rho)](z_2(\theta_t\omega)\nabla\delta,\nabla v)| \\
&\quad \le \frac{\mu-\rho(\nu-\rho)}{2}\|\nabla v\|^2 + 8\epsilon^2[\mu-\rho(\nu-\rho)]\|z_2(\theta_t\omega)\|^2\|\nabla\delta\|^2,
\end{aligned}$$

$$\begin{aligned}&\left|2\epsilon\left[\mu-\rho(\nu-\rho)\right]\left(\delta\nabla z_2(\theta_t\omega),\nabla\nu\right)\right|\\&\quad\le\frac{\mu-\rho(\nu-\rho)}{2}\|\nabla\nu\|^2+8\epsilon^2\left[\mu-\rho(\nu-\rho)\right]\left\|\nabla z_2(\theta_t\omega)\right\|^2\|\delta\|^2,\end{aligned}$$

which along with (3.23), Remark 3.1 and assumption gives that there exists a positive constant

$$C_6=\min\left\{\alpha-\frac{12M_1\beta^2}{\nu-\rho},\nu-\rho,\rho,\mu-\rho(\nu-\rho)-\frac{8M_1}{\alpha}\right\}>0$$

such that for $t>T_1$

$$\frac{d}{dt}E_2(\tilde{u},\nu,\tilde{\psi})+C_6E_2(\tilde{u},\nu,\tilde{\psi})\le F\left(z_1(\theta_t\omega),z_2(\theta_t\omega)\right),\tag{3.24}$$

where

$$E_2(\tilde{u},\nu,\tilde{\psi})=\|\nabla\tilde{u}\|^2+\|\nabla\tilde{\psi}\|^2+\|\Delta\nu\|^2+\left(\mu-\rho(\nu-\rho)\right)\|\nabla\nu\|^2,$$

and

$$\begin{aligned}&F\left(z_1(\theta_t\omega),z_2(\theta_t\omega)\right)\\&\quad=\frac{8}{\alpha}\left(\|\nabla f\|^2+\epsilon^2\|f\|^2\left\|\nabla z_1(\theta_t\omega)\right\|^2\right)\\&\qquad+24\epsilon^2(\nu-\rho)\left(\|\delta\|^2\left\|\nabla z_2(\theta_t\omega)\right\|^2+\left\|z_2(\theta_t\omega)\right\|^2\|\nabla\delta\|^2\right)+\frac{12}{\nu-\rho}\|\nabla g\|^2\\&\qquad+\frac{12\epsilon^2}{\rho}\left(\|\Delta\delta\|^2\left\|z_2(\theta_t\omega)\right\|^2+\|\delta\|^2\left\|\Delta z_2(\theta_t\omega)\right\|^2\right)\\&\qquad+8\epsilon^2\left[\mu-\rho(\nu-\rho)\right]\left(\left\|z_2(\theta_t\omega)\right\|^2\|\nabla\delta\|^2+\left\|\nabla z_2(\theta_t\omega)\right\|^2\|\delta\|^2\right).\end{aligned}$$

Let

$$P_2(\theta_t\omega)=C\left(\|\nabla z_1\|^4+\|z_2\|^2+\|\nabla z_2\|^4+\|\Delta z_2\|^4\right).$$

Since $z_1(\theta_t\omega)=y_1(\theta_t\omega_1)$ and $z_2(\theta_t\omega)=\delta y_2(\theta_t\omega_2)$, $\delta\in\mathbb{H}^2(\mathbb{R}^n)$. Therefore, by Lemma 3.1, we have

$$P_2(\theta_t\omega)\le C\sum_{i=1}^{2}\left(\left|y_i(\theta_t\omega_i)\right|^2+\left|y_i(\theta_t\omega_i)\right|_p^p\right)+C,$$

and

$$P_2(\theta_t\omega)\le Ce^{\frac{C}{2}|t|}r(\omega)+C\quad\text{for all }t\in\mathbb{R}.$$

By replacing ω by $\theta_{-t}\omega$, it follows from Gronwall's lemma that

$$\begin{aligned}&E_2\left(\tilde{u}\left(t,\theta_{-t}\omega,\tilde{u}_0(\theta_{-t}\omega)\right),\nu\left(t,\theta_{-t}\omega,\nu_0(\theta_{-t}\omega)\right),\tilde{\psi}\left(t,\theta_{-t}\omega,\tilde{\psi}_0(\theta_{-t}\omega)\right)\right)\\&\quad\le e^{-C_6t}E_2\left(\tilde{u}_0(\theta_{-t}\omega),\nu_0(\theta_{-t}\omega),\tilde{\psi}_0(\theta_{-t}\omega)\right)+\epsilon\int_0^t e^{C_6(s-t)}P_2(\theta_{s-t}\omega)\,ds+C_7\end{aligned}$$

$$\le e^{-C_6 t} E_2\big(\tilde{u}_0(\theta_{-t}\omega), v_0(\theta_{-t}\omega), \tilde{\psi}_0(\theta_{-t}\omega)\big) + \epsilon \int_{-t}^{0} e^{C_6 s} P_2(\theta_s\omega)\, ds + C_7$$
$$\le e^{-C_6 t} E_2\big(\tilde{u}_0(\theta_{-t}\omega), v_0(\theta_{-t}\omega), \tilde{\psi}_0(\theta_{-t}\omega)\big) + \epsilon C_6 \int_{-t}^{0} e^{\frac{C_6}{2} s} r(\omega)\, ds + C_7$$
$$\le e^{-C_6 t} E_2\big(\tilde{u}_0(\theta_{-t}\omega), v_0(\theta_{-t}\omega), \tilde{\psi}_0(\theta_{-t}\omega)\big) + \epsilon C_6 r(\omega) + C_7,$$

which implies that

$$\begin{aligned}&\big\|\tilde{u}\big(t,\theta_{-t}\omega,\tilde{u}_0(\theta_{-t}\omega)\big)\big\|_{\mathbb{H}^1}^2 + \big\|v\big(t,\theta_{-t}\omega,v_0(\theta_{-t}\omega)\big)\big\|_{\mathbb{H}^2}^2 + \big\|\tilde{\psi}\big(t,\theta_{-t}\omega,\tilde{\psi}_0(\theta_{-t}\omega)\big)\big\|_{\mathbb{H}^1}^2 \\ &\quad \le e^{-C_6 t} E_2\big(\tilde{u}_0(\theta_{-t}\omega), v_0(\theta_{-t}\omega), \tilde{\psi}_0(\theta_{-t}\omega)\big) + \epsilon C_6 r(\omega) + C_7,\end{aligned} \tag{3.25}$$

where $r(\omega)$ is a tempered function.

Note that $\tilde{u} = uz$, $\tilde{\psi} = \psi - \epsilon\delta z_2(\theta_t\omega)$, by (3.25) we obtain

$$\begin{aligned}&\big\|u\big(t,\theta_{-t}\omega,u_0(\theta_{-t}\omega)\big)\big\|_{\mathbb{H}^1}^2 + \big\|v\big(t,\theta_{-t}\omega,v_0(\theta_{-t}\omega)\big)\big\|_{\mathbb{H}^2}^2 + \big\|\psi\big(t,\theta_{-t}\omega,\psi_0(\theta_{-t}\omega)\big)\big\|_{\mathbb{H}^1}^2 \\
&\quad = \big\|\tilde{u}\big(t,\theta_{-t}\omega,\tilde{u}_0(\theta_{-t}\omega)e^{-i\epsilon z_1(\theta_{-t}\omega)}\big)e^{-i\epsilon z_1(\omega)}\big\|_{\mathbb{H}^1}^2 + \big\|v\big(t,\theta_{-t}\omega,v_0(\theta_{-t}\omega)\big)\big\|_{\mathbb{H}^2}^2 \\
&\qquad + \big\|\tilde{\psi}\big(t,\theta_{-t}\omega,\tilde{\psi}_0(\theta_{-t}\omega) + \epsilon\delta z_2(\theta_{-t}\omega)\big) + \epsilon\delta z_2(\omega)\big\|_{\mathbb{H}^1}^2 \\
&\quad \le \big\|\tilde{u}\big(t,\theta_{-t}\omega,\tilde{u}_0(\theta_{-t}\omega)e^{-i\epsilon z_1(\theta_{-t}\omega)}\big)\big\|_{\mathbb{H}^1}^2 + \big\|\nabla v\big(t,\theta_{-t}\omega,v_0(\theta_{-t}\omega)\big)\big\|_{\mathbb{H}^2}^2 \\
&\qquad + \big\|\tilde{\psi}\big(t,\theta_{-t}\omega,\tilde{\psi}_0(\theta_{-t}\omega) + \epsilon\delta z_2(\theta_{-t}\omega)\big)\big\|_{\mathbb{H}^1}^2 + 2\epsilon^2\|\delta\|_{\mathbb{H}^1}^2\big\|z_2(\omega)\big\|_{\mathbb{H}^1}^2 \\
&\quad \le e^{-C_6 t} E_2\big(\tilde{u}_0(\theta_{-t}\omega), v_0(\theta_{-t}\omega), \tilde{\psi}_0(\theta_{-t}\omega)\big) + 2\epsilon^2\|\delta\|_{\mathbb{H}^1}^2\big\|z_2(\theta_{-t}\omega)\big\|_{\mathbb{H}^1}^2 \\
&\qquad + \epsilon C_6 r(\omega) + C_7 + 2\epsilon^2\|\delta\|_{\mathbb{H}^1}^2\big\|z_2(\omega)\big\|_{\mathbb{H}^1}^2.\end{aligned}$$

Since $\|\nabla z_2(\omega)\|^2$ is tempered, there exists $T'_{\mathbb{B}}(\omega) > T_1 > 0$, for all $t \ge T'_{\mathbb{B}}$,

$$\big\|u\big(t,\theta_{-t}\omega,u_0(\theta_{-t}\omega)\big)\big\|_{\mathbb{H}^1}^2 + \big\|v\big(t,\theta_{-t}\omega,v_0(\theta_{-t}\omega)\big)\big\|_{\mathbb{H}^2}^2 + \big\|\psi\big(t,\theta_{-t}\omega,\psi_0(\theta_{-t}\omega)\big)\big\|_{\mathbb{H}^1}^2 \le R(\omega),$$

where $R(\omega) = \epsilon C_6 r(\omega) + C_7 + 2\epsilon^2\|\delta\|_{\mathbb{H}^1}^2\|z_2(\omega)\|_{\mathbb{H}^1}^2$. This completes the proof. □

In what follows, we will use the method of [11, 16] to derive uniform estimates on the tails of solution when x exists on unbounded domains.

Lemma 3.6 *Let $f \in \mathbb{L}^4$, $g \in \mathbb{L}^2$, $\delta \in \mathbb{H}^1 \cap \mathbb{W}^{1,4}$, $\mathbb{B} = \{B(\omega)\}_{\omega\in\Omega} \in \mathbb{D}$ and $(u_0, v_0, \psi_0) \in \mathbb{B}(\omega)$. Assume that $\mu - \rho(\nu - \rho) > 0$. Then, for every $\varepsilon > 0$ and P-a.e. $\omega \in \Omega$, there exist $T' = T'(\omega,\varepsilon) < 0$ and $m = m(\omega,\varepsilon) > 0$ such that the solution of system* (3.1) *satisfies, for all $t \ge T'(\omega,\varepsilon)$,*

$$\int_{|x|\ge m} \big|\phi\big(t,\theta_{-t}\omega,(u_0,v_0,\phi_0)(\theta_{-t}\omega)\big)\big|^2 dx \le \varepsilon.$$

Proof Let $\eta(x) \in \mathbb{C}(\mathbb{R}^+, [0,1])$ be a cut-off function satisfying

$$\eta(x) = 0, \quad \text{for all } x \in [0,1]; \qquad \eta(x) = 1, \quad \text{for all } x \in [2,+\infty),$$

and $|\eta'(x)| \le \eta_0$ (a positive constant).

Taking the imaginary part of the inner product of the first equation of (3.6) with $\eta(\frac{|x|^2}{m^2})\tilde{u}$ in $\mathbb{L}^2(\mathbb{R}^n)$, we have

$$\frac{d}{dt}\int_{\mathbb{R}^n}\eta\left(\frac{|x|^2}{m^2}\right)|\tilde{u}|^2\,dx + 2\alpha\int_{\mathbb{R}^n}\eta\left(\frac{|x|^2}{m^2}\right)|\tilde{u}|^2\,dx = 2\,\mathrm{Im}\int_{\mathbb{R}^n}\eta\left(\frac{|x|^2}{m^2}\right)fz\tilde{u}\,dx. \tag{3.26}$$

Taking the inner product of the third equation of (3.6) with $\eta(\frac{|x|^2}{m^2})\tilde{\psi}$ in $\mathbb{L}^2(\mathbb{R}^n)$, we get

$$\begin{aligned}
&\frac{d}{dt}\int_{\mathbb{R}^n}\eta\left(\frac{|x|^2}{m^2}\right)|\tilde{\psi}|^2\,dx + 2(\nu-\rho)\int_{\mathbb{R}^n}\eta\left(\frac{|x|^2}{m^2}\right)|\tilde{\psi}|^2\,dx \\
&\quad = -2[\mu-\rho(\nu-\rho)]\int_{\mathbb{R}^n}\eta\left(\frac{|x|^2}{m^2}\right)\nu\tilde{\psi}\,dx + 2\int_{\mathbb{R}^n}\eta\left(\frac{|x|^2}{m^2}\right)(\triangle\nu)\tilde{\psi}\,dx \\
&\quad + 2\beta\int_{\mathbb{R}^n}\eta\left(\frac{|x|^2}{m^2}\right)|\tilde{u}z^{-1}|^2\tilde{\psi}\,dx - 2\epsilon(\nu-\rho)\int_{\mathbb{R}^n}\eta\left(\frac{|x|^2}{m^2}\right)\delta z_2(\theta_t\omega)\tilde{\psi}\,dx \\
&\quad + 2\int_{\mathbb{R}^n}\eta\left(\frac{|x|^2}{m^2}\right)g\tilde{\psi}\,dx.
\end{aligned} \tag{3.27}$$

Note that

$$\begin{aligned}
&-2[\mu-\rho(\nu-\rho)]\int_{\mathbb{R}^n}\eta\left(\frac{|x|^2}{m^2}\right)\nu\tilde{\psi}\,dx \\
&\quad = -[\mu-\rho(\nu-\rho)]\frac{d}{dt}\int_{\mathbb{R}^n}\eta\left(\frac{|x|^2}{m^2}\right)|\nu|^2\,dx - 2\rho[\mu-\rho(\nu-\rho)]\int_{\mathbb{R}^n}\eta\left(\frac{|x|^2}{m^2}\right)|\nu|^2\,dx \\
&\quad + 2\epsilon[\mu-\rho(\nu-\rho)]\int_{\mathbb{R}^n}\eta\left(\frac{|x|^2}{m^2}\right)\delta z_2(\theta_t\omega)\nu\,dx,
\end{aligned} \tag{3.28}$$

and

$$\begin{aligned}
&2\int_{\mathbb{R}^n}\eta\left(\frac{|x|^2}{m^2}\right)(\triangle\nu)\tilde{\psi}\,dx \\
&\quad = -2\int_{\mathbb{R}^n}\eta'\left(\frac{|x|^2}{m^2}\right)\left(\frac{2x}{m^2}\right)(\nabla\nu)\tilde{\psi}\,dx - \frac{d}{dt}\int_{\mathbb{R}^n}\eta\left(\frac{|x|^2}{m^2}\right)|\nabla\nu|^2\,dx \\
&\quad - 2\rho\int_{\mathbb{R}^n}\eta\left(\frac{|x|^2}{m^2}\right)|\nabla\nu|^2\,dx - 2\rho\int_{\mathbb{R}^n}\eta'\left(\frac{|x|^2}{m^2}\right)\left(\frac{2x}{m^2}\right)\nu\nabla\nu\,dx \\
&\quad + 2\epsilon\int_{\mathbb{R}^n}\eta\left(\frac{|x|^2}{m^2}\right)z_2(\theta_t\omega)\nabla\nu\nabla\delta\,dx + 2\epsilon\int_{\mathbb{R}^n}\eta\left(\frac{|x|^2}{m^2}\right)\delta\nabla\nu\nabla z_2(\theta_t\omega)\,dx \\
&\quad + 2\epsilon\int_{\mathbb{R}^n}\eta'\left(\frac{|x|^2}{m^2}\right)\left(\frac{2x}{m^2}\right)\delta z_2(\theta_{-t}\omega)\nabla\nu.
\end{aligned} \tag{3.29}$$

Summing up (3.26)-(3.29), we have

$$\begin{aligned}
&\frac{d}{dt}\int_{\mathbb{R}^n}\eta\left(\frac{|x|^2}{m^2}\right)\left(|\tilde{u}|^2 + (\mu-\rho(\nu-\rho))|\nu|^2 + |\nabla\nu|^2 + |\tilde{\psi}|^2\right)dx \\
&\quad + 2\alpha\int_{\mathbb{R}^n}\eta\left(\frac{|x|^2}{m^2}\right)|\tilde{u}|^2\,dx + 2\rho(\mu-\rho(\nu-\rho))\int_{\mathbb{R}^n}\eta\left(\frac{|x|^2}{m^2}\right)|\nu|^2\,dx \\
&\quad + 2\rho\int_{\mathbb{R}^n}\eta\left(\frac{|x|^2}{m^2}\right)|\nabla\nu|^2\,dx + 2(\nu-\rho)\int_{\mathbb{R}^n}\eta\left(\frac{|x|^2}{m^2}\right)|\tilde{\psi}|^2\,dx
\end{aligned}$$

$$= 2\,\mathrm{Im}\int_{\mathbb{R}^n}\eta\Big(\frac{|x|^2}{m^2}\Big)fz\tilde{u}\,dx + 2\epsilon\big(\mu-\rho(\nu-\rho)\big)\int_{\mathbb{R}^n}\eta\Big(\frac{|x|^2}{m^2}\Big)\delta v z_2(\theta_t\omega)\,dx$$
$$-2\int_{\mathbb{R}^n}\eta'\Big(\frac{|x|^2}{m^2}\Big)\Big(\frac{2x}{m^2}\Big)(\nabla v)\tilde{\psi}\,dx + 2\epsilon\int_{\mathbb{R}^n}\eta\Big(\frac{|x|^2}{m^2}\Big)z_2(\theta_t\omega)\nabla v\nabla\delta\,dx$$
$$+2\epsilon\int_{\mathbb{R}^n}\eta\Big(\frac{|x|^2}{m^2}\Big)\delta\nabla v\nabla z_2(\theta_t\omega)\,dx - 2\rho\int_{\mathbb{R}^n}\eta'\Big(\frac{|x|^2}{m^2}\Big)\Big(\frac{2x}{m^2}\Big)v\nabla v\,dx$$
$$+2\epsilon\int_{\mathbb{R}^n}\eta'\Big(\frac{|x|^2}{m^2}\Big)\Big(\frac{2x}{m^2}\Big)\delta z_2(\theta_{-t}\omega)\nabla v\,dx + 2\beta\int_{\mathbb{R}^n}\eta\Big(\frac{|x|^2}{m^2}\Big)\big|\tilde{u}z^{-1}\big|^2\tilde{\psi}\,dx$$
$$+2\int_{\mathbb{R}^n}\eta\Big(\frac{|x|^2}{m^2}\Big)g\tilde{\psi}\,dx - 2\epsilon(\nu-\rho)\int_{\mathbb{R}^n}\eta\Big(\frac{|x|^2}{m^2}\Big)\delta z_2(\theta_t\omega)\tilde{\psi}\,dx. \tag{3.30}$$

We now estimate the right-hand side term in (3.30) as follows. Firstly, by the definition of η, we have

$$-2\int_{\mathbb{R}^n}\eta'\Big(\frac{|x|^2}{m^2}\Big)\Big(\frac{2x}{m^2}\Big)(\nabla v)\tilde{\psi}\,dx \le 4\eta_0\int_{m\le x\le\sqrt{2}m}\Big(\frac{|x|}{m^2}\Big)|\nabla v||\tilde{\psi}|\,dx$$
$$\le \frac{4\sqrt{2}\eta_0}{m}\big(\|\nabla v\|^2 + \|\tilde{\psi}\|^2\big), \tag{3.31}$$
$$-2\rho\int_{\mathbb{R}^n}\eta'\Big(\frac{|x|^2}{m^2}\Big)\Big(\frac{2x}{m^2}\Big)v\nabla v\,dx \le 4\rho\eta_0\int_{m\le x\le\sqrt{2}m}\Big(\frac{|x|}{m^2}\Big)|v||\nabla v|\,dx$$
$$\le \frac{4\rho\sqrt{2}\eta_0}{m}\big(\|\nabla v\|^2 + \|v\|^2\big), \tag{3.32}$$
$$2\epsilon\int_{\mathbb{R}^n}\eta'\Big(\frac{|x|^2}{m^2}\Big)\Big(\frac{2x}{m^2}\Big)\delta z_2(\theta_t\omega)\nabla v\,dx$$
$$\le 4\epsilon\eta_0\int_{m\le x\le\sqrt{2}m}\Big(\frac{|x|}{m^2}\Big)|\delta|\big|z_2(\theta_t\omega)\big||\nabla v|\,dx$$
$$\le \frac{4\epsilon\sqrt{2}\eta_0}{m}\big(\|\nabla v\|^2 + \|\delta\|^2\big\|z_2(\theta_{-t}\omega)\big\|^2\big). \tag{3.33}$$

Then, by Young's inequality, we get

$$2\,\mathrm{Im}\int_{\mathbb{R}^n}\eta\Big(\frac{|x|^2}{m^2}\Big)fz\tilde{u}\,dx \le \alpha\int_{\mathbb{R}^n}\eta\Big(\frac{|x|^2}{m^2}\Big)|\tilde{u}|^2\,dx + \frac{4}{\alpha}\int_{\mathbb{R}^n}\eta\Big(\frac{|x|^2}{m^2}\Big)|f|^2\,dx; \tag{3.34}$$
$$2\epsilon\big(\mu-\rho(\nu-\rho)\big)\int_{\mathbb{R}^n}\eta\Big(\frac{|x|^2}{m^2}\Big)\delta v z_2(\theta_t\omega)\,dx$$
$$\le \rho\big(\mu-\rho(\nu-\rho)\big)\int_{\mathbb{R}^n}\eta\Big(\frac{|x|^2}{m^2}\Big)|v|^2\,dx$$
$$+\frac{4\epsilon^2}{\rho\big(\mu-\rho(\nu-\rho)\big)}\int_{\mathbb{R}^n}\eta\Big(\frac{|x|^2}{m^2}\Big)|\delta|^2\big|z_2(\theta_t\omega)\big|^2\,dx; \tag{3.35}$$
$$2\epsilon\int_{\mathbb{R}^n}\eta\Big(\frac{|x|^2}{m^2}\Big)z_2(\theta_t\omega)\nabla v\nabla\delta\,dx$$
$$\le \frac{\rho}{2}\int_{\mathbb{R}^n}\eta\Big(\frac{|x|^2}{m^2}\Big)|\nabla v|^2\,dx + \frac{8\epsilon^2}{\rho}\int_{\mathbb{R}^n}\eta\Big(\frac{|x|^2}{m^2}\Big)\big|z_2(\theta_t\omega)\big|^2|\nabla\delta|^2\,dx; \tag{3.36}$$

$$2\epsilon \int_{\mathbb{R}^n} \eta\left(\frac{|x|^2}{m^2}\right)\delta \nabla z_2(\theta_t\omega)\nabla v\, dx$$
$$\leq \frac{\rho}{2}\int_{\mathbb{R}^n} \eta\left(\frac{|x|^2}{m^2}\right)|\nabla v|^2\, dx + \frac{8\epsilon^2}{\rho}\int_{\mathbb{R}^n} \eta\left(\frac{|x|^2}{m^2}\right)\left|\nabla z_2(\theta_t\omega)\right|^2|\delta|^2\, dx; \tag{3.37}$$
$$2\beta \int_{\mathbb{R}^n} \eta\left(\frac{|x|^2}{m^2}\right)\left|\tilde{u}z^{-1}\right|^2\tilde{\psi}\, dx$$
$$\leq \frac{v-\rho}{2}\int_{\mathbb{R}^n} \eta\left(\frac{|x|^2}{m^2}\right)|\tilde{\psi}|^2\, dx + \frac{8\beta^2}{v-\rho}\int_{\mathbb{R}^n} \eta\left(\frac{|x|^2}{m^2}\right)|\tilde{u}|^4\, dx; \tag{3.38}$$
$$-2\epsilon(v-\rho)\int_{\mathbb{R}^n} \eta\left(\frac{|x|^2}{m^2}\right)\delta z_2(\theta_t\omega)\tilde{\psi}\, dx$$
$$\leq \frac{v-\rho}{2}\int_{\mathbb{R}^n} \eta\left(\frac{|x|^2}{m^2}\right)|\tilde{\psi}|^2\, dx + 8\epsilon^2(v-\rho)\int_{\mathbb{R}^n} \eta\left(\frac{|x|^2}{m^2}\right)\left|z_2(\theta_t\omega)\right|^2|\delta|^2\, dx; \tag{3.39}$$
$$2\int_{\mathbb{R}^n} \eta\left(\frac{|x|^2}{m^2}\right)g\tilde{\psi}\, dx$$
$$\leq \frac{v-\rho}{2}\int_{\mathbb{R}^n} \eta\left(\frac{|x|^2}{m^2}\right)|\tilde{\psi}|^2\, dx + \frac{8}{v-\rho}\int_{\mathbb{R}^n} \eta\left(\frac{|x|^2}{m^2}\right)|g|^2\, dx. \tag{3.40}$$

Finally, by (3.30)-(3.40), we obtain

$$\begin{aligned}
&\frac{d}{dt}\int_{\mathbb{R}^n} \eta\left(\frac{|x|^2}{m^2}\right)\left(|\tilde{u}|^2 + \left(\mu - \rho(v-\rho)\right)|v|^2 + |\nabla v|^2 + |\tilde{\psi}|^2\right) dx \\
&\quad + \alpha\int_{\mathbb{R}^n} \eta\left(\frac{|x|^2}{m^2}\right)|\tilde{u}|^2\, dx + \rho\left(\mu - \rho(v-\rho)\right)\int_{\mathbb{R}^n} \eta\left(\frac{|x|^2}{m^2}\right)|v|^2\, dx \\
&\quad + \rho\int_{\mathbb{R}^n} \eta\left(\frac{|x|^2}{m^2}\right)|\nabla v|^2\, dx + \frac{v-\rho}{2}\int_{\mathbb{R}^n} \eta\left(\frac{|x|^2}{m^2}\right)|\tilde{\psi}|^2\, dx \\
&\leq \frac{4}{\alpha}\int_{\mathbb{R}^n} \eta\left(\frac{|x|^2}{m^2}\right)|f|^2\, dx + \frac{4\epsilon^2}{\rho\left(\mu - \rho(v-\rho)\right)}\int_{\mathbb{R}^n} \eta\left(\frac{|x|^2}{m^2}\right)|\delta|^2\left|z_2(\theta_t\omega)\right|^2 dx \\
&\quad + \frac{8\epsilon^2}{\rho}\int_{\mathbb{R}^n} \eta\left(\frac{|x|^2}{m^2}\right)\left(\left|z_2(\theta_t\omega)\right|^2|\nabla\delta|^2 + \left|\nabla z_2(\theta_t\omega)\right|^2|\delta|^2\right) dx \\
&\quad + \frac{8\beta^2}{v-\rho}\int_{\mathbb{R}^n} \eta\left(\frac{|x|^2}{m^2}\right)|\tilde{u}|^4\, dx + 8\epsilon^2(v-\rho)\int_{\mathbb{R}^n} \eta\left(\frac{|x|^2}{m^2}\right)\left|z_2(\theta_t\omega)\right|^2|\delta|^2\, dx \\
&\quad + \frac{8}{v-\rho}\int_{\mathbb{R}^n} \eta\left(\frac{|x|^2}{m^2}\right)|g|^2\, dx + \frac{4\sqrt{2}\eta_0}{m}\left(\|\nabla v\|^2 + \|\tilde{\psi}\|^2\right) \\
&\quad + \frac{4\rho\sqrt{2}\eta_0}{m}\left(\|\nabla v\|^2 + \|v\|^2\right) + \frac{4\epsilon\sqrt{2}\eta_0}{m}\left(\|\nabla v\|^2 + \|\delta\|^2\left\|z_2(\theta_{-t}\omega)\right\|^2\right).
\end{aligned} \tag{3.41}$$

By (3.41) and assumption, there exist positive constants $C_8 := \min\{\alpha, \rho, \frac{v-\rho}{2}\}$ and $C_9 := C_9(\alpha, \beta, \mu, v, \rho)$ such that

$$\begin{aligned}
&\frac{d}{dt}\int_{\mathbb{R}^n} \eta\left(\frac{|x|^2}{m^2}\right)\left(|\tilde{u}|^2 + \left(\mu - \rho(v-\rho)\right)|v|^2 + |\nabla v|^2 + |\tilde{\psi}|^2\right) dx \\
&\quad + C_8\int_{\mathbb{R}^n} \eta\left(\frac{|x|^2}{m^2}\right)\left(|\tilde{u}|^2 + \rho\left(\mu - \rho(v-\rho)\right)|v|^2 + |\nabla v|^2 + |\tilde{\psi}|^2\right) dx \\
&\leq \epsilon C_9\int_{\mathbb{R}^n} \eta\left(\frac{|x|^2}{m^2}\right)\left(\left|z_2(\theta_t\omega)\right|^4 + \left|\nabla z_2(\theta_t\omega)\right|^4\right) dx
\end{aligned}$$

$$+ C_9 \int_{\mathbb{R}^n} \eta\left(\frac{|x|^2}{m^2}\right)\left(|f|^4 + |g|^2 + |\delta|^4 + |\nabla\delta|^4\right) dx + C_9 \int_{\mathbb{R}^n} \eta\left(\frac{|x|^2}{m^2}\right)|\tilde{u}|^4 dx$$
$$+ \frac{4\sqrt{2}\eta_0}{m}\left(\|\nabla v\|^2 + \|\tilde{\psi}\|^2\right) + \frac{4\epsilon\sqrt{2}\eta_0}{m}\left(\|\nabla v\|^2 + \|\delta\|^2\left\|z_2(\theta_{-t}\omega)\right\|^2\right). \tag{3.42}$$

Now, replacing ω by $\theta_{-t}\omega$, and then integrating (3.42) over (T_2, t) with $t \geq T_2$, we have

$$\int_{\mathbb{R}^n} \eta\left(\frac{|x|^2}{m^2}\right)\Big(\left|\tilde{u}(t,\theta_{-t}\omega,\tilde{u}_0(\theta_{-t}\omega))\right|^2 + \left(\mu - \rho(\nu-\rho)\right)\left|v(t,\theta_{-t}\omega,v_0(\theta_{-t}\omega))\right|^2$$
$$+ \left|\nabla v(t,\theta_{-t}\omega,v_0(\theta_{-t}\omega))\right|^2 + \left|\tilde{\psi}(t,\theta_{-t}\omega,\tilde{\psi}_0(\theta_{-t}\omega))\right|^2\Big) dx$$
$$\leq e^{-C_8(t-T_2)} \int_{\mathbb{R}^n} \eta\left(\frac{|x|^2}{m^2}\right)\Big(\left|\tilde{u}(T_2,\theta_{-t}\omega,\tilde{u}_0(\theta_{-t}\omega))\right|^2 + \left|\nabla v(T_2,\theta_{-t}\omega,v_0(\theta_{-t}\omega))\right|^2$$
$$+ \rho\left(\mu - \rho(\nu-\rho)\right)\left|v(T_2,\theta_{-t}\omega,v_0(\theta_{-t}\omega))\right|^2 + \left|\tilde{\psi}(T_2,\theta_{-t}\omega,\tilde{\psi}_0(\theta_{-t}\omega))\right|^2\Big) dx$$
$$+ \epsilon C_9 \int_{T_2}^{t} e^{C_8(s-t)} \int_{\mathbb{R}^n} \eta\left(\frac{|x|^2}{m^2}\right)\left(\left|z_2(\theta_{s-t}\omega)\right|^4 + \left|\nabla z_2(\theta_{s-t}\omega)\right|^4\right) dx\, ds$$
$$+ C_9 \int_{T_2}^{t} e^{C_8(s-t)} \int_{\mathbb{R}^n} \eta\left(\frac{|x|^2}{m^2}\right)\left(|f|^4 + |g|^2 + |\delta|^4 + |\nabla\delta|^4 + |\tilde{u}|^4\right) dx\, ds$$
$$+ \frac{C^*}{m}\int_{T_2}^{t} e^{C_8(s-t)}\Big(\left\|v(s,\theta_{-t}\omega,v_0(\theta_{-t}\omega))\right\|^2 + \left\|\nabla v(s,\theta_{-t}\omega,v_0(\theta_{-t}\omega))\right\|^2$$
$$+ \left\|\tilde{\psi}(s,\theta_{-t}\omega,\tilde{\psi}_0(\theta_{-t}\omega))\right\|^2\Big) ds + \frac{4\epsilon\sqrt{2}\eta_0\|\delta\|^2}{m}\int_{T_2}^{t} e^{C_8(s-t)}\left\|z_2(\theta_{-t}\omega)\right\|^2 ds, \tag{3.43}$$

where C^* is a fixed constant.

In what follows, we estimate the terms in (3.43). First replacing t by T_2 and then replacing ω by $\theta_{-t}\omega$ in (3.18), we have the following bounds for the first term on the right-hand side of (3.43):

$$e^{-C_8(t-T_2)} \int_{\mathbb{R}^n} \eta\left(\frac{|x|^2}{m^2}\right)\Big(\left|\tilde{u}(T_2,\theta_{-t}\omega,\tilde{u}_0(\theta_{-t}\omega))\right|^2 + \left|\nabla v(T_2,\theta_{-t}\omega,v_0(\theta_{-t}\omega))\right|^2$$
$$+ \rho\left(\mu - \rho(\nu-\rho)\right)\left|v(T_2,\theta_{-t}\omega,v_0(\theta_{-t}\omega))\right|^2 + \left|\tilde{\psi}(T_2,\theta_{-t}\omega,\tilde{\psi}_0(\theta_{-t}\omega))\right|^2\Big) dx$$
$$\leq e^{-C_8(t-T_2)}\Big(e^{-C_8T_2}\left\|\tilde{u}_0(\theta_{-t}\omega)\right\|^2 + \rho\left(\mu - \rho(\nu-\rho)\right)\left\|v_0(\theta_{-t}\omega)\right\|^2$$
$$+ \left\|\tilde{\psi}_0(\theta_{-t}\omega)\right\|^2\Big) + e^{-C_8(t-T_2)}\int_0^{T_2} e^{C_8(s-T_2)}\left(P_1(\theta_{s-t}\omega) + C_9\right) ds$$
$$\leq e^{-C_8t}\Big(\left\|\tilde{u}_0(\theta_{-t}\omega)\right\|^2 + \rho\left(\mu - \rho(\nu-\rho)\right)\left\|v_0(\theta_{-t}\omega)\right\|^2 + \left\|\tilde{\psi}_0(\theta_{-t}\omega)\right\|^2\Big)$$
$$+ \int_{-t}^{T_2-t} e^{C_8s}P_1(\theta_s\omega)\, ds + C_9e^{C_8(T_2-t)}$$
$$\leq e^{-C_8t}\Big(\left\|\tilde{u}_0(\theta_{-t}\omega)\right\|^2 + \rho\left(\mu - \rho(\nu-\rho)\right)\left\|v_0(\theta_{-t}\omega)\right\|^2 + \left\|\tilde{\psi}_0(\theta_{-t}\omega)\right\|^2\Big)$$
$$+ C_9\int_{-t}^{T_2-t} e^{\frac{C_8}{2}s}r(\omega)\, ds + C_9e^{C_8(T_2-t)}$$
$$\leq e^{-C_8t}\Big(\left\|\tilde{u}_0(\theta_{-t}\omega)\right\|^2 + \rho\left(\mu - \rho(\nu-\rho)\right)\left\|v_0(\theta_{-t}\omega)\right\|^2 + \left\|\tilde{\psi}_0(\theta_{-t}\omega)\right\|^2\Big)$$
$$+ \frac{2C_9}{C_8}e^{\frac{C_8}{2}(T_2-t)}r(\omega) + C_9e^{C_8(T_2-t)}. \tag{3.44}$$

By (3.44) we find that, given $\varepsilon > 0$, there is $T_3 = T_3(\mathbb{B}, \omega, \varepsilon) > T_2$ such that for all $t > T_3$,

$$\begin{aligned} & e^{-C(t-T_2)} \int_{\mathbb{R}^n} \eta\left(\frac{|x|^2}{m^2}\right)\left(\left|\tilde{u}\left(T_2, \theta_{-t}\omega, \tilde{u}_0(\theta_{-t}\omega)\right)\right|^2 + \left|\nabla v\left(T_2, \theta_{-t}\omega, v_0(\theta_{-t}\omega)\right)\right|^2 \right. \\ & \quad \left. + \rho\left(\mu - \rho(v - \rho)\right)\left|v\left(T_2, \theta_{-t}\omega, v_0(\theta_{-t}\omega)\right)\right|^2 + \left|\tilde{\psi}\left(T_2, \theta_{-t}\omega, \tilde{\psi}_0(\theta_{-t}\omega)\right)\right|^2\right) dx \\ & \leq \varepsilon. \end{aligned} \tag{3.45}$$

Note that $\delta \in \mathbb{H}^1 \cap \mathbb{W}^{1,4}$, hence there is $R_1 = R_1(\omega, \varepsilon)$ such that for all $m \geq R_1$,

$$\int_{|x|\geq m} \left(|\nabla \delta(x)|^4 + |\delta(x)|^4\right) dx \leq \frac{\varepsilon}{r(\omega)}, \tag{3.46}$$

where $r(\omega)$ is a tempered function. By (3.46), we have the following estimate:

$$\begin{aligned} & C_9 \int_{T_2}^{t} e^{C_8(s-t)} \int_{\mathbb{R}^n} \eta\left(\frac{|x|^2}{m^2}\right)\left(|z_2(\theta_{s-t}\omega)|^4 + |\nabla z_2(\theta_{s-t}\omega)|^4\right) dx\, ds \\ & \leq C_9 \int_{T_2}^{t} e^{C_8(s-t)} \int_{|x|\geq m} \left(|z_2(\theta_{s-t}\omega)|^4 + |\nabla z_2(\theta_{s-t}\omega)|^4\right) dx\, ds \\ & \leq C_9 \int_{T_2}^{t} e^{C_8(s-t)} \int_{|x|\geq m} \left(|\delta|^4 |y_2(\theta_{s-t}\omega_2)|^4 + |\nabla \delta|^4 |\nabla y_2(\theta_{s-t}\omega_2)|^4\right) dx\, ds \\ & \leq \frac{C_9\varepsilon}{r(\omega)} \int_{T_2}^{t} e^{C_8(s-t)} \sum_{i=1}^{2}\left(|y_i(\theta_{s-t}\omega_i)|^2 + |y_i(\theta_{s-t}\omega_i)|^4\right) ds \\ & \leq \frac{C_9\varepsilon}{r(\omega)} \int_{T_2}^{t} e^{C_8(s-t)} r(\theta_{s-t}\omega)\, ds \leq \frac{C_9\varepsilon}{r(\omega)} \int_{T_2-t}^{0} e^{C_8 s} r(\theta_s\omega)\, ds \\ & \leq \frac{C_9\varepsilon}{r(\omega)} \int_{0}^{T_2-t} e^{\frac{C_8}{2}s} r(\omega)\, ds \leq \varepsilon. \end{aligned} \tag{3.47}$$

By Lemma 3.3, we have

$$\begin{aligned} & C_9 \int_{T_2}^{t} e^{C_8(s-t)} \int_{\mathbb{R}^n} \eta\left(\frac{|x|^2}{m^2}\right) |\tilde{u}|^4\, dx\, ds \\ & \leq C_9 \int_{T_2}^{t} e^{C_8(s-t)} e^{-\alpha s} \int_{|x|\geq m} |\tilde{u}_0|^4\, dx\, ds + \frac{64 C_9}{\alpha^4} \int_{T_2}^{t} e^{C_8(s-t)} \int_{|x|\geq m} |f|^4\, dx\, ds \\ & \leq \frac{C_9}{C_9 + \alpha} e^{-\alpha t} \int_{|x|\geq m} |\tilde{u}_0|^4\, dx\, ds + \frac{64 C_9}{\alpha^4} \int_{T_2}^{t} e^{C_8(s-t)} \int_{|x|\geq m} |f|^4\, dx\, ds. \end{aligned} \tag{3.48}$$

Note that $f \in \mathbb{L}^4$, $g \in \mathbb{L}^2$ and $\delta \in \mathbb{H}^1 \cap \mathbb{W}^{1,4}$, there is $R_2 = R_2(\varepsilon)$ such that for all $m \geq R_2$,

$$\int_{|x|\geq m} \left(|f|^4 + |g|^2 + |\delta|^4 + |\nabla \delta|^2\right) dx \leq \varepsilon. \tag{3.49}$$

Then, by (3.48) and (3.49), there is $T_4 = T_4(\mathbb{B}, \omega, \varepsilon) > 0$ such that for all $t > T_4$,

$$\begin{aligned} & C_9 \int_{T_1}^{t} e^{C_8(s-t)} \int_{\mathbb{R}^n} \eta\left(\frac{|x|^2}{m^2}\right)\left(|f|^4 + |g|^2 + |\delta|^2 + |\nabla \delta|^4 + |\tilde{u}|^4\right) dx\, ds \\ & \leq C_9 \int_{T_1}^{t} e^{C_8(s-t)} \int_{|x|\geq m} \left(|f|^4 + |g|^2 + |\delta|^2 + |\nabla \delta|^4\right) dx\, ds \end{aligned}$$

$$
\begin{aligned}
&+ C_9 e^{-\alpha t} \int_{|x|\geq m} |\tilde{u}_0|^4 \, dx \, ds \\
&\leq C_9 \varepsilon \int_{T_1}^{t} e^{C_8(s-t)} \, ds + C_9 e^{-\alpha t} \int_{|x|\geq m} |\tilde{u}_0|^4 \, dx \, ds \leq \varepsilon. \qquad (3.50)
\end{aligned}
$$

Now, we estimate the last term on the right-hand side of (3.43). Denote $E_3(v, \tilde{\psi}) = \|v\|^2 + \|\nabla v\|^2 + \|\tilde{\psi}\|^2$. Replacing t by s and then replacing ω by $\theta_{-t}\omega$ in (3.18), we have the following estimate:

$$
\begin{aligned}
&\frac{C^*}{m} \int_{T_2}^{t} e^{C_8(s-t)} E_3\big(v(s, \theta_{-t}\omega, v_0(\theta_{-t}\omega)), \tilde{\psi}(s, \theta_{-t}\omega, \tilde{\psi}_0(\theta_{-t}\omega))\big) \, ds \\
&\leq \frac{C^*}{m} \int_{T_2}^{t} e^{-C_8 t} E_1\big(\tilde{u}_0(\theta_{-t}\omega), v_0(\theta_{-t}\omega), \tilde{\psi}_0(\theta_{-t}\omega)\big) \, ds \\
&\quad + \frac{C^*}{m} \int_{T_2}^{t} e^{C_8(s-t)} \int_0^s e^{C_8(\tau-s)} P_1(\theta_{\tau-t}\omega) \, d\tau \, ds + \frac{C^*}{m} \int_{T_2}^{t} e^{C_8(s-t)} \, ds \\
&\leq \frac{C^*}{m} e^{-C_8 t}(t - T_2) E_1\big(\tilde{u}_0(\theta_{-t}\omega), v_0(\theta_{-t}\omega), \tilde{\psi}_0(\theta_{-t}\omega)\big) \\
&\quad + \frac{C^*}{m} \int_{T_2}^{t} \int_0^s e^{C_8(\tau-t)} P_1(\theta_{\tau-t}\omega) \, d\tau \, ds + \frac{C^*}{m} \\
&\leq \frac{C^*}{m} e^{-C_8 t}(t - T_2) E_1\big(\tilde{u}_0(\theta_{-t}\omega), v_0(\theta_{-t}\omega), \tilde{\psi}_0(\theta_{-t}\omega)\big) \\
&\quad + \frac{C^*}{m} \int_{T_2}^{t} \int_{-t}^{s-t} e^{C_8 \tau} P_1(\theta_{\tau}\omega) \, d\tau \, ds + \frac{C^*}{m} \\
&\leq \frac{C^*}{m} e^{-C_8 t}(t - T_2) E_1\big(\tilde{u}_0(\theta_{-t}\omega), v_0(\theta_{-t}\omega), \tilde{\psi}_0(\theta_{-t}\omega)\big) \\
&\quad + \frac{C^*}{m} r(\omega) \int_{T_2}^{t} \int_{-t}^{s-t} e^{\frac{C_8}{2}\tau} \, d\tau \, ds + \frac{C^*}{m} \\
&\leq \frac{C^*}{m} e^{-C_8 t}(t - T_2) E_1\big(\tilde{u}_0(\theta_{-t}\omega), v_0(\theta_{-t}\omega), \tilde{\psi}_0(\theta_{-t}\omega)\big) + \frac{C^*}{m} r(\omega) + \frac{C^*}{m}, \qquad (3.51)
\end{aligned}
$$

which implies that there exist $T_5 = T_5(\mathbb{B}, \omega, \varepsilon) > T_2$ and $R_3 = R_3(\omega, \varepsilon)$ such that for all $t \geq T_5$ and $m \geq R_3$,

$$
\frac{C^*}{m} \int_{T_2}^{t} e^{C_8(s-t)} E_3\big(v(s, \theta_{-t}\omega, v_0(\theta_{-t}\omega)), \tilde{\psi}(s, \theta_{-t}\omega, \tilde{\psi}_0(\theta_{-t}\omega))\big) \, ds \leq \varepsilon. \qquad (3.52)
$$

Obviously, there exists $R_4 = R_4(\omega, \epsilon)$ such that for all $m > R_4$, we have

$$
\frac{4\epsilon\sqrt{2}\eta_0\|\delta\|^2}{m} \int_{T_2}^{t} e^{C_8(s-t)} \left\|z_2(\theta_{-t}\omega)\right\|^2 ds \leq \varepsilon. \qquad (3.53)
$$

By (3.43), (3.45), (3.47), (3.50), (3.52) and (3.53), there exist $R = \max\{R_1, R_2, R_3, R_4\}$ and $T = \min\{T_2, T_3, T_4, T_5\}$ such that for all $m \geq R$ and $t \geq T$,

$$
\begin{aligned}
&\int_{\mathbb{R}^n} \eta\left(\frac{|x|^2}{m^2}\right)\Big(\left|\tilde{u}(t, \theta_{-t}\omega, \tilde{u}_0(\theta_{-t}\omega))\right|^2 + \big(\mu - \rho(v - \rho)\big)\left|v(t, \theta_{-t}\omega, v_0(\theta_{-t}\omega))\right|^2 \\
&\quad + \left|\tilde{\psi}(t, \theta_{-t}\omega, \tilde{\psi}_0(\theta_{-t}\omega))\right|^2\Big) \leq 5\varepsilon,
\end{aligned}
$$

which shows that for all $m \geq R$ and $t \geq T$,

$$\int_{|x|\geq\sqrt{2}m} \left(\left|\tilde{u}\left(t,\theta_{-t}\omega,\tilde{u}_0(\theta_{-t}\omega)\right)\right|^2 + \left|v\left(t,\theta_{-t}\omega,v_0(\theta_{-t}\omega)\right)\right|^2 + \left|\tilde{\psi}\left(t,\theta_{-t}\omega,\tilde{\psi}_0(\theta_{-t}\omega)\right)\right|^2\right) \leq 5\varepsilon.$$

Note that

$$\int_{|x|\geq\sqrt{2}m} \left|z_2(\omega)\right|^2 dx = \int_{|x|\geq\sqrt{2}m} |\delta|^2 \left|y_2(\omega_2)\right|^2 dx \leq \frac{\varepsilon}{r(\omega)} \left|y_2(\omega_2)\right|^2 \leq \varepsilon.$$

Therefore

$$\begin{aligned}
&\int_{|x|\geq\sqrt{2}m} \left|\phi\left(t,\theta_{-t}\omega,(u_0,v_0,\phi_0)(\theta_{-t}\omega)\right)\right|^2 dx \\
&= \int_{|x|\geq\sqrt{2}m} \left(\left|u\left(t,\theta_{-t}\omega,u_0(\theta_{-t}\omega)\right)\right|^2 + \left|v\left(t,\theta_{-t}\omega,v_0(\theta_{-t}\omega)\right)\right|^2 \right.\\
&\quad \left. + \left|\psi\left(t,\theta_{-t}\omega,\psi_0(\theta_{-t}\omega)\right)\right|^2\right) dx \\
&= \int_{|x|\geq\sqrt{2}m} \left(\left|\tilde{u}\left(t,\theta_{-t}\omega,\tilde{u}_0(\theta_{-t}\omega)\right)e^{i\epsilon z_1(\omega)}\right|^2 + \left|v\left(t,\theta_{-t}\omega,v_0(\theta_{-t}\omega)\right)\right|^2 \right.\\
&\quad \left. + \left|\tilde{\psi}\left(t,\theta_{-t}\omega,\tilde{\psi}_0(\theta_{-t}\omega)\right) + \epsilon\delta z_2(\omega)\right|^2\right) \\
&\leq 2\int_{|x|\geq\sqrt{2}m} \left(\left|\tilde{u}\left(t,\theta_{-t}\omega,\tilde{u}_0(\theta_{-t}\omega)\right)\right|^2 + \left|v\left(t,\theta_{-t}\omega,v_0(\theta_{-t}\omega)\right)\right|^2 \right.\\
&\quad \left. + \left|\tilde{\psi}\left(t,\theta_{-t}\omega,\tilde{\psi}_0(\theta_{-t}\omega)\right)\right|^2 + 2\epsilon^2|\delta|^2\left|z_2(\omega)\right|^2\right) \\
&\leq C'\varepsilon,
\end{aligned}$$

where C' is a fixed positive constant. This completes the proof. □

Theorem 3.7 *Let $f \in \mathbb{H}^2$, $g \in \mathbb{L}^2$ and $\delta \in \mathbb{H} \cap \mathbb{W}^{2,4}$. Assume that $\alpha(\nu - \rho) > 12M_1\beta^2$, $\alpha(\mu - \rho(\nu - \rho)) > 8M_1$ and $\mu - \rho(\nu - \rho) > 0$. Then the random dynamical system ϕ_ϵ possesses a unique $\mathbb{D}$-random attractor in $\mathbb{I}$.*

Proof From Proposition 2.8, we only need to prove the asymptotic compactness of RDS ϕ. By Lemma 3.4, for any sequence $t_n \to \infty$ and $x_n(\theta_{-t_n}\omega) \in \mathbb{K}(\omega)$, we have

$$\left\{\phi(t_n,\theta_{-t_n}\omega,x_n)\right\}_{n=1}^{\infty} \text{ is bounded in } \mathbb{I}. \tag{3.54}$$

Hence, by (3.54), there exists $\phi_0 \in L^2(\mathbb{R}^n)$ such that

$$\left\{\phi\left(t_n,\theta_{-t_n}\omega,x_n(\theta_{-t_n}\omega)\right)\right\}_{n=1}^{\infty} \to \phi_0 \text{ weakly in } \mathbb{I}. \tag{3.55}$$

Next, we prove that (3.55) is actually strong convergence. Define the set S by $S = \{x \in \mathbb{R}^n : |x| \leq m\}$, where m will be specified latter. Notice the compactness of embedding $H^1(S), H^2(S) \hookrightarrow L^2(S)$, by Lemma 3.5 it follows that

$$\phi\left(t_n,\theta_{-t_n}\omega,x_n(\theta_{-t_n}\omega)\right) \to \phi_0 \text{ strongly in } \mathbb{I}(S),$$

which shows that for any given $\varepsilon > 0$, there exists $N_1 = N_1(\mathbb{K}(\omega), \omega, \varepsilon)$ such that for all $n \geq N_1$,

$$\left\| \phi\left(t_n, \theta_{-t_n}\omega, x_n(\theta_{-t_n}\omega)\right) - \phi_0 \right\|_{\mathbb{I}(S)}^2 \leq \varepsilon. \tag{3.56}$$

On the other hand, by Lemma 3.6, there exist $T_1 = T_1(\mathbb{K}(\omega), \omega, \varepsilon)$ and $N_2 = N_2(\omega, \varepsilon)$ (large enough) such that $t_n \geq T_1$ for every $n \geq N_2$, we have

$$\int_{|x|>m_1} \left| \phi\left(t_n, \theta_{-t_n}\omega, x_n(\theta_{-t_n}\omega)\right)(x) \right|^2 dx \leq \varepsilon. \tag{3.57}$$

Since $\delta \in \mathbb{H} \cap \mathbb{W}^{2,4}$, there exists $m_2 = m_2(\varepsilon)$ such that

$$\int_{|x|>m_2} \left| \phi_0(x) \right|^2 dx \leq \varepsilon. \tag{3.58}$$

Let $m = \max\{m_1, m_2\}$ and $N = \max\{N_1, N_2\}$. By (3.56)-(3.58) we find that for all $n \geq N$,

$$\begin{aligned}
&\left\| \phi\left(t_n, \theta_{-t_n}\omega, x_n(\theta_{-t_n}\omega)\right)(x) - \phi_0(x) \right\|_{\mathbb{I}}^2 \\
&\quad \leq \int_{|x|\leq m} \left| \phi\left(t_n, \theta_{-t_n}\omega, x_n(\theta_{-t_n}\omega)\right)(x) - \phi_0(x) \right|^2 dx \\
&\qquad + \int_{|x|>m} \left| \phi\left(t_n, \theta_{-t_n}\omega, x_n(\theta_{-t_n}\omega)\right)(x) - \phi_0(x) \right|^2 dx \\
&\quad \leq 3\varepsilon.
\end{aligned}$$

This completes the proof. □

Competing interests
The authors declare that they have no competing interests.

Authors' contributions
All authors contributed equally to the writing of this paper. All authors read and approved the final manuscript.

Author details
[1]College of Information Technology, Jilin Agricultural University, Changchun, 130118, P.R. China. [2]School of Mathematics and Statistics, Shaoguan University, Shaoguan, 512005, P.R. China. [3]Department of Mathematics, Sun Yat-sen (Zhongshan) University, Guangzhou, 510275, P.R. China.

Acknowledgements
The authors sincerely thank Prof. Yong Li for his instructions and many invaluable suggestions. This work is supported financially by the NSFJP Grant (201215184), Guangdong Provincial Culture of Seedling of China (no. 2013LYM0081), the Shaoguan Science and Technology Foundation (no. 313140546).

References
1. Crauel, H, Flandoli, F: Attractors for random dynamical systems. Probab. Theory Relat. Fields **100**(3), 365-393 (1994)
2. Flandoli, F, Schmalfuß, B: Random attractors for the 3D stochastic Navier-Stokes equation with multiplicative noise. Stoch. Int. J. Probab. Stoch. Process. **59**(1), 21-45 (1996)
3. Ruelle, D: Characteristic exponents for a viscous fluid subjected to time dependent forces. Commun. Math. Phys. **93**(3), 285-300 (1984)
4. Schmalfuß, B: Backward cocycles and attractors of stochastic differential equations. In: Reitmann, V, Riedrich, T, Koksch, N (eds.) International Seminar on Applied Mathematics-Nonlinear Dynamics: Attractor Approximation and Global Behaviour, Dresden, Technische Universität, pp. 185-192 (1992)
5. Arnold, L: Random Dynamical Systems. Springer, Berlin (1995)

6. Crauel, H, Debussche, A, Flandoli, F: Random attractors. J. Dyn. Differ. Equ. **9**(2), 307-341 (1997)
7. Fan, X: Random attractor for a damped sine-Gordon equation with white noise. Pac. J. Math. **216**(1), 63-76 (2004)
8. Zhou, S, Yin, F, Ouyang, Z: Random attractor for damped nonlinear wave equations with white noise. SIAM J. Appl. Dyn. Syst. **4**(4), 883-903 (2005)
9. Shen, Z, Zhou, S, Shen, W: One-dimensional random attractor and rotation number of the stochastic damped sine-Gordon equation. J. Differ. Equ. **248**(6), 1432-1457 (2010)
10. Zhao, C, Duan, J: Random attractor for the Ladyzhenskaya model with additive noise. J. Math. Anal. Appl. **362**(1), 241-251 (2010)
11. Bates, PW, Lu, K, Wang, B: Random attractors for stochastic reaction-diffusion equations on unbounded domains. J. Differ. Equ. **246**(2), 845-869 (2009)
12. Wang, B: Random attractors for the stochastic Benjamin-Bona-Mahony equation on unbounded domains. J. Differ. Equ. **246**(6), 2506-2537 (2009)
13. Karachalios, NI, Stavrakakis, NM, Xanthopoulos, P: Parametric exponential energy decay for dissipative electron-ion plasma waves. Z. Angew. Math. Phys. **56**(2), 218-238 (2005)
14. Fukuda, I, Tsutsumi, M: On coupled Klein-Gordon-Schródinger equations. II. J. Math. Anal. Appl. **66**(2), 358-378 (1978)
15. Biler, P: Attractors for the system of Schrödinger and Klein-Gordon equations with Yukawa coupling. SIAM J. Math. Anal. **21**(5), 1190-1212 (1990)
16. Wang, B, Lange, H: Attractors for the Klein-Gordon-Schrödinger equation. J. Math. Phys. **40**(5), 2445-2457 (1999)
17. Guo, BL, Li, YS: Attractor for dissipative Klein-Gordon-Schrödinger equations in $\mathbb{R}^3$. J. Differ. Equ. **136**(2), 356-377 (1997)
18. Lu, K, Wang, B: Global attractors for the Klein-Gordon-Schrödinger equation in unbounded domains. J. Differ. Equ. **170**(2), 281-316 (2001)
19. Yan, WP, Ji, SG, Li, Y: Random attractors for stochastic discrete Klein-Gordon-Schrödinger equations. Phys. Lett. A **373**(14), 1268-1275 (2009)
20. Yan, WP, Li, Y, Ji, SG: Random attractors for first order stochastic retarded lattice dynamical systems. J. Math. Phys. **51**(3), 032702 (2010)
21. Xu, L, Yan, WP: Stochastic Fitzhugh-Nagumo systems with delay. Taiwan. J. Math. **16**(3), 1079-1103 (2012)
22. Flandoli, F, Schmalfuß, B: Random attractors for the 3D stochastic Navier-Stokes equation with multiplicative white noise. Stoch. Int. J. Probab. Stoch. Process. **59**(1-2), 21-45 (1996)
23. Duan, J, Lu, K, Schmalfuß, B: Invariant manifolds for stochastic partial differential equations. Ann. Probab. **31**(4), 2109-2135 (2003)

2

Recursive stochastic linear-quadratic optimal control and nonzero-sum differential game problems with random jumps

Na Li[1,2] and Zhiyong Yu[2*]

*Correspondence:
yuzhiyong@sdu.edu.cn
[2]School of Mathematics, Shandong University, Jinan, 250100, China
Full list of author information is available at the end of the article

Abstract

This paper presents an existence and uniqueness result for a kind of forward-backward stochastic differential equations (FBSDEs for short) driven by Brownian motion and Poisson process under some monotonicity conditions. By virtue of the conclusion of FBSDEs, we solve a linear-quadratic stochastic optimal control problem for forward-backward stochastic systems with random jumps. Moreover, we also solve a linear-quadratic nonzero-sum stochastic differential game problem. We obtain explicit forms of the unique optimal control and the unique Nash equilibrium point, respectively.

MSC: 60H10; 93E20; 91A15

Keywords: forward-backward stochastic differential equation; Poisson process; stochastic optimal control; linear-quadratic problem; nonzero-sum stochastic differential game; Nash equilibrium

1 Introduction

This paper is concerned with a kind of linear-quadratic stochastic optimal control (LQ SOC) problems and linear-quadratic nonzero-sum stochastic differential game (LQ NZSSDG) problems for forward-backward systems with random jumps. In detail, for the LQ NZSSDG problems, the controlled system is given by the following controlled linear forward-backward stochastic differential equation with Poisson process (FBSDE):

$$\begin{cases} dx_t = (A_t x_{t-} + B_t^1 v_t^1 + B_t^2 v_t^2 + \mu_t)\,dt + (C_t x_{t-} + D_t^1 v_t^1 + D_t^2 v_t^2 + \nu_t)\,dW_t \\ \qquad + \int_{\mathcal{E}} (E_t(e) x_{t-} + F_t^1(e) v_t^1 + F_t^2(e) v_t^2 + \psi_t(e)) \widetilde{N}(dt, de), \\ -dy_t = (G_t x_{t-} + H_t y_{t-} + I_t z_t + \int_{\mathcal{E}} J_t(e) k_t(e) \pi(de) + K_t^1 v_t^1 + K_t^2 v_t^2 + \phi_t)\,dt \\ \qquad - z_t\,dW_t - \int_{\mathcal{E}} k_t(e) \widetilde{N}(dt, de), \\ x_0 = a, \qquad y_T = \Upsilon x_T + \eta. \end{cases} \tag{1}$$

For simplicity of notation, here we only consider the case of two players. The corresponding conclusions for the case of n players can be obtained in the same way. In addition, we are given two cost (or utility, performance criterion) functionals for each player:

$$\mathcal{J}^i(v^1(\cdot),v^2(\cdot)) = \frac{1}{2}\mathbb{E}\int_0^T \Big[\langle L_t^i x_t, x_t\rangle + \langle M_t^i y_t, y_t\rangle + \langle N_t^i z_t, z_t\rangle + \int_{\mathcal{E}} \langle O_t^i(e)k_t(e), k_t(e)\rangle \pi(de)$$
$$+ \langle \Gamma_t^i v_t^i, v_t^i\rangle\Big] dt + \frac{1}{2}\mathbb{E}\langle R^i x_T, x_T\rangle + \frac{1}{2}\langle S^i y_0, y_0\rangle \quad (i = 1,2). \tag{2}$$

Under suitable conditions, the controlled system (1) and the cost functionals (2) are well defined. Our final objective is to find the unique Nash equilibrium point explicitly. It is well known that a stochastic optimal control problem can be regarded as a game problem with only one player. From this point of view, the LQ SOC problem is a special case of the LQ NZSSDG problem. We also study its unique optimal control.

We notice that in the controlled system (1), besides a forward stochastic differential equation (SDE), there is another backward stochastic differential equation (BSDE). The linear BSDE was introduced by Bismut [1] to solve some stochastic optimal control problems, and the general nonlinear BSDEs were introduced by Pardoux and Peng [2] and Duffie and Epstein [3] independently. It worth noting that in [3] and the subsequent paper [4] by El Karoui *et al.*, BSDEs were used to characterize a kind of stochastic differential recursive utility, which is an extension of the standard additive utility with the instantaneous utility depending not only on the instantaneous consumption rate but also on the future utility. From this viewpoint, the LQ SOC problem and LQ NZSSDG problem for the forward-backward system can be regarded as an extension of the recursive stochastic optimal control problem and the recursive differential game problem. On the other side, in the classical stochastic optimal control and differential game theory, the performance criterion is described by the linear expectation $\mathbb{E}$. However, sometimes the classical linear expectation $\mathbb{E}$ does not quite represent people's preferences (see Allais [5], Ellsberg [6]). A remedial way is to use the so-called *generalized expectation* introduced by Peng [7, 8] instead of the linear expectation. From the theory of the generalized expectation, it is also defined by some BSDEs. For research of stochastic optimization problems with generalized expectation, we refer to Yong [9] and Hui and Xiao [10]. Therefore, the LQ SOC problem and LQ NZSSDG problem studied in this paper can also be regarded as an extension of stochastic optimization problems with generalized expectations.

Recently, researchers paid much attention to the so-called *principal-agent problem* due to its wide applications in finance. Williams [11] constructed a model of dynamic principal-agent problems in continuous time. The actions of the principal and the agent affect the same controlled system, which is described by a forward SDE. By virtue of the classical stochastic maximum principle in the optimal control theory, the optimal control of the agent is given by a coupled FBSDE (called the stochastic Hamiltonian system). Then the principal's problem is formulated as a stochastic optimization problem for a coupled forward-backward system. The LQ SOC problem and the LQ NZSSDG problem (with forward-backward systems in the decoupled form) studied in this paper can be regarded as a simple case of those complicated problems.

We also notice that the controlled system (1) is in a linear form, and the cost functionals (2) are in quadratic forms. Such a kind of optimization problem is called a linear-quadratic (LQ) problem. The LQ problem is an important class of stochastic optimization problems, which provides a basic knowledge for more general problems since lots of nonlinear problems can be approximated by LQ problem reasonably. The LQ problem for forward system was firstly studied by Wonham [12]. Then Bismut [13] proved an existence result for optimal controls. By virtue of a square complete technique, Chen *et al.* [14] studied stochastic LQ problems with indefinite control weight costs. They gave the optimal feedback control by the solution of a stochastic Riccati equation. For the forward controlled system, Hamadène [15] linked an LQ nonzero-sum stochastic differential game problem to a linear coupled FBSDE by virtue of changing variables, and he also gave an existence result for Nash equilibrium points of the game problem. Wu [16] and Wu and Yu [17] improved and extended the result of [15]. Recently, Yu [18] considered the LQ problem for the forward-backward system, which is linked closely to a new type of FBSDEs: the linear coupled FBSDEs with double dimensions. The solution of this kind of FBSDEs played an important role in the construction of optimal controls and Nash equilibrium points.

In this paper, we adopt the model driven by both Brownian motion and Poisson process (see (1) and (2)), since jump diffusion processes characterize stochastic phenomena better than just Brownian motion diffusion processes, providing us more realistic models in practice. For example, in finance stock prices often exhibit some jump behavior. Moreover, financial markets with jump stock prices provide a rich family of incomplete financial models. For more information as regards jump diffusion models, interested readers may refer to Cont and Tankov [19] and Øksendal and Sulem [20]. We also refer to Tang and Li [21], Situ [22], and Wu and Wang [23] for some research on BSDEs driven by both Brownian motions and Poisson processes.

Inspired by the idea of stochastic maximum principle in optimal control theory and Hamadène's transform (see [15]), both the LQ SOC problem and the LQ NZSSDG problem are closed linked to a kind of coupled FBSDEs involving Poisson jumps, which is out of scope of the existing results in the FBSDEs theory. Historically, fully coupled FBSDEs were firstly studied by Antonelli [24]. He obtained a local solvability result by the fixed point theorem. For the global existence and uniqueness results, there exist two main methods. One concerns a kind of *four step scheme* approach introduced by Ma *et al.* [25], which can be regarded as a method combining partial differential equations and probability theory. In this method, the forward diffusion is required to be non-degenerate and the coefficients are required to be deterministic. The other one, named the *method of continuation*, is probabilistic; it was introduced originally by Hu and Peng [26], and then developed by Peng and Wu [27] and Yong [28, 29]. In this method, the conditions in [25] are relaxed, but some monotonicity assumption is proposed. However, the monotonicity assumption is naturally satisfied by many coupled FBSDEs arising from stochastic optimal control problems. Recently, Ma *et al.* [30] proposed a *unified approach*, which can be regarded as a combination of existing methods.

In the present paper, we use the method of continuation to obtain an existence and uniqueness result for a kind of FBSDEs with Poisson jumps under some monotonicity conditions. The solvability of FBSDEs is important for the following optimization problems and is also interesting in its own right. Then we use the solution of FBSDE to construct the unique optimal control for the LQ SOC problem. We develop Hamadène's transform to study the LQ NZSSDG problem. Under suitable conditions, we prove an existence and uniqueness result for the Nash equilibrium point. Moreover, we give an explicit form for it. Our method used in this paper is effective in studying the uniqueness of Nash equilibrium points, which is not considered in [15–17]. It is also worthy to point out that for the existence of Nash equilibrium point our conclusion is more general than the corresponding results in [15–18].

The rest of this paper is organized as follows. In Section 2, we give some notations. Section 3 is devoted to the proof of an existence and uniqueness theorem for adapted solutions of a new kind of FBSDEs driven by Brownian motion and a Poisson process. In Section 4, we give an explicit form of the unique optimal control for LQ SOC problem. In the last section we study the LQ NZSSDG problem and construct the unique Nash equilibrium point explicitly.

2 Notations

Let $\mathbb{R}^n$ be the n-dimensional Euclidean space with the usual norm $\|\cdot\|$ and inner product $\langle\cdot,\cdot\rangle$. Let $\mathbb{R}^{m\times n}$ be the collection of $m\times n$ matrices with the inner product:

$$\langle A,B\rangle=\mathrm{tr}\{AB^\top\},\quad \text{for any } A,B\in\mathbb{R}^{m\times n},$$

where $\top$ denotes the transpose of matrices.

Let $T>0$ is a constant and $[0,T]$ denote the finite time span. Let $(\Omega,\mathcal{F},\mathbb{F},\mathbb{P})$ be a complete filtered probability space. The filtration $\mathbb{F}=\{\mathcal{F}_t;0\le t\le T\}$ is generated by two mutually independent stochastic processes. One is a 1-dimensional Brownian motion W, and the other one is a Poisson random measure N defined on $\mathbb{R}_+\times\mathcal{E}$, where $\mathcal{E}=\mathbb{R}^d-\{0\}$ is a nonempty Borel subset of some Euclidean space. The compensator of N is $\overline{N}(dt,de)=\pi(de)\,dt$, which makes $\{\widetilde{N}((0,t]\times A)=(N-\overline{N})((0,t]\times A)\}_{t\ge0}$ a martingale for any A belonging to the Borel field $\mathcal{B}(\mathcal{E})$ with $\pi(A)<\infty$. Here π is a given σ-finite measure on the measurable space $(\mathcal{E},\mathcal{B}(\mathcal{E}))$ satisfying $\int_{\mathcal{E}}(1\wedge|e|^2)\pi(de)<\infty$. Then $\mathcal{F}_t$ is defined by

$$\sigma\{W_s:0\le s\le t\}\vee\sigma\left\{\int_{\mathcal{E}}\widetilde{N}(s,de):0\le s\le t\right\}\vee\mathcal{N},\quad 0\le t\le T,$$

where $\mathcal{N}$ denotes the totality of $\mathbb{P}$-null sets, and $\mathcal{F}=\mathcal{F}_T$. We note that in this paper we assume the dimension of Brownian motion $d_1=1$ and the number of Poisson random measure $d_2=1$ just for the simplicity of notations. Actually, all of the following conclusions still hold true for the case of $d_1>1$ and/or $d_2>1$.

For any Euclidean space $\mathbb{R}^m$, we introduce the following notations:

- $L^2(\mathcal{F}_T;\mathbb{R}^m)=\{\varsigma|\varsigma$ is an $\mathbb{R}^m$-valued $\mathcal{F}_T$-measurable random variable such that $\mathbb{E}[|\varsigma|^2]<\infty\}$,
- $L^2_{\mathbb{F}}(0,T;\mathbb{R}^m)=\{g:[0,T]\times\Omega\to\mathbb{R}^m|g(\cdot)$ is an $\mathbb{R}^m$-valued $\mathbb{F}$-adapted stochastic process such that $\|g\|^2=\mathbb{E}\int_0^T|g(t)|^2\,dt<\infty\}$,
- $C^2_{\mathbb{F}}(0,T;\mathbb{R}^m)=\{g:[0,T]\times\Omega\to\mathbb{R}^m|g(\cdot)$ is a càdlàg process in $L^2_{\mathbb{F}}(0,T;\mathbb{R}^m)$ such that $\|g\|^2=\mathbb{E}[\sup_{0\le t\le T}|g(t)|^2]<\infty\}$,
- $M^2_{\mathbb{F}}(0,T;\mathbb{R}^m)=\{r:[0,T]\times\mathcal{E}\times\Omega\to\mathbb{R}^m|\,r(\cdot,\cdot)$ is an $\mathbb{R}^m$-valued $\mathbb{F}$-adapted process such that $\|r\|^2=\mathbb{E}\int_0^T\int_{\mathcal{E}}|r(t,e)|^2\pi(de)\,dt<\infty\}$,

where $\|\cdot\|$ is the norm in each subspace.

Moreover, we define the space

$$\mathcal{R}=\mathbb{R}^n\times\mathbb{R}^m\times\mathbb{R}^m\times\mathbb{R}^m\times\mathbb{R}^m\times\mathbb{R}^n\times\mathbb{R}^n\times\mathbb{R}^n,$$

and the inner product in $\mathcal{R}$ is

$$\left\langle(\lambda,\xi),(\bar{\lambda},\bar{\xi})\right\rangle=\langle x,\bar{x}\rangle+\langle y,\bar{y}\rangle+\langle z,\bar{z}\rangle+\langle k,\bar{k}\rangle+\langle p,\bar{p}\rangle+\langle q,\bar{q}\rangle+\langle r,\bar{r}\rangle+\langle\theta,\bar{\theta}\rangle,$$

for any $\lambda = (x, y, z, k)$, $\xi = (p, q, r, \theta)$, $\bar{\lambda} = (\bar{x}, \bar{y}, \bar{z}, \bar{k})$ and $\bar{\xi} = (\bar{p}, \bar{q}, \bar{r}, \bar{\theta})$. Then the norm in $\mathcal{R}$ is deduced by $\|(\lambda, \xi)\| = \sqrt{\langle(\lambda, \xi), (\lambda, \xi)\rangle}$. We also define

$$\begin{aligned}\mathcal{L}[0,T] = {} & C^2_{\mathbb{F}}(0,T;\mathbb{R}^n) \times C^2_{\mathbb{F}}(0,T;\mathbb{R}^m) \times L^2_{\mathbb{F}}(0,T;\mathbb{R}^m) \times M^2_{\mathbb{F}}(0,T;\mathbb{R}^m) \\ & \times C^2_{\mathbb{F}}(0,T;\mathbb{R}^m) \times C^2_{\mathbb{F}}(0,T;\mathbb{R}^n) \times L^2_{\mathbb{F}}(0,T;\mathbb{R}^n) \times M^2_{\mathbb{F}}(0,T;\mathbb{R}^n)\end{aligned}$$

with the norm

$$\begin{aligned}\|(\lambda(\cdot), \xi(\cdot))\| = {} & \left\{\mathbb{E}\Big(\sup_{t\in[0,T]} |x_t|^2 + \sup_{t\in[0,T]} |y_t|^2 + \sup_{t\in[0,T]} |p_t|^2 + \sup_{t\in[0,T]} |q_t|^2\Big)\right. \\ & \left. + \mathbb{E}\int_0^T \big(|z_t|^2 + |r_t|^2\big)\,dt + \mathbb{E}\int_0^T \int_{\mathcal{E}} \big(|k_t(e)|^2 + |\theta_t(e)|^2\big)\pi(de)\,dt\right\}^{\frac{1}{2}}\end{aligned}$$

for any $(\lambda(\cdot), \xi(\cdot)) = (x(\cdot), y(\cdot), z(\cdot), k(\cdot,\cdot), p(\cdot), q(\cdot), r(\cdot), \theta(\cdot,\cdot)) \in \mathcal{L}[0,T]$ and

$$\begin{aligned}\mathcal{G}[0,T] = {} & L^2_{\mathbb{F}}(0,T;\mathbb{R}^n) \times L^2_{\mathbb{F}}(0,T;\mathbb{R}^n) \times M^2_{\mathbb{F}}(0,T;\mathbb{R}^n) \times L^2_{\mathbb{F}}(0,T;\mathbb{R}^m) \\ & \times L^2(\mathcal{F}_T;\mathbb{R}^m) \times L^2_{\mathbb{F}}(0,T;\mathbb{R}^m) \times L^2_{\mathbb{F}}(0,T;\mathbb{R}^m) \times M^2_{\mathbb{F}}(0,T;\mathbb{R}^m) \\ & \times L^2_{\mathbb{F}}(0,T;\mathbb{R}^n) \times L^2(\mathcal{F}_T;\mathbb{R}^n) \times \mathbb{R}^m.\end{aligned}$$

The element of $\mathcal{G}[0,T]$ is often denoted by

$$\kappa = \big(\mu_1(\cdot), \nu_1(\cdot), \psi_1(\cdot,\cdot), \varphi_1(\cdot), \eta_1, \mu_2(\cdot), \nu_2(\cdot), \psi_2(\cdot,\cdot), \varphi_2(\cdot), \eta_2, b\big).$$

3 Solvability of forward-backward stochastic differential equation with Poisson jump

In this section, we consider the solvability of a new kind of forward-backward stochastic differential equation (FBSDE) with a Poisson jump, which is obtained by adjoint equations deduced from the stochastic maximum principle. The solution of FBSDE plays a key role to solve the LQ stochastic optimal control problems in the next section and the LQ nonzero-sum stochastic differential game problems in Section 5.

We consider the following FBSDE:

$$\begin{cases} dx_t = \alpha_1(t, x_{t-}, B_t p_{t-}, C_t q_{t-}, D_t r_t, \int_{\mathcal{E}} E_t(e)\theta_t(e)\pi(de))\,dt \\ \qquad + \beta_1(t, x_{t-}, B_t p_{t-}, C_t q_{t-}, D_t r_t, \int_{\mathcal{E}} E_t(e)\theta_t(e)\pi(de))\,dW_t \\ \qquad + \int_{\mathcal{E}} \gamma_1(t, e, x_{t-}, B_t p_{t-}, C_t q_{t-}, D_t r_t, \int_{\mathcal{E}} E_t(e)\theta_t(e)\pi(de))\widetilde{N}(dt, de), \\ -dy_t = f_1(t, x_{t-}, y_{t-}, z_t, k_t, B_t p_{t-}, C_t q_{t-}, D_t r_t, \int_{\mathcal{E}} E_t(e)\theta_t(e)\pi(de))\,dt - z_t\,dW_t \\ \qquad - \int_{\mathcal{E}} k_t(e)\widetilde{N}(dt, de), \\ dp_t = \alpha_2(t, x_{t-}, y_{t-}, z_t, k_t, p_{t-}, q_{t-}, r_t, \theta_t)\,dt + \beta_2(t, x_{t-}, y_{t-}, z_t, k_t, p_{t-}, q_{t-}, r_t, \theta_t)\,dW_t \\ \qquad + \int_{\mathcal{E}} \gamma_2(t, e, x_{t-}, y_{t-}, z_t, k_t, p_{t-}, q_{t-}, r_t, \theta_t)\widetilde{N}(dt, de), \\ -dq_t = f_2(t, x_{t-}, y_{t-}, z_t, k_t, p_{t-}, q_{t-}, r_t, \theta_t)\,dt - r_t\,dW_t - \int_{\mathcal{E}} \theta_t(e)\widetilde{N}(dt, de), \\ x_0 = a, \qquad q_T = \Phi_2(x_T, p_T), \\ p_0 = \Psi(y_0), \qquad y_T = \Phi_1(x_T), \end{cases} \tag{3}$$

where $(x, y, z, k, p, q, r, \theta)$ takes values in $\mathcal{R}$, B, C, D, E are $\mathbb{F}$-adapted matrix-valued bounded processes with appropriate dimensions, for each fixed $(x, y, z, k, p, q, r, \theta)$, α_1, β_1,

$\gamma_1, f_1, \alpha_2, \beta_2, \gamma_2, f_2, \Phi_1, \Phi_2, \Psi$ are $\mathbb{F}$-adapted mappings with appropriate dimensions, and $a \in \mathbb{R}^n$. For convenience, we use the following notations:

$$\begin{aligned} &\lambda = (x, y, z, k), \qquad \xi = (p, q, r, \theta), \\ &\xi_t^M = \left(B_t p, C_t q, D_t r, \int_{\mathcal{E}} E_t(e)\theta(e)\pi(de) \right), \\ &\xi_t^\Sigma = B_t p + C_t q + D_t r + \int_{\mathcal{E}} E_t(e)\theta(e)\pi(de). \end{aligned} \tag{4}$$

Definition 3.1 A process $(x(\cdot), y(\cdot), z(\cdot), k(\cdot,\cdot), p(\cdot), q(\cdot), r(\cdot), \theta(\cdot,\cdot)) \in \mathcal{L}[0,T]$ is called an adapted solution of FBSDE (3) if it satisfies (3). If there exists a unique adapted solution, then (3) is said to be uniquely solvable.

Similar to Hu and Peng [26] and Peng and Wu [27], for each $t \in [0,T]$, $(\lambda, \xi) \in \mathcal{R}$, we introduce the following notation:

$$\begin{aligned} A(t,\lambda,\xi) = \big(&-f_2(t,\lambda,\xi), \alpha_2(t,\lambda,\xi), \beta_2(t,\lambda,\xi), \gamma_2(t,\lambda,\xi), \\ &-f_1\left(t,\lambda,\xi_t^M\right), \alpha_1\left(t,x,\xi_t^M\right), \beta_1\left(t,x,\xi_t^M\right), \gamma_1\left(t,x,\xi_t^M\right)\big). \end{aligned}$$

Now we give assumptions on the coefficients of (3).

Assumption 1 (Lipschitz condition)

(i) For any $(\lambda,\xi) \in \mathcal{R}$, $A(\cdot,\lambda,\xi) \in L^2_{\mathbb{F}}(0,T;\mathcal{R})$. For any $x \in \mathbb{R}^n$, $\Phi_1(x) \in L^2(\mathcal{F}_T;\mathbb{R}^m)$. For any $(x,p) \in \mathbb{R}^n \times \mathbb{R}^m$, $\Phi_2(x,p) \in L^2(\mathcal{F}_T;\mathbb{R}^n)$. For any $y \in \mathbb{R}^m$, $\Psi(y) \in \mathbb{R}^m$.

(ii) The mappings A, Φ_1, Φ_2, and Ψ are uniformly Lipschitz continuous with respect to (λ,ξ), x, (x,p), and y, respectively.

Assumption 2 (Domination condition) There exists a constant $L > 0$ such that, for any $\lambda = (x,y,z,k)$, any $\xi = (p,q,r,\theta)$ and any $\bar{\xi} = (\bar{p},\bar{q},\bar{r},\bar{\theta})$,

$$\left|g_1\left(t,x,\xi_t^M\right) - g_1\left(t,x,\bar{\xi}_t^M\right)\right| \leq L\left|B_t\hat{p} + C_t\hat{q} + D_t\hat{r} + \int_{\mathcal{E}} E_t(e)\hat{\theta}(e)\pi(de)\right|, \quad \text{a.s. a.e.} \tag{5}$$

and

$$\left|f_1\left(t,\lambda,\xi_t^M\right) - f_1\left(t,\lambda,\bar{\xi}_t^M\right)\right| \leq L\left|B_t\hat{p} + C_t\hat{q} + D_t\hat{r} + \int_{\mathcal{E}} E_t(e)\hat{\theta}(e)\pi(de)\right|, \quad \text{a.s. a.e.}, \tag{6}$$

where ξ_t^M and $\bar{\xi}_t^M$ are defined by (4), $\hat{\xi} = (\hat{p},\hat{q},\hat{r},\hat{\theta}) = (p-\bar{p}, q-\bar{q}, r-\bar{r}, \theta-\bar{\theta})$, and $g_1 = \alpha_1, \beta_1, \gamma_1$.

Assumption 3 (Monotonicity condition)

(i) There exists a constant $l > 0$ such that, for any $\lambda = (x,y,z,k)$, any $\bar{\lambda} = (\bar{x},\bar{y},\bar{z},\bar{k})$, any $\xi = (p,q,r,\theta)$, and any $\bar{\xi} = (\bar{p},\bar{q},\bar{r},\bar{\theta})$,

$$\begin{aligned} &\left\langle A(t,\lambda,\xi) - A(t,\bar{\lambda},\bar{\xi}), (\hat{\lambda},\hat{\xi}) \right\rangle \\ &\quad \leq -l\left|B_t\hat{p} + C_t\hat{q} + D_t\hat{r} + \int_{\mathcal{E}} E_t(e)\hat{\theta}(e)\pi(de)\right|^2, \quad \text{a.s. a.e.,} \end{aligned}$$

where $(\hat{\lambda},\hat{\xi}) = (\hat{x},\hat{y},\hat{z},\hat{k},\hat{p},\hat{q},\hat{r},\hat{\theta}) = (x-\bar{x}, y-\bar{y}, z-\bar{z}, k-\bar{k}, p-\bar{p}, q-\bar{q}, r-\bar{r}, \theta-\bar{\theta})$.

(ii) For any $x, \bar{x} \in \mathbb{R}^n$ and any $p, \bar{p} \in \mathbb{R}^m$,

$$\left\langle \Phi_1(x) - \Phi_1(\bar{x}), \hat{p} \right\rangle + \left\langle \Phi_2(x,p) - \Phi_2(\bar{x},\bar{p}), \hat{x} \right\rangle \geq 0, \quad \text{a.s.}$$

(iii) For any $y, \bar{y} \in \mathbb{R}^m$,

$$\left\langle \Psi(y) - \Psi(\bar{y}), \hat{y} \right\rangle \leq 0.$$

Now we are in the position to give the main result of this section.

Theorem 3.2 *Under Assumptions* 1, 2, *and* 3, *FBSDE* (3) *admits a unique solution,*

$$\big(\lambda(\cdot), \xi(\cdot)\big) = \big(x(\cdot), y(\cdot), z(\cdot), k(\cdot,\cdot), p(\cdot), q(\cdot), r(\cdot), \theta(\cdot,\cdot)\big) \in \mathcal{L}[0,T].$$

We shall employ the 'method of continuation' introduced by [26, 27] to prove Theorem 3.2. To this aim, we consider a family of FBSDEs parameterized by $\rho \in [0,1]$ as follows:

$$\begin{cases} dx_t^\rho = [-(1-\rho)lC_t^\top \xi_{t-}^{\rho\Sigma} + \rho\alpha_1(t, x_{t-}^\rho, \xi_{t-}^{\rho M}) + \mu_1(t)]\, dt \\ \qquad + [-(1-\rho)lD_t^\top \xi_{t-}^{\rho\Sigma} + \rho\beta_1(t, x_{t-}^\rho, \xi_{t-}^{\rho M}) + \nu_1(t)]\, dW_t \\ \qquad + \int_{\mathcal{E}} [-(1-\rho)lE_t^\top(e) \xi_{t-}^{\rho\Sigma} + \rho\gamma_1(t, e, x_{t-}^\rho, \xi_{t-}^{\rho M}) + \psi_1(t,e)] \widetilde{N}(dt, de), \\ -dy_t^\rho = [-(1-\rho)lB_t^\top \xi_{t-}^{\rho\Sigma} + \rho f_1(t, \lambda_{t-}^\rho, \xi_{t-}^{\rho M}) + \varphi_1(t)]\, dt - z_t^\rho\, dW_t \\ \qquad - \int_{\mathcal{E}} k_t^\rho(e) \widetilde{N}(dt, de), \\ dp_t^\rho = [\rho\alpha_2(t, \lambda_{t-}^\rho, \xi_{t-}^\rho) + \mu_2(t)]\, dt + [\rho\beta_2(t, \lambda_{t-}^\rho, \xi_{t-}^\rho) + \nu_2(t)]\, dW_t \\ \qquad + [\int_{\mathcal{E}} \rho\gamma_2(t, e, \lambda_{t-}^\rho, \xi_{t-}^\rho) + \psi_2(t,e)] \widetilde{N}(dt, de), \\ -dq_t^\rho = [\rho f_2(t, \lambda_{t-}^\rho, \xi_{t-}^\rho) + \varphi_2(t)]\, dt - r_t^\rho\, dW_t - \int_{\mathcal{E}} \theta_t^\rho(e) \widetilde{N}(dt, de), \\ x_0^\rho = a, \qquad q_T^\rho = \rho\Phi_2(x_T^\rho, q_T^\rho) + \eta_2, \\ p_0^\rho = \rho\Psi(y_0^\rho) + b, \qquad y_T^\rho = \rho\Phi_1(x_T^\rho) + \eta_1. \end{cases} \tag{7}$$

Clearly, when $\rho = 0$ the FBSDE (7) is in a decoupled form. From the classical theory of stochastic differential equations (SDEs) and backward stochastic differential equations (BSDEs), it is uniquely solvable. When $\rho = 1$ and $\kappa = (\mu_1(\cdot), \nu_1(\cdot), \psi_1(\cdot), \varphi_1(\cdot), \eta_1, \mu_2(\cdot), \nu_2(\cdot), \psi_2(\cdot), \varphi_2(\cdot), \eta_2, b)$ vanishes, FBSDE (7) coincides with FBSDE (3). The following lemma, which provides a path from the solvable case $\rho = 0$ to the desired unsolvable case $\rho = 1$, plays a key role.

Lemma 3.3 *Let Assumptions* 1, 2, *and* 3 *hold true. There exists a positive constant* δ_0 *such that, if for some* $\rho_0 \in [0,1)$, *there exists a unique solution* $(\lambda_t^{\rho_0}, \xi_t^{\rho_0})$ *of FBSDE* (7) *with any* $\kappa \in \mathcal{G}[0,T]$, *then there exists a unique solution* $(\lambda_t^{\rho_0+\delta}, \xi_t^{\rho_0+\delta})$ *of FBSDE* (7) *for* $\rho = \rho_0 + \delta$ *with* $\delta \in [0, \delta_0]$ *and* $\rho_0 + \delta \leq 1$.

Proof Let $\delta_0 > 0$ be undetermined, and $\delta \in [0, \delta_0]$. We consider the following FBSDE:

$$\begin{cases} dX_t = [-(1-\rho_0)lC_t^\top \Xi_{t-}^\Sigma + \rho_0\alpha_1(t,X_{t-},\Xi_{t-}^M) + \delta(lC_t^\top\xi_{t-}^\Sigma + \alpha_1(t,x_{t-},\xi_{t-}^M)) + \mu_1(t)]\,dt \\ \qquad + [-(1-\rho_0)lD_t^\top \Xi_{t-}^\Sigma + \rho_0\beta_1(t,X_{t-},\Xi_{t-}^M) + \delta(lD_t^\top\xi_{t-}^\Sigma + \beta_1(t,x_{t-},\xi_{t-}^M)) + \nu_1(t)]\,dW_t \\ \qquad + \int_{\mathcal{E}}[-(1-\rho_0)lE_t^\top(e)\Xi_{t-}^\Sigma + \rho_0\gamma_1(t,e,X_{t-},\Xi_{t-}^M) + \delta(lE_t^\top(e)\xi_{t-}^\Sigma + \gamma_1(t,e,x_{t-},\xi_{t-}^M)) \\ \qquad + \psi_1(t,e)]\widetilde{N}(dt,de), \\ -dY_t = [-(1-\rho_0)lB_t^\top \Xi_{t-}^\Sigma + \rho_0 f_1(t,\Lambda_t,\Xi_{t-}^M) + \delta(lB_t^\top\xi_{t-}^\Sigma + f_1(t,\lambda_t,\xi_{t-}^M)) + \varphi_1(t)]\,dt \\ \qquad - Z_t\,dW_t - \int_{\mathcal{E}} K_t(e)\widetilde{N}(dt,de), \\ dP_t = [\rho_0\alpha_2(t,\Lambda_{t-},\Xi_{t-}) + \delta\alpha_2(t,\lambda_{t-},\xi_{t-}) + \mu_2(t)]\,dt \\ \qquad + [\rho_0\beta_2(t,\Lambda_{t-},\Xi_{t-}) + \delta\beta_2(t,\lambda_{t-},\xi_{t-}) + \nu_2(t)]\,dW_t \\ \qquad + [\int_{\mathcal{E}}\rho_0\gamma_2(t,e,\Lambda_{t-},\Xi_{t-}) + \delta\gamma_2(t,e,\lambda_{t-},\xi_{t-}) + \psi_2(t,e)]\widetilde{N}(dt,de), \\ -dQ_t = [\rho_0 f_2(t,\Lambda_{t-},\Xi_{t-}) + \delta f_2(t,\lambda_{t-},\xi_{t-}) + \varphi_2(t)]\,dt - R_t\,dW_t - \int_{\mathcal{E}}\Theta_t(e)\widetilde{N}(dt,de), \\ X_0 = a, \qquad Q_T = \rho_0\Phi_2(X_T,P_T) + \delta\Phi_2(x_T,p_T) + \eta_2, \\ P_0 = \rho_0\Psi(Y_0) + \delta\Psi(y_0) + b, \qquad Y_T = \rho_0\Phi_1(X_T^\rho) + \delta\Phi_1(x_T^\rho) + \eta_1, \end{cases} \tag{8}$$

where, similar to (4), we denote

$$\Lambda = (X,Y,Z,K), \qquad \Xi = (P,Q,R,\Theta),$$

$$\Xi_t^M = \left(B_tP, C_tQ, D_tR, \int_{\mathcal{E}} E_t(e)\Theta(e)\pi(de)\right),$$

$$\Xi_t^\Sigma = B_tP + C_tQ + D_tR + \int_{\mathcal{E}} E_t(e)\theta(e)\pi(de).$$

Our assumption says, when $\rho = \rho_0$, for any $\kappa \in \mathcal{G}[0,T]$, FBSDE (7) admits a unique solution. Applying it to FBSDE (8), we have the result: for each $(\lambda(\cdot),\xi(\cdot)) = (x(\cdot),y(\cdot),z(\cdot),k(\cdot,\cdot),$ $p(\cdot),q(\cdot),r(\cdot),\theta(\cdot,\cdot)) \in \mathcal{L}[0,T]$, FBSDE (8) admits a unique solution $(\Lambda(\cdot),\Xi(\cdot)) = (X(\cdot),Y(\cdot),$ $Z(\cdot),K(\cdot,\cdot),P(\cdot),Q(\cdot),R(\cdot),\Theta(\cdot,\cdot)) \in \mathcal{L}[0,T]$. Consequently, this result implies a mapping:

$$(\Lambda,\Xi) = I_{\rho_0+\delta}(\lambda,\xi) : \mathcal{L}[0,T] \to \mathcal{L}[0,T]. \tag{9}$$

Next, we shall prove the above mapping is a contraction.

Let $(\bar{\lambda},\bar{\xi}) = (\bar{x},\bar{y},\bar{z},\bar{k},\bar{p},\bar{k},\bar{q},\bar{\theta}) \in \mathcal{L}[0,T]$ and $(\bar{\Lambda},\overline{\Xi}) = (\bar{X},\bar{Y},\bar{Z},\bar{K},\bar{P},\bar{Q},\bar{R},\bar{\Theta}) = I_{\rho_0+\delta}(\bar{\lambda},\bar{\xi})$. We set

$$\begin{aligned}(\hat{\lambda},\hat{\xi}) &= (\lambda-\bar{\lambda},\xi-\bar{\xi}) = (\hat{x},\hat{y},\hat{z},\hat{k},\hat{p},\hat{q},\hat{r},\hat{\theta}) \\ &= (x-\bar{x},y-\bar{y},z-\bar{z},k-\bar{k},p-\bar{p},q-\bar{q},r-\bar{r},\theta-\bar{\theta})\end{aligned}$$

and

$$\begin{aligned}(\hat{\Lambda},\hat{\Xi}) &= (\Lambda-\bar{\Lambda},\Xi-\overline{\Xi}) = (\hat{X},\hat{Y},\hat{Z},\hat{K},\hat{P},\hat{Q},\hat{R},\hat{\Theta}) \\ &= (X-\bar{X},Y-\bar{Y},Z-\bar{Z},K-\bar{K},P-\bar{P},Q-\bar{Q},R-\bar{R},\Theta-\bar{\Theta}).\end{aligned}$$

Applying Itô's formula to $\langle\hat{X}_t,\hat{Q}_t\rangle + \langle\hat{Y}_t,\hat{P}_t\rangle$ yields

$$\begin{aligned}&\rho_0\mathbb{E}\{\langle\Phi_2(X_T,P_T)-\Phi_2(\bar{X}_T,\bar{P}_T),\hat{X}_T\rangle + \langle\Phi_1(X_T)-\Phi_1(\bar{X}_T),\hat{P}_T\rangle\} \\ &\quad - \rho_0\langle\Psi(Y_0)-\Psi(\bar{Y}_0),\hat{Y}_0\rangle + \delta\big[\mathbb{E}\{\langle\Phi_2(x_T,p_T)-\Phi_2(\bar{x}_T,\bar{p}_T),\hat{X}_T\rangle \\ &\quad + \langle\Phi_1(x_T)-\Phi_1(\bar{x}_T),\hat{P}_T\rangle\}\big] - \delta\langle\Psi(y_0)-\Psi(\bar{y}_0),\hat{Y}_0\rangle\end{aligned}$$

$$= -(1-\rho_0)l\mathbb{E}\int_0^T |\hat{\Xi}_t^{\Sigma}|^2\,dt + \rho_0\mathbb{E}\int_0^T \langle A(t,\Lambda_t,\Xi_t) - A(t,\bar{\Lambda}_t,\overline{\Xi}_t), (\hat{\Lambda}_t,\hat{\Xi}_t)\rangle\,dt$$

$$+ \delta l\mathbb{E}\int_0^T \Big[\langle \hat{P}_t, B_t^{\top}\hat{\xi}_{t-}^{\Sigma}\rangle + \langle \hat{Q}_t, C_t^{\top}\hat{\xi}_{t-}^{\Sigma}\rangle + \langle \hat{R}_t, D_t^{\top}\hat{\xi}_{t-}^{\Sigma}\rangle + \int_{\mathcal{E}}\langle \hat{\Theta}_t(e), E_t^{\top}(e)\hat{\xi}_{t-}^{\Sigma}\rangle\pi(de)\Big]\,dt$$

$$+ \delta\mathbb{E}\int_0^T \Big[\langle \hat{X}_t, -\hat{f}_2(t)\rangle + \langle \hat{Y}_t, \hat{\alpha}_2(t)\rangle + \langle \hat{Z}_t, \hat{\beta}_2(t)\rangle + \int_{\mathcal{E}}\langle \hat{K}_t(e), \hat{\gamma}_2(t,e)\rangle\pi(de)$$

$$+ \langle \hat{P}_t, -\hat{f}_1(t)\rangle + \langle \hat{Q}_t, \hat{\alpha}_1(t)\rangle + \langle \hat{R}_t, \hat{\beta}_1(t)\rangle + \int_{\mathcal{E}}\langle \Theta_t(e), \hat{\gamma}_1(t,e)\rangle\pi(de)\Big]\,dt,$$

where

$$\hat{g}_1(t) = g_1(t,x,\xi_t^M) - g_1(t,\bar{x},\bar{\xi}_t^M), \qquad \hat{f}_1(t) = f_1(t,\lambda,\xi) - f_1(t,\bar{\lambda},\bar{\xi}),$$
$$\hat{g}_2(t) = g_2(t,\lambda,\xi) - g_2(t,\bar{\lambda},\bar{\xi}),$$

with $g_1 = \alpha_1, \beta_1, \gamma_1$ and $g_2 = \alpha_2, \beta_2, \gamma_2, f_2$. By Assumptions 1, 2, and 3, for any $\varepsilon > 0$ we have

$$\mathbb{E}\int_0^T \Big| B_t\hat{P}_{t-} + C_t\hat{Q}_{t-} + D_t\hat{R}_t + \int_{\mathcal{E}} E_t(e)\hat{\Theta}_t\pi(de)\Big|^2\,dt$$
$$\le \varepsilon C\big\|(\hat{\Lambda}_t,\hat{\Xi}_t)\big\|^2 + \frac{\delta C}{\varepsilon}\big\|(\hat{\lambda}_t,\hat{\xi}_t)\big\|^2, \tag{10}$$

where $C > 0$ is a constant. Here and hereafter, C represents some generic constant which can be changed from line to line. We also apply Itô's formula to $|\hat{X}_t|^2$. By Assumptions 1 and 2, combining with Gronwall's inequality and Burkholder-Davis-Gundy's inequality, we have

$$\mathbb{E}\sup_{t\in[0,T]}|\hat{X}_t|^2 \le C\mathbb{E}\int_0^T \Big| B_t\hat{P}_{t-} + C_t\hat{Q}_{t-} + D_t\hat{R}_t + \int_{\mathcal{E}} E_t(e)\hat{\Theta}_t\pi(de)\Big|^2\,dt$$
$$+ \delta C\Big[\mathbb{E}\sup_{t\in[0,T]}|\hat{x}_t|^2 + \|\hat{\xi}_t\|^2\Big]. \tag{11}$$

Similarly, by applying Itô's formula to $|\hat{Y}_t|^2$, $|\hat{P}_t|^2$ and $|\hat{Q}_t|^2$, respectively, we obtain three other estimates:

$$\mathbb{E}\sup_{t\in[0,T]}|\hat{Y}_t|^2 + \mathbb{E}\int_0^T |\hat{Z}_t|^2\,dt + \mathbb{E}\int_0^T\int_{\mathcal{E}}|\hat{K}_t(e)|^2\pi(de)\,dt$$
$$\le C\mathbb{E}\int_0^T \Big| B_t\hat{P}_{t-} + C_t\hat{Q}_{t-} + D_t\hat{R}_t + \int_{\mathcal{E}} E_t(e)\hat{\Theta}_t\pi(de)\Big|^2\,dt$$
$$+ C\mathbb{E}\sup_{t\in[0,T]}|\hat{X}_t|^2 + \delta C\big\|(\hat{\lambda}_t,\hat{\xi}_t)\big\|^2, \tag{12}$$

$$\mathbb{E}\sup_{t\in[0,T]}|\hat{P}_t|^2 \le C\|\hat{\Lambda}_t\|^2 + \delta C\big\|(\hat{\lambda}_t,\hat{\xi}_t)\big\|^2, \tag{13}$$

and

$$\mathbb{E}\sup_{t\in[0,T]}|\hat{Q}_t|^2 + \mathbb{E}\int_0^T |\hat{R}_t|^2\,dt + \mathbb{E}\int_0^T\int_{\mathcal{E}}|\hat{\Theta}_t(e)|^2\pi(de)\,dt$$
$$\le C\|\hat{\Lambda}_t\|^2 + C\mathbb{E}\sup_{t\in[0,T]}|\hat{P}_t|^2 + \delta C\big\|(\hat{\lambda}_t,\hat{\xi}_t)\big\|^2. \tag{14}$$

By the combination of (10), (11), (12), (13), and (14) and taking $\varepsilon = 1/(2C)$, we have

$$\left\|(\hat{\Lambda}_t, \hat{\Xi}_t)\right\|^2 \leq \delta C \left\|(\hat{\lambda}_t, \hat{\xi}_t)\right\|^2. \tag{15}$$

We emphasize that the constant $C > 0$ appearing in the above inequality is independent of ρ_0 and δ. Now, we choose $\delta_0 > 0$ such that $C\delta_0 \leq 1/2$; then, for any $\delta \in [0, \delta_0]$,

$$\left\|(\hat{\Lambda}_t, \hat{\Xi}_t)\right\|^2 \leq \frac{1}{2} \left\|(\hat{\lambda}_t, \hat{\xi}_t)\right\|^2. \tag{16}$$

This implies that the mapping $I_{\rho_0+\delta}$ is a contraction. It follows immediately that the mapping admits a unique fixed point which is exactly the unique solution of (7) for $\rho = \rho_0 + \delta$. We thus complete the proof. □

Proof of Theorem 3.2 By Lemma 3.3, there exists a fixed step $\delta_0 > 0$ such that, if for a given $\rho_0 \in [0, 1)$, FBSDE (7) is uniquely solvable for any $\kappa \in \mathcal{G}[0, T]$ and any $a \in \mathbb{R}^n$, then (7) is also uniquely solvable for $\rho = \rho_0 + \delta$ with $\delta \in [0, \delta_0]$. Obviously, for $\rho = 0$, FBSDE (7) is in a decoupled form and then can be uniquely solved. So we can increase the parameter ρ step by step from $\rho = 0$ to $\rho = 1$. Especially, taking $\kappa = 0$ and $\rho = 1$, we get the desired conclusion: FBSDE (3) is uniquely solvable. □

4 LQ SOC problem

In this section, we apply the solvability result of FBSDEs studied in the above section to deal with a linear-quadratic stochastic optimal control (LQ SOC) problem for a forward-backward system with Poisson random jumps, in which the controlled system is given by

$$\begin{cases} dx_t = (A_t x_{t-} + B_t v_t + \mu_t)\, dt + (C_t x_{t-} + D_t v_t + \nu_t)\, dW_t \\ \qquad + \int_{\mathcal{E}} (E_t(e) x_{t-} + F_t(e) v_t + \psi_t(e)) \widetilde{N}(dt, de), \\ -dy_t = (G_t x_{t-} + H_t y_{t-} + I_t z_t + \int_{\mathcal{E}} J_t(e) k_t(e) \pi(de) + K_t v_t + \phi_t)\, dt, \\ \qquad - z_t\, dW_t - \int_{\mathcal{E}} k_t(e) \widetilde{N}(dt, de), \\ x_0 = a, \qquad y_T = \Upsilon x_T + \eta, \end{cases} \tag{17}$$

where A, B, C, D, E, F, G, H, I, J, K are $\mathbb{F}$-adapted matrix-valued bounded processes with appropriate dimensions, Υ is an $\mathcal{F}_T$-measurable $m \times n$ matrix-valued bounded random variable, $\mu, \nu \in L^2_{\mathbb{F}}(0, T; \mathbb{R}^n)$, $\psi \in M^2_{\mathbb{F}}(0, T; \mathbb{R}^n)$, $\phi \in L^2_{\mathbb{F}}(0, T; \mathbb{R}^n)$, $a \in \mathbb{R}^n$, and $\eta \in L^2(\mathbb{F}_T; \mathbb{R}^m)$. The admissible control set is denoted by $\mathcal{V} = L^2_{\mathbb{F}}(0, T; \mathbb{R}^k)$, in which each element $v(\cdot)$ is called an admissible control. Clearly, for any admissible control $v(\cdot)$, FBSDE (17) admits a unique adapted solution

$$\begin{aligned} \left(x^v(\cdot), y^v(\cdot), z^v(\cdot), k^v(\cdot, \cdot)\right) \in C^2_{\mathbb{F}}\left(0, T; \mathbb{R}^n\right) \times C^2_{\mathbb{F}}\left(0, T; \mathbb{R}^m\right) \\ \times L^2_{\mathbb{F}}\left(0, T; \mathbb{R}^m\right) \times M^2_{\mathbb{F}}\left(0, T; \mathbb{R}^m\right), \end{aligned}$$

and we called it the state trajectory corresponding to $v(\cdot)$. Additionally, we give a cost functional associated with $v(\cdot)$ in quadratic form as follows:

$$\begin{aligned} \mathcal{J}(v(\cdot)) = \frac{1}{2} \mathbb{E} \int_0^T \Big[\langle L_t x_t, x_t \rangle + \langle M_t y_t, y_t \rangle + \langle N_t z_t, z_t \rangle + \int_{\mathcal{E}} \langle O_t(e) k_t(e), k_t(e) \rangle \pi(de) \\ + \langle \Gamma_t v_t, v_t \rangle \Big] dt + \frac{1}{2} \mathbb{E} \langle R x_T, x_T \rangle + \frac{1}{2} \langle S y_0, y_0 \rangle. \end{aligned} \tag{18}$$

Here L, M, N, O are $\mathbb{F}$-adapted symmetric and nonnegative definite matrix-valued bounded processes with appropriate dimensions, Γ is an $\mathbb{F}$-adapted $k \times k$ symmetric and positive definite matrix-valued bounded process, and there exists a constant $\delta > 0$ such that

$$\langle \Gamma_t v, v \rangle \geq \delta |v|^2, \quad \text{a.s.}$$

for any $v \in \mathbb{R}^k$ and for almost every $t \in [0, T]$, R is an $\mathcal{F}_T$-measurable $n \times n$ symmetric and nonnegative definite matrix-valued bounded random variable, and S is an $m \times m$ symmetric and nonnegative definite matrix.

Problem (LQ SOC) The problem is to look for an admissible control $u(\cdot) \in \mathcal{V}$ which satisfies

$$\mathcal{J}\big(u(\cdot)\big) = \inf_{v(\cdot) \in \mathcal{V}} \mathcal{J}\big(v(\cdot)\big). \tag{19}$$

Such an admissible control $u(\cdot)$ is called an optimal control, and $(x(\cdot), y(\cdot), z(\cdot), k(\cdot,\cdot)) = (x^u(\cdot), y^u(\cdot), z^u(\cdot), k^u(\cdot,\cdot))$ is called the corresponding optimal trajectory.

The following theorem gives an explicit characterization of the unique optimal control.

Theorem 4.1 *Problem (LQ SOC) admits a unique optimal control which is in the following form:*

$$u_t = -\Gamma_t^{-1}\left(B_t^\top q_t + D_t^\top r_t + \int_{\mathcal{E}} F_t^\top(e)\theta_t(e)\pi(de) - K_t^\top p_t\right), \quad t \in [0, T], \tag{20}$$

where $(x(\cdot), y(\cdot), z(\cdot), k(\cdot,\cdot), p(\cdot), q(\cdot), r(\cdot), \theta(\cdot,\cdot)) \in \mathcal{L}[0, T]$ is the unique solution of the following FBSDE (for the sake of simplicity, we denote it by FBSDE-LQ-SOC):

$$\begin{cases} dx_t = (A_t x_{t-} - B_t\Gamma_t^{-1}(B_t^\top q_{t-} + D_t^\top r_t + \int_{\mathcal{E}} F_t^\top(e)\theta_t(e)\pi(de) - K_t^\top p_t) + \mu_t)\,dt \\ \qquad\quad + (C_t x_{t-} - D_t\Gamma_t^{-1}(B_t^\top q_{t-} + D_t^\top r_t + \int_{\mathcal{E}} F_t^\top(e)\theta_t(e)\pi(de) - K_t^\top p_t) + \nu_t)\,dW_t \\ \qquad\quad + \int_{\mathcal{E}}(E_t(e)x_{t-} - F_t(e)\Gamma_t^{-1}(B_t^\top q_{t-} + D_t^\top r_t + \int_{\mathcal{E}} F_t^\top(e)\theta_t(e)\pi(de) - K_t^\top p_t) \\ \qquad\quad + \psi_t(e))\widetilde{N}(dt, de), \\ -dy_t = (G_t x_{t-} + H_t y_{t-} + I_t z_t + \int_{\mathcal{E}} J_t(e)k_t(e)\pi(de) - K_t\Gamma_t^{-1}(B_t^\top q_{t-} + D_t^\top r_t \\ \qquad\quad + \int_{\mathcal{E}} F_t^\top(e)\theta_t(e)\pi(de) - K_t^\top p_{t-}) + \phi_t)\,dt - z_t\,dW_t - \int_{\mathcal{E}} k_t(e)\widetilde{N}(dt, de), \\ dp_t = (H_t^\top p_{t-} - M_t y_t)\,dt + (I_t^\top p_{t-} - N_t z_t)\,dW_t \\ \qquad\quad + \int_{\mathcal{E}}(J_t^\top(e)p_{t-} - O_t(e)k_t(e))\widetilde{N}(dt, de), \\ -dq_t = (A_t^\top q_{t-} + C_t^\top r_t + \int_{\mathcal{E}} E_t^\top(e)\theta_t(e)\pi(de) - G_t^\top p_{t-} + L_t x_t)\,dt - r_t\,dW_t \\ \qquad\quad - \int_{\mathcal{E}} \theta_t(e)\widetilde{N}(dt, de), \\ x_0 = a, \qquad p_0 = -Sy_0, \\ y_T = \Upsilon x_T + \eta, \qquad q_T = Rx_T - \Upsilon^\top p_T. \end{cases} \tag{21}$$

Proof By Theorem 3.2, FBSDE-LQ-SOC (21) admits a unique solution $(x(\cdot), y(\cdot), z(\cdot), k(\cdot,\cdot), p(\cdot), q(\cdot), r(\cdot), \theta(\cdot,\cdot)) \in \mathcal{L}[0, T]$.

Next, we prove that $u(\cdot)$ defined by (20) is an optimal control. For any given $v(\cdot) \in \mathcal{V}$, we denote the corresponding state trajectory by $(x^v(\cdot), y^v(\cdot), z^v(\cdot), k^v(\cdot,\cdot))$. By applying Itô's formula to $\langle x_t^v - x_t, q_t\rangle + \langle p_t, y_t^v - y_t\rangle$, we have

$$
\begin{aligned}
&\mathbb{E}\langle Rx_T, x_T^v - x_T\rangle + \langle Sy_0, y_0^v - y_0\rangle \\
&= -\mathbb{E}\int_0^T \Big[\langle L_t x_t, x_t^v - x_t\rangle + \langle M_t y_t, y_t^v - y_t\rangle + \langle N_t z_t, z_t^v - z_t\rangle \\
&\quad + \int_{\mathcal{E}} \langle O_t(e)k_t(e), k_t^v(e) - k_t(e)\rangle \pi(de) + \langle \Gamma_t u_t, v_t - u_t\rangle \Big] dt.
\end{aligned} \tag{22}
$$

Then we analyze the difference between $\mathcal{J}(v(\cdot))$ and $\mathcal{J}(u(\cdot))$:

$$
\begin{aligned}
&\mathcal{J}(v(\cdot)) - \mathcal{J}(u(\cdot)) \\
&= \frac{1}{2}\mathbb{E}\int_0^T \Big[\langle L_t x_t^v, x_t^v\rangle - \langle L_t x_t, x_t\rangle + \langle M_t y_t^v, y_t^v\rangle - \langle M_t y_t, y_t\rangle \\
&\quad + \langle N_t z_t^v, z_t^v\rangle - \langle N_t z_t, z_t\rangle + \int_{\mathcal{E}} \langle O_t(e)k_t^v(e), k_t^v(e)\rangle \pi(de) - \int_{\mathcal{E}} \langle O_t(e)k_t(e), k_t(e)\rangle \pi(de) \\
&\quad + \langle \Gamma_t v_t, v_t\rangle - \langle \Gamma_t u_t, u_t\rangle \Big] dt + \frac{1}{2}\mathbb{E}\big[\langle Rx_T^v, x_T^v\rangle - \langle Rx_T, x_T\rangle\big] \\
&\quad + \frac{1}{2}\big[\langle Sy_0^v, y_0^v\rangle - \langle Sy_0, y_0\rangle\big] \\
&= \frac{1}{2}\mathbb{E}\int_0^T \Big[\langle L_t(x_t^v - x_t), x_t^v - x_t\rangle + \langle M_t(y_t^v - y_t), y_t^v - y_t\rangle + \langle N_t(z_t^v - z_t), z_t^v - z_t\rangle \\
&\quad + \int_{\mathcal{E}} \langle O_t(e)(k_t^v(e) - k_t(e)), k_t^v(e) - k_t(e)\rangle \pi(de) + \langle \Gamma_t(v_t - u_t), v_t - u_t\rangle \Big] dt \\
&\quad + \frac{1}{2}\mathbb{E}\langle R(x_T^v - x_T), x_T^v - x_T\rangle + \frac{1}{2}\langle S(y_0^v - y_0), y_0^v - y_0\rangle + \triangle,
\end{aligned} \tag{23}
$$

where

$$
\begin{aligned}
\triangle &= \mathbb{E}\int_0^T \Big[\langle L_t x_t, x_t^v - x_t\rangle + \langle M_t y_t, y_t^v - y_t\rangle + \langle N_t z_t, z_t^v - z_t\rangle \\
&\quad + \int_{\mathcal{E}} \langle O_t(e)k_t(e), k_t^v(e) - k_t(e)\rangle \pi(de) + \langle \Gamma_t u_t, v_t - u_t\rangle \Big] dt \\
&\quad + \mathbb{E}\langle Rx_T, x_T^v - x_T\rangle + \langle Sy_0, y_0^v - y_0\rangle.
\end{aligned}
$$

From (22), we have $\triangle = 0$. Moreover, since L, M, N, O, R, S are nonnegative definite and Γ_t is positive definite, from (23), we get

$$
\mathcal{J}(v(\cdot)) - \mathcal{J}(u(\cdot)) \geq 0.
$$

This implies $u(\cdot)$ defined by (20) is an optimal control.

Finally, we show the uniqueness of optimal control. As we proved above, u defined by (20) is an optimal control of Problem (LQ SOC). Let $\bar{u}$ be another optimal control, and denote the corresponding state trajectory by $(x^{\bar{u}}(\cdot), y^{\bar{u}}(\cdot), z^{\bar{u}}(\cdot), k^{\bar{u}}(\cdot,\cdot))$. Obviously, $\mathcal{J}(\bar{u}(\cdot)) =$

$\mathcal{J}(u(\cdot))$. Coming back to (23), we have

$$
\begin{aligned}
0 &= \mathcal{J}(\bar{u}(\cdot)) - \mathcal{J}(u(\cdot)) \\
&= \frac{1}{2}\mathbb{E}\int_0^T \Big[\langle L_t(x_t^{\bar{u}} - x_t), x_t^{\bar{u}} - x_t\rangle + \langle M_t(y_t^{\bar{u}} - y_t), y_t^{\bar{u}} - y_t\rangle + \langle N_t(z_t^{\bar{u}} - z_t), z_t^{\bar{u}} - z_t\rangle \\
&\quad + \int_{\mathcal{E}} \langle O_t(e)(k_t^{\bar{u}}(e) - k_t(e)), k_t^{\bar{u}}(e) - k_t(e)\rangle \pi(de) + \langle \Gamma_t(\bar{u}_t - u_t), \bar{u}_t - u_t\rangle \Big] dt \\
&\quad + \frac{1}{2}\mathbb{E}\langle R(x_T^{\bar{u}} - x_T), x_T^{\bar{u}} - x_T\rangle + \frac{1}{2}\langle S(y_0^{v} - y_0), y_0^{v} - y_0\rangle \\
&\geq \mathbb{E}\int_0^T \langle \Gamma_t(\bar{u}_t - u_t), \bar{u}_t - u_t\rangle dt.
\end{aligned}
$$

Because Γ_t is positive, we get $\bar{u}(\cdot) = u(\cdot)$. Due to the arbitrariness of $\bar{u}$, we obtain the uniqueness of the optimal control. □

5 LQ NZSSDG problem

Now we extend the LQ SOC problem to a linear-quadratic nonzero-sum stochastic differential game (LQ NZSSDG) problem for a forward-backward system with random jumps. Without loss of generality, we only consider the case of two players in this paper. The case of n (> 2) players can be treated in the same way. In detail, the game system is described by

$$
\begin{cases}
dx_t = (A_t x_{t-} + B_t^1 v_t^1 + B_t^2 v_t^2 + \mu_t)\,dt + (C_t x_{t-} + D_t^1 v_t^1 + D_t^2 v_t^2 + \nu_t)\,dW_t \\
\qquad + \int_{\mathcal{E}} (E_t(e) x_{t-} + F_t^1(e) v_t^1 + F_t^2(e) v_t^2 + \psi_t(e)) \widetilde{N}(dt, de), \\
-dy_t = (G_t x_{t-} + H_t y_{t-} + I_t z_t + \int_{\mathcal{E}} J_t(e) k_t(e) \pi(de) + K_t^1 v_t^1 + K_t^2 v_t^2 + \phi_t)\,dt, \\
\qquad - z_t\,dW_t - \int_{\mathcal{E}} k_t(e) \widetilde{N}(dt, de), \\
x_0 = a, \qquad y_T = \Upsilon x_T + \eta,
\end{cases}
\tag{24}
$$

where A, B^1, B^2, C, D^1, D^2, E, F^1, F^2, G, H, I, J, K^1, K^2 are $\mathbb{F}$-adapted matrix-valued bounded processes with appropriate dimensions, Υ is an $\mathcal{F}_T$-measurable $m \times n$ matrix-valued bounded random variable, $\mu, \nu \in L^2_{\mathbb{F}}(0, T; \mathbb{R}^n)$, $\psi \in M^2_{\mathbb{F}}(0, T; \mathbb{R}^n)$, $\phi \in L^2_{\mathbb{F}}(0, T; \mathbb{R}^n)$, $a \in \mathbb{R}^n$, and $\eta \in L^2(\mathcal{F}_T; \mathbb{R}^m)$. Different from Problem (LQ SOC), there exist two control processes $v^1(\cdot)$ and $v^2(\cdot)$ belonging to two players, respectively, to affect the state of the game system at the same time. Let $\mathcal{V}^i = L^2_{\mathbb{F}}(0, T; \mathbb{R}^{k_i})$ $(i = 1, 2)$ denote the set of admissible controls. Each element $v^i(\cdot) \in \mathcal{V}^i$ is called an admissible control for Player i $(i = 1, 2)$. Moreover, $\mathcal{V}^1 \times \mathcal{V}^2$ is called the set of admissible controls for the players. For any given pair of admissible controls $(v^1(\cdot), v^2(\cdot)) \in \mathcal{V}^1 \times \mathcal{V}^2$, FBSDE (24) admits a unique solution

$$
\begin{aligned}
(x^{v_1,v_2}(\cdot), y^{v_1,v_2}(\cdot), z^{v_1,v_2}(\cdot), k^{v_1,v_2}(\cdot,\cdot)) \in C^2_{\mathbb{F}}(0, T; \mathbb{R}^n) \times C^2_{\mathbb{F}}(0, T; \mathbb{R}^m) \times L^2_{\mathbb{F}}(0, T; \mathbb{R}^m) \\
\times M^2_{\mathbb{F}}(0, T; \mathbb{R}^m),
\end{aligned}
$$

which is called the state trajectory corresponding to $(v^1(\cdot), v^2(\cdot))$. Additionally, each player has his own cost functional:

$$\mathcal{J}^i(v^1(\cdot), v^2(\cdot))$$
$$= \frac{1}{2}\mathbb{E}\int_0^T \Big[\langle L_t^i x_t, x_t\rangle + \langle M_t^i y_t, y_t\rangle + \langle N_t^i z_t, z_t\rangle + \int_{\mathcal{E}} \langle O_t^i(e)k_t(e), k_t(e)\rangle \pi(de)$$
$$+ \langle \Gamma_t^i v_t^i, v_t^i\rangle\Big] dt + \frac{1}{2}\mathbb{E}\langle R^i x_T, x_T\rangle + \frac{1}{2}\langle S^i y_0, y_0\rangle \quad (i = 1, 2). \tag{25}$$

Here for $i = 1, 2$, L^i, M^i, N^i, O^i are $\mathbb{F}$-adapted symmetric and nonnegative definite matrix-valued bounded processes with appropriate dimensions, Γ^i is an $\mathbb{F}$-adapted $k \times k$ symmetric and positive definite matrix-valued bounded process, and there exists a constant $\delta > 0$ such that

$$\langle \Gamma_t^i v, v\rangle \geq \delta |v|^2, \quad \text{a.s.}$$

for any $v \in \mathbb{R}^k$ and for almost every $t \in [0, T]$, R^i is an $\mathcal{F}_T$-measurable $n \times n$ symmetric and nonnegative definite matrix-valued bounded random variable, and S^i is an $m \times m$ symmetric and nonnegative definite matrix.

Suppose each player hopes to minimize his cost functional $\mathcal{J}^i(v^1(\cdot), v^2(\cdot))$ by selecting an appropriate admissible control $v^i(\cdot)$ $(i = 1, 2)$, then the game problem (24)-(25) is formulated as follows.

Problem (LQ NZSSDG) The problem is to find a pair of admissible controls $(u^1(\cdot), u^2(\cdot)) \in \mathcal{V}^1 \times \mathcal{V}^2$ such that

$$\begin{aligned} \mathcal{J}^1(u^1(\cdot), u^2(\cdot)) &= \inf_{v^1(\cdot)\in\mathcal{V}^1} \mathcal{J}^1(v^1(\cdot), u^2(\cdot)), \\ \mathcal{J}^2(u^1(\cdot), u^2(\cdot)) &= \inf_{v^2(\cdot)\in\mathcal{V}^2} \mathcal{J}^2(u^1(\cdot), v^2(\cdot)). \end{aligned} \tag{26}$$

Such a pair of admissible controls $(u^1(\cdot), u^2(\cdot))$ is called a Nash equilibrium point. For notational convenience, we denote by $(x(\cdot), y(\cdot), z(\cdot), k(\cdot,\cdot)) = (x^{u^1,u^2}(\cdot), y^{u^1,u^2}(\cdot), z^{u^1,u^2}(\cdot), k^{u^1,u^2}(\cdot,\cdot))$ the state trajectory corresponding to a Nash equilibrium point $(u^1(\cdot), u^2(\cdot))$.

Similar to Problem (LQ SOC) studied in the above section, we will link Nash equilibrium points of Problem (LQ NZSSDG) to solutions of some FBSDE.

Theorem 5.1 *$(u^1(\cdot), u^2(\cdot))$ is a Nash equilibrium point of Problem (LQ NZSSDG), if and only if $(u^1(\cdot), u^2(\cdot))$ has the form*

$$\begin{pmatrix} u_t^1 \\ u_t^2 \end{pmatrix} = \begin{pmatrix} -(\Gamma_t^1)^{-1}[(B_t^1)^\top q_t^1 + (D_t^1)^\top r_t^1 + \int_{\mathcal{E}}(F_t^1)^\top(e)\theta_t^1(e)\pi(de) - (K_t^1)^\top p_t^1] \\ -(\Gamma_t^2)^{-1}[(B_t^2)^\top q_t^2 + (D_t^2)^\top r_t^2 + \int_{\mathcal{E}}(F_t^2)^\top(e)\theta_t^2(e)\pi(de) - (K_t^2)^\top p_t^2] \end{pmatrix}, \tag{27}$$

$t \in [0, T]$, where $(x(\cdot), y(\cdot), z(\cdot), k(\cdot,\cdot), p^1(\cdot), q^1(\cdot), r^1(\cdot), \theta^1(\cdot,\cdot), p^2(\cdot), q^2(\cdot), r^2(\cdot), \theta^2(\cdot,\cdot))$ satisfies the following FBSDE (for simplicity, we denote it by FBSDE-LQ-NZSSDG):

$$\begin{cases} dx_t = \{A_t x_{t-} - B_t^1(\Gamma_t^1)^{-1}[(B_t^1)^\top q_{t-}^1 + (D_t^1)^\top r_t^1 + \int_{\mathcal{E}}(F_t^1)^\top(e)k_t^1(e)\pi(de) - (K_t^1)^\top p_{t-}^1] \\ \quad - B_t^2(\Gamma_t^2)^{-1}[(B_t^2)^\top q_{t-}^2 + (D_t^2)^\top r_t^2 + \int_{\mathcal{E}}(F_t^2)^\top(e)k_t^2(e)\pi(de) - (K_t^2)^\top p_{t-}^2] + \mu_t\}\,dt \\ \quad + \{C_t x_{t-} - D_t^1(\Gamma_t^1)^{-1}[(B_t^1)^\top q_{t-}^1 + (D_t^1)^\top r_t^1 + \int_{\mathcal{E}}(F_t^1)^\top(e)k_t^1(e)\pi(de) - (K_t^1)^\top p_{t-}^1] \\ \quad - D_t^2(\Gamma_t^2)^{-1}[(B_t^2)^\top q_{t-}^2 + (D_t^2)^\top r_t^2 + \int_{\mathcal{E}}(F_t^2)^\top(e)k_t^2(e)\pi(de) - (K_t^2)^\top p_{t-}^2] + \nu_t\}\,dW_t \\ \quad + \int_{\mathcal{E}}\{E_t(e)x_{t-} - F_t^1(e)(\Gamma_t^1)^{-1}[(B_t^1)^\top q_{t-}^1 + (D_t^1)^\top r_t^1 + \int_{\mathcal{E}}(F_t^1)^\top(e)k_t^1(e)\pi(de) \\ \quad - (K_t^1)^\top p_{t-}^1] - F_t^2(e)(\Gamma_t^2)^{-1}((B_t^2)^\top q_{t-}^2 + (D_t^2)^\top r_t^2 + \int_{\mathcal{E}}(F_t^2)^\top(e)k_t^2(e)\pi(de) \\ \quad - (K_t^2)^\top p_{t-}^2) + \psi_t(e)\}\widetilde{N}(dt,de), \\ -dy_t = \{G_t x_{t-} + H_t y_{t-} + I_t z_t + \int_{\mathcal{E}} J_t(e)k_t(e)\pi(de) - K_t^1(\Gamma_t^1)^{-1}[(B_t^1)^\top q_{t-}^1 + (D_t^1)^\top r_t^1 \\ \quad + \int_{\mathcal{E}}(F_t^1)^\top(e)k_t^1(e)\pi(de) - (K_t^1)^\top p_{t-}^1] - K_t^2(\Gamma_t^2)^{-1}[(B_t^2)^\top q_{t-}^2 + (D_t^2)^\top r_t^2 \\ \quad + \int_{\mathcal{E}}(F_t^2)^\top(e)k_t^2(e)\pi(de) - (K_t^2)^\top p_{t-}^2] + \phi_t\}\,dt - z_t\,dW_t - \int_{\mathcal{E}} k_t(e)\widetilde{N}(dt,de), \\ dp_t^1 = (H_t^\top p_{t-}^1 - M_t^1 y_{t-})\,dt + (I_t^\top p_{t-}^1 - N_t^1 z_t)\,dW_t + \int_{\mathcal{E}}(J_t^\top(e)p_{t-}^1 - O_t^1(e)k_t(e))\widetilde{N}(dt,de), \\ dp_t^2 = (H_t^\top p_{t-}^2 - M_t^2 y_{t-})\,dt + (I_t^\top p_{t-}^2 - N_t^2 z_t)\,dW_t \\ \quad + \int_{\mathcal{E}}(J_t^\top(e)p_{t-}^2 - O_t^2(e)k_t(e))\widetilde{N}(dt,de), \\ -dq_t^1 = (A_t^\top q_{t-}^1 + C_t^\top r_t^1 + \int_{\mathcal{E}} E_t^\top(e)\theta_t^1(e)\pi(de) - G_t^\top p_{t-}^1 + L_t^1 x_{t-})\,dt - r_t^1\,dW_t \\ \quad - \int_{\mathcal{E}}\theta_t^1(e)\widetilde{N}(dt,de), \\ -dq_t^2 = (A_t^\top q_{t-}^2 + C_t^\top r_t^2 + \int_{\mathcal{E}} E_t^\top(e)\theta_t^2(e)\pi(de) - G_t^\top p_{t-}^2 + L_t^2 x_{t-})\,dt - r_t^2\,dW_t \\ \quad - \int_{\mathcal{E}}\theta_t^2(e)\widetilde{N}(dt,de), \\ x_0 = a, \qquad y_T = \Upsilon x_T + \eta, \\ p_0^1 = -S^1 y_0, \qquad q_T^1 = R^1 x_T - \Upsilon^\top p_T^1, \\ p_0^2 = -S^2 y_0, \qquad q_T^2 = R^2 x_T - \Upsilon^\top p_T^2. \end{cases} \tag{28}$$

Proof Noticing the definition of Nash equilibrium point, we link Problem (LQ NZSSDG) with two Problems (LQ SOC) studied in the above section. Precisely, for $i = 1, 2$, we denote $j = 3 - i$. Fix $u^j(\cdot)$, to minimize the following cost functional:

$$\mathcal{J}^i(v^i(\cdot), u^j(\cdot)) = \frac{1}{2}\mathbb{E}\int_0^T \Big[\langle L_t^i x_t, x_t\rangle + \langle M_t^i y_t, y_t\rangle + \langle N_t^i z_t, z_t\rangle + \int_{\mathcal{E}}\langle O_t^i(e)k_t(e), k_t(e)\rangle\pi(de) + \langle \Gamma_t^i v_t^i, v_t^i\rangle\Big]\,dt + \frac{1}{2}\mathbb{E}\langle R^i x_T, x_T\rangle + \frac{1}{2}\langle S^i y_0, y_0\rangle, \tag{29}$$

subject to

$$\begin{cases} dx_t = [A_t x_{t-} + B_t^i v_t^i + (B_t^j v_t^j + \mu_t)]\,dt + [C_t x_{t-} + D_t^i v_t^i + (D_t^j v_t^j + \nu_t)]\,dW_t \\ \quad + \int_{\mathcal{E}}[E_t(e)x_{t-} + F_t^i(e)v_t^i + (F_t^j(e)v_t^j + \psi_t(e))]\widetilde{N}(dt,de), \\ -dy_t = [G_t x_{t-} + H_t y_{t-} + I_t z_t + \int_{\mathcal{E}} J_t(e)k_t(e)\pi(de) + K_t^i v_t^i + (K_t^j v_t^j + \phi_t)]\,dt \\ \quad - z_t\,dW_t - \int_{\mathcal{E}} k_t(e)\widetilde{N}(dt,de), \\ x_0 = a, \qquad y_T = \Upsilon x_T + \eta, \end{cases} \tag{30}$$

over $\mathcal{V}^i$ is an LQ SOC problem. By Theorem 4.1, the unique optimal control of LQ SOC problem (29)-(30) has the following form:

$$u_t^i = -(\Gamma_t^i)^{-1}\Big[(B_t^i)^\top q_t^i + (D_t^i)^\top r_t^i + \int_{\mathcal{E}}(F_t^i)^\top(e)\theta_t^i(e)\pi(de) + (K_t^i)^\top p_t^i\Big], \quad t \in [0, T],$$

where $(x(\cdot), y(\cdot), z(\cdot), k(\cdot,\cdot), p^i(\cdot), q^i(\cdot), r^i(\cdot), \theta^i(\cdot,\cdot))$ is the unique solution of the following FBSDEF-LQ-SOC ($i = 1, 2$):

$$
\begin{cases}
dx_t = \{A_t x_{t-} - B_t^i(\Gamma_t^i)^{-1}[(B_t^i)^\top q_{t-}^i + (D_t^i)^\top r_t^i + \int_{\mathcal{E}}(F_t^i)^\top(e)\theta_t^i(e)\pi(de) - (K_t^i)^\top p_{t-}^i] \\
\qquad + (B_t^j u_t^j + \mu_t)\}\, dt + \{C_t x_{t-} - D_t^i(\Gamma_t^i)^{-1}[(B_t^i)^\top q_{t-}^i + (D_t^i)^\top r_t^i \\
\qquad + \int_{\mathcal{E}}(F_t^i)^\top(e)\theta_t^i(e)\pi(de) \\
\qquad - (K_t^i)^\top p_{t-}^i] + (D_t^j u_t^j + \nu_t)\}\, dW_t + \int_{\mathcal{E}}\{E_t(e)x_{t-} - F_t^i(e)(\Gamma_t^i)^{-1}[(B_t^i)^\top q_{t-}^i + (D_t^i)^\top r_t^i \\
\qquad + \int_{\mathcal{E}}(F_t^i)^\top(e)k_t^i(e)\pi(de) - (K_t^i)^\top p_{t-}^i] + (F_t^j(e)u_t^j + \psi_t(e))\}\widetilde{N}(dt, de), \\
-dy_t = \{G_t x_{t-} + H_t y_{t-} + I_t z_t + \int_{\mathcal{E}} J_t(e)k_t(e)\pi(de) - K_t^i(\Gamma_t^i)^{-1}[(B_t^i)^\top q_{t-}^i + (D_t^i)^\top r_t^1 \\
\qquad + \int_{\mathcal{E}}(F_t^i)^\top(e)k_t^i(e)\pi(de) - (K_t^i)^\top p_{t-}^i] + (K_t^j u_t^j + \phi_t)\}\, dt - z_t\, dW_t \\
\qquad - \int_{\mathcal{E}} k_t(e)\widetilde{N}(dt, de), \\
dp_t^i = (H_t^\top p_{t-}^i - M_t^i y_{t-})\, dt + (I_t^\top p_{t-}^i - N_t^i z_t)\, dW_t + \int_{\mathcal{E}}(J_t^\top(e)p_{t-}^i - O_t^i(e)k_t(e))\widetilde{N}(dt, de), \\
-dq_t^i = (A_t^\top q_{t-}^i + C_t^\top r_t^i + \int_{\mathcal{E}} E_t^\top(e)\theta_t^i(e)\pi(de) - G_t^\top p_{t-}^i + L_t^i x_{t-})\, dt - r_t^i\, dW_t \\
\qquad - \int_{\mathcal{E}} \theta_t^i(e)\widetilde{N}(dt, de), \\
p_0^i = -S^i y_0, \qquad q_T^i = R^i x_T - \Upsilon^\top p_T^i.
\end{cases}
\tag{31}
$$

Combining the two FBSDE-LQ-SOCs (31) for the cases $i = 1$ and $i = 2$, we get FBSDE-LQ-NZSSDG (28), and we thus finish the proof. □

Next, we would like to employ and extend the linear transform introduce by Hamadène [15] to discuss the solvability of FBSDE-LQ-NZSSDG (28). To this aim, we introduce the following assumptions.

Assumption 4

(i) The dimension of x is equal to that of y, i.e. $n = m$.

(ii) There exist four constants $\zeta_B, \zeta_D, \zeta_F, \zeta_K \in \mathbb{R}$ and two $\mathbb{F}$-adapted matrix-valued bounded processes $\mathcal{Q}_t^1$ and $\mathcal{Q}_t^2$ with appropriate dimensions, such that

$$
\begin{aligned}
&B_t^i = \zeta_B \mathcal{Q}_t^i, \qquad D_t^i = \zeta_D \mathcal{Q}_t^i, \\
&F_t^i(e) = \zeta_F \mathcal{Q}_t^i, \qquad K_t^i = \zeta_K \mathcal{Q}_t^i, \quad t \in [0, T], i = 1, 2.
\end{aligned}
\tag{32}
$$

(iii) The matrix-valued processes $\mathcal{Q}_t^i(\Gamma_t^i)^{-1}(\mathcal{Q}_t^i)^\top$ $(i = 1, 2)$, are independent of t.

(iv) The following commutation relations among matrices hold true:

$$
\mathcal{Q}_t^i\left(\Gamma_t^i\right)^{-1}\left(\mathcal{Q}_t^i\right)^\top \mathcal{P} = \mathcal{P}\mathcal{Q}_t^i\left(\Gamma_t^i\right)^{-1}\left(\mathcal{Q}_t^i\right)^\top, \quad t \in [0, T], i = 1, 2,
\tag{33}
$$

where $\mathcal{P} = A_t^\top, C_t^\top, E_t^\top, G_t^\top, H_t^\top, I_t^\top, J_t^\top, L_t^i, M_t^i, N_t^i, O_t^i, R^i, S^i, K_t^i, \Gamma_t^i$.

We note that Assumption 4 is easy to satisfy when the matrix-valued processes appearing in the game system (24) and the cost functionals (25) are valued in $\mathbb{R}$ and independent of t.

Theorem 5.2 *Under Assumption* 4, *FBSDE-LQ-NZSSDG* (28) *is uniquely solvable. Moreover,* $(u^1(\cdot), u^2(\cdot))$ *defined by* (27) *is the unique Nash equilibrium point of Problem* (*LQ NZSSDG*).

Proof Due to Theorem 5.1, we only need to prove the existence and uniqueness of FBSDE-LQ-NZSSDG (28). To this aim, we introduce another FBSDE:

$$\begin{cases} d\tilde{x}_t = [A_t\tilde{x}_{t-} - (\zeta_B)^2\tilde{q}_{t-} - \zeta_B\zeta_D\tilde{r}_t - \zeta_B\zeta_F\int_{\mathcal{E}}\tilde{\theta}_t(e)\pi(de) + \zeta_B\zeta_K\tilde{p}_{t-} + \mu_t]\,dt \\ \qquad + [C_t\tilde{x}_{t-} - \zeta_B\zeta_D\tilde{q}_{t-} - (\zeta_D)^2\tilde{r}_t - \zeta_D\zeta_F\int_{\mathcal{E}}\tilde{\theta}_t(e)\pi(de) + \zeta_D\zeta_K\tilde{p}_{t-} + \nu_t]\,dW_t \\ \qquad + \int_{\mathcal{E}}[E_t(e)\tilde{x}_{t-} - \zeta_B\zeta_F\tilde{q}_{t-} - \zeta_D\zeta_F\tilde{r}_t \\ \qquad - (\zeta_F)^2\int_{\mathcal{E}}\tilde{\theta}_t(e)\pi(de) + \zeta_F\zeta_K\tilde{p}_{t-} + \psi_t(e)]\widetilde{N}(dt,de), \\ -d\tilde{y}_t = [G_t\tilde{x}_{t-} + H_t\tilde{y}_{t-} + I_t\tilde{z}_t + \int_{\mathcal{E}}J_t(e)\tilde{k}_t(e)\pi(de) \\ \qquad - \zeta_B\zeta_K\tilde{q}_{t-} - \zeta_D\zeta_K\tilde{r}_t - \zeta_F\zeta_K\int_{\mathcal{E}}\tilde{\theta}_t(e)\pi(de) \\ \qquad + (\zeta_K)^2\tilde{p}_{t-} + \phi_t]\,dt - \tilde{z}_t\,dW_t - \int_{\mathcal{E}}\tilde{k}_t(e)\widetilde{N}(dt,de), \\ d\tilde{p}_t = \{H_t^\top\tilde{p}_{t-} - [Q_t^1(\Gamma_t^1)^{-1}(Q_t^1)^\top M_t^1 + Q_t^2(\Gamma_t^2)^{-1}(Q_t^2)^\top M_t^2]\tilde{y}_{t-}\}\,dt \\ \qquad + \{I_t^\top\tilde{p}_{t-} - [Q_t^1(\Gamma_t^1)^{-1}(Q_t^1)^\top N_t^1 + Q_t^2(\Gamma_t^2)^{-1}(Q_t^2)^\top N_t^2]\tilde{z}_t\}\,dW_t \\ \qquad + \int_{\mathcal{E}}\{J_t^\top(e)\tilde{p}_{t-} - [Q_t^1(\Gamma_t^1)^{-1}(Q_t^1)^\top O_t^1(e) + Q_t^2(\Gamma_t^2)^{-1}(Q_t^2)^\top O_t^2(e)]\tilde{k}_t(e)\}\widetilde{N}(dt,de), \\ -d\tilde{q}_t = \{A_t^\top\tilde{q}_{t-} + C_t^\top\tilde{r}_t + \int_{\mathcal{E}}E_t^\top(e)\tilde{\theta}_t(e)\pi(de) - G_t^\top\tilde{p}_{t-} + [Q_t^1(\Gamma_t^1)^{-1}(Q_t^1)^\top L_t^1 \\ \qquad + Q_t^2(\Gamma_t^2)^{-1}(Q_t^2)^\top L_t^2]\tilde{x}_{t-}\}\,dt - \tilde{r}_t\,dW_t - \int_{\mathcal{E}}\tilde{\theta}_t(e)\widetilde{N}(dt,de), \\ \tilde{x}_0 = a, \qquad \tilde{p}_0 = -[Q_t^1(\Gamma_t^1)^{-1}(Q_t^1)^\top S^1 + Q_t^2(\Gamma_t^2)^{-1}(Q_t^2)^\top S^2]\tilde{y}_0, \\ \tilde{y}_T = \Upsilon\tilde{x}_T + \eta, \qquad \tilde{q}_T = [Q_t^1(\Gamma_t^1)^{-1}(Q_t^1)^\top R^1 + Q_t^2(\Gamma_t^2)^{-1}(Q_t^2)^\top R^2]\tilde{x}_T - \Upsilon^\top\tilde{p}_T, \end{cases} \tag{34}$$

and we shall show that the unique solvability of (28) is equivalent to that of (34). On the one hand, based on the commutation relations among matrices (see (33)), if $(x, y, z, k, p^1, q^1, r^1, \theta^1, p^2, q^2, r^2, \theta^2)$ is a solution of FBSDE-LQ-NZSSDG (28), then $(\tilde{x}, \tilde{y}, \tilde{z}, \tilde{k}, \tilde{p}, \tilde{q}, \tilde{r}, \tilde{\theta})$ solves FBSDE (34), where

$$\begin{cases} \tilde{x}_t = x_t, \qquad \tilde{y}_t = y_t, \qquad \tilde{z}_t = z_t, \qquad \tilde{k}_t = k_t, \\ \tilde{p}_t = Q_t^1(\Gamma_t^1)^{-1}(Q_t^1)^\top p_t^1 + Q_t^2(\Gamma_t^2)^{-1}(Q_t^2)^\top p_t^2, \\ \tilde{q}_t = Q_t^1(\Gamma_t^1)^{-1}(Q_t^1)^\top q_t^1 + Q_t^2(\Gamma_t^2)^{-1}(Q_t^2)^\top q_t^2, \\ \tilde{r}_t = Q_t^1(\Gamma_t^1)^{-1}(Q_t^1)^\top r_t^1 + Q_t^2(\Gamma_t^2)^{-1}(Q_t^2)^\top r_t^2, \\ \tilde{\theta}_t = Q_t^1(\Gamma_t^1)^{-1}(Q_t^1)^\top \theta_t^1 + Q_t^2(\Gamma_t^2)^{-1}(Q_t^2)^\top \theta_t^2. \end{cases}$$

On the other hand, when $(\tilde{x}, \tilde{y}, \tilde{z}, \tilde{k}, \tilde{p}, \tilde{q}, \tilde{r}, \tilde{\theta})$ is a solution of FBSDE (34), we let $x(\cdot) = \tilde{x}(\cdot)$, $y(\cdot) = \tilde{y}(\cdot)$, $z(\cdot) = \tilde{z}(\cdot)$, $k(\cdot,\cdot) = \tilde{k}(\cdot,\cdot)$. From the existence and uniqueness results of SDEs and BSDEs, the equation

$$\begin{cases} dp_t^1 = (H_t^\top p_{t-}^1 - M_t^1\tilde{y}_{t-})\,dt + (I_t^\top p_{t-}^1 - N_t^1\tilde{z}_t)\,dW_t \\ \qquad + \int_{\mathcal{E}}(J_t^\top(e)p_{t-}^1 - O_t^1(e)\tilde{k}_t(e))\widetilde{N}(dt,de), \\ dp_t^2 = (H_t^\top p_{t-}^2 - M_t^2\tilde{y}_{t-})\,dt + (I_t^\top p_{t-}^2 - N_t^2\tilde{z}_t)\,dW_t \\ \qquad + \int_{\mathcal{E}}(J_t^\top(e)p_{t-}^2 - O_t^2(e)\tilde{k}_t(e))\widetilde{N}(dt,de), \\ -dq_t^1 = (A_t^\top q_{t-}^1 + C_t^\top r_t^1 + \int_{\mathcal{E}}E_t^\top(e)\theta_t^1(e)\pi(de) - G_t^\top p_{t-}^1 + L_t^1\tilde{x}_{t-})\,dt - r_t^1\,dW_t \\ \qquad - \int_{\mathcal{E}}\theta_t^1(e)\widetilde{N}(dt,de), \\ -dq_t^2 = (A_t^\top q_{t-}^2 + C_t^\top r_t^2 + \int_{\mathcal{E}}E_t^\top(e)\theta_t^2(e)\pi(de) - G_t^\top p_{t-}^2 + L_t^2\tilde{x}_{t-})\,dt - r_t^2\,dW_t \\ \qquad - \int_{\mathcal{E}}\theta_t^2(e)\widetilde{N}(dt,de), \\ p_0^1 = -S^1\tilde{y}_0, \qquad q_T^1 = R^1\tilde{x}_T - \Upsilon^\top p_T^1, \\ p_0^2 = -S^2\tilde{y}_0, \qquad q_T^2 = R^2\tilde{x}_T - \Upsilon^\top p_T^2, \end{cases}$$

admits a unique solution denoted by $(p^1(\cdot),q^1(\cdot),r^1(\cdot),\theta^1(\cdot,\cdot),p^2(\cdot),q^2(\cdot),r^2(\cdot),\theta^2(\cdot,\cdot))$. It is easy to check $(\tilde{x}(\cdot),\tilde{y}(\cdot),\tilde{z}(\cdot),p^1(\cdot),q^1(\cdot),r^1(\cdot),\theta^1(\cdot,\cdot),p^2(\cdot),q^2(\cdot),r^2(\cdot),\theta^2(\cdot,\cdot))$ defined above is a solution of FBSDE-LQ-NZSSDG (28), *i.e.* the existence of (28) is equivalent to that of (34). Moreover, in a similar way, one can prove that the uniqueness of (28) is also equivalent to that of (34).

It is easy to check the coefficients of FBSDE (34) satisfy Assumption 1, Assumption 2, and Assumption 3. By Theorem 3.2, FBSDE (34) admits a unique solution, then the same is true for FBSDE-LQ-NZSSDG (34). The proof is completed. □

Remark 5.3 In [15–18], when they studied nonzero-sum game problems, some stronger assumptions, such as the diffusion of the forward equation in the game system (24) does not depend on the controls $u_1(\cdot)$ and $u_2(\cdot)$ (*i.e.*, $D_i = 0$, $i = 1,2$) and so on, are imposed. Here we relax this kind of conditions to (32).

Competing interests

The authors declare that there are no competing interest regarding the publication of this paper.

Authors' contributions

All authors contributed equally to the writing of this paper. All authors read and approved the final manuscript.

Author details

[1]Department of Mathematics, QiLu Normal University, Jinan, 250013, China. [2]School of Mathematics, Shandong University, Jinan, 250100, China.

Acknowledgements

The authors would like to thank the associated editor and the referees for their constructive and insightful comments, which helped to improve the quality of this work, and Prof. Zhen Wu for helpful discussions. This work is supported by the National Natural Science Foundation of China (11471192), and the Nature Science Foundation of Shandong Province (JQ201401).

References

1. Bismut, JM: Théorie probabiliste du contrôle des diffusions. Mem. Am. Math. Soc. **4**, 167 (1976)
2. Pardoux, É, Peng, S: Adapted solution of a backward stochastic differential equation. Syst. Control Lett. **14**(1), 55-61 (1990)
3. Duffie, D, Epstein, L: Stochastic differential utility. With an appendix by the authors and, Skiadas, C. Econometrica **60**(2), 353-394 (1992)
4. El Karoui, N, Peng, S, Quenez, MC: Backward stochastic differential equations in finance. Math. Finance **7**(1), 1-71 (1997)
5. Allais, M: Le comportement de l'homme rationnel devant le risque: critique des postulats et axiomes de l'école americaine. Econometrica **21**, 503-546 (1953) (in French)
6. Ellsberg, D: Risk, ambiguity and the savage axioms. Q. J. Econ. **75**, 643-669 (1961)
7. Peng, S: Backward SDE and related *g*-expectation. In: Backward Stochastic Differential Equations (Paris, 1995-1996). Pitman Res. Notes Math. Ser., vol. 364, pp. 141-159. Longman, Harlow (1997)
8. Peng, S: Nonlinear expectations, nonlinear evaluations and risk measures. In: Stochastic Methods in Finance. Lecture Notes in Math., vol. 1856, pp. 165-253. Springer, Berlin (2004)
9. Yong, J: A stochastic linear quadratic optimal control problem with generalized expectation. Stoch. Anal. Appl. **26**(6), 1136-1160 (2008)
10. Hui, ECM, Xiao, H: Maximum principle for differential games of forward-backward stochastic systems with applications. J. Math. Anal. Appl. **386**(1), 412-427 (2012)
11. Williams, N: On dynamic principal-agent problems in continuous time. Working Paper (2009)
12. Wonham, WM: On the separation theorem of stochastic control. SIAM J. Control **6**, 312-326 (1968)
13. Bismut, JM: Contrôle des systèmes linéaires quadratiques: applications de l'intégrale stochastique. In: Séminaire de Probabilités, XII (Univ. Strasbourg, Strasbourg, 1976/1977). Lecture Notes in Math., vol. 649, pp. 180-264. Springer, Berlin (1978) (in French)
14. Chen, S, Li, X, Zhou, X: Stochastic linear quadratic regulators with indefinite control weight costs. SIAM J. Control Optim. **36**(5), 1685-1702 (1998)
15. Hamadène, S: Backward-forward SDE's and stochastic differential games. Stoch. Process. Appl. **77**(1), 1-15 (1998)
16. Wu, Z: Forward-backward stochastic differential equations, linear quadratic stochastic optimal control and nonzero sum differential games. J. Syst. Sci. Complex. **18**(2), 179-192 (2005)
17. Wu, Z, Yu, Z: Linear quadratic nonzero-sum differential games with random jumps. Appl. Math. Mech. **26**(8), 1034-1039 (2005)

18. Yu, Z: Linear-quadratic optimal control and nonzero-sum differential game of forward-backward stochastic system. Asian J. Control **14**(1), 173-185 (2012)
19. Cont, R, Tankov, P: Financial Modelling with Jump Processes. Chapman & Hall/CRC Financial Mathematics Series. Chapman & Hall/CRC, Boca Raton (2004)
20. Øksendal, B, Sulem, A: Applied Stochastic Control of Jump Diffusions, 2nd edn. Springer, Berlin (2007)
21. Tang, S, Li, X: Necessary conditions for optimal control of stochastic systems with random jumps. SIAM J. Control Optim. **32**(5), 1447-1475 (1994)
22. Situ, R: On solutions of backward stochastic differential equations with jumps and applications. Stoch. Process. Appl. **66**(2), 209-236 (1997)
23. Wu, Z, Wang, X: FBSDE with Poisson process and its application to linear quadratic stochastic optimal control problem with random jumps. Acta Autom. Sin. **29**(6), 821-826 (2003)
24. Antonelli, F: Backward-forward stochastic differential equations. Ann. Appl. Probab. **3**(3), 777-793 (1993)
25. Ma, J, Protter, P, Yong, J: Solving forward-backward stochastic differential equations explicitly-a four step scheme. Probab. Theory Relat. Fields **98**(3), 339-359 (1994)
26. Hu, Y, Peng, S: Solution of forward-backward stochastic differential equations. Probab. Theory Relat. Fields **103**(2), 273-283 (1995)
27. Peng, S, Wu, Z: Fully coupled forward-backward stochastic differential equations and applications to optimal control. SIAM J. Control Optim. **37**(3), 825-843 (1999)
28. Yong, J: Finding adapted solutions of forward-backward stochastic differential equations: method of continuation. Probab. Theory Relat. Fields **107**(4), 537-572 (1997)
29. Yong, J: Forward-backward stochastic differential equations with mixed initial-terminal conditions. Trans. Am. Math. Soc. **362**(2), 1047-1096 (2010)
30. Ma, J, Wu, Z, Zhang, D, Zhang, J: On wellposedness of forward-backward SDEs - a unified approach. Ann. Appl. Probab. (to appear)

3

Approximate controllability of fractional impulsive partial neutral integrodifferential inclusions with infinite delay in Hilbert spaces

Zuomao Yan* and Xiumei Jia

*Correspondence:
yanzuomao@163.com
Department of Mathematics, Hexi University, Zhangye, Gansu 734000, P.R. China

Abstract

In this paper, the approximate controllability of fractional impulsive partial neutral integrodifferential inclusions with infinite delay in a Hilbert space is studied. By using the nonlinear alternative of Leray-Schauder type for multivalued maps due to O'Regan and properties of the α-resolvent operator combined with approximation techniques, we derive a new set of sufficient conditions for the approximate controllability of fractional impulsive evolution system under the assumption that the corresponding linear system is approximately controllable. An example is provided to illustrate the obtained theory.

MSC: 34A37; 34K05; 93B05; 26A33

Keywords: approximate controllability; fractional impulsive partial neutral integrodifferential inclusions; multivalued map; infinite delay; α-resolvent operator

1 Introduction

The study of impulsive functional differential equations is linked to their utility in simulating processes and phenomena subject to short-time perturbations during their evolution. The perturbations are performed discretely and their duration is negligible in comparison with the total duration of the processes and phenomena. One may refer to [1–3] and the references therein.

The nonlinear fractional differential equations has in recent years been an object of increasing interest because of its wide applicability in nonlinear oscillations of earthquakes, and many physical phenomena such as seepage flow in porous media and in fluid dynamic traffic models; see [4–6]. Recently, the existence of solutions for fractional semilinear differential or integrodifferential equations has been extensively studied by many authors [7–12]. Functional differential equations with infinite delay appear frequently in applications as model equations and for this reason the study of such equations has received great attention in the last few years. Many authors [13–15] were interested in the existence of solutions for fractional functional differential equations with infinite delay in Banach spaces. Further, the existence, uniqueness and other quantitative and qualitative properties of solutions to various fractional impulsive semilinear differential and integrodifferential systems have received considerable interest among researchers. With regard to this matter, we refer the reader to Mophou [16], Shu *et al.* [17], Dabas and Chauhan [18], Balachandran *et al.* [19], and so on.

Controllability is one of the fundamental concepts in mathematical control theory and plays an important role in control systems. The exact controllability of nonlinear systems represented by fractional evolution equations or inclusions in abstract spaces has been considered in many publications; see [20–23]. There exist only a limited number of papers on the exact controllability of the fractional impulsive evolution systems. For example, Tai and Wang [24] studied the exact controllability of fractional-order impulsive neutral functional integrodifferential systems by using fractional calculus, a semigroup of operators and Krasnoselskii's fixed point theorem. Debbouche and Baleanu [25] established the exact controllability result of a class of fractional evolution nonlocal impulsive quasilinear delay integrodifferential systems. Liu and Li [26] discussed the exact controllability of impulsive fractional evolution inclusions in Banach spaces. However, in infinite-dimensional spaces the concept of exact controllability is usually too strong and, indeed, has limited applicability (see [27] and references therein). Approximate controllable systems are more prevalent and very often approximate controllability is completely adequate in applications (see [28, 29] and references therein). Therefore, it is important, in fact necessary, to study the weaker concept of controllability, namely approximate controllability for nonlinear systems. In recent years, for fractional semilinear control systems in Banach spaces, several papers were devoted to the approximate controllability. For example, Sakthivel *et al.* [30, 31] discussed the approximate controllability of semilinear fractional differential systems. The approximate controllability problem for nonlinear fractional stochastic system in Hilbert spaces has been investigated [32]. Kumar and Sukavanam [33] proved some sufficient conditions for the approximate controllability of fractional-order semilinear systems with bounded delay. Sukavanam and Kumar [34] discussed the approximate controllability of a fractional-order system in which the nonlinear term depends on both state and control variables. Yan [35] studied the approximate controllability of partial neutral functional differential systems of fractional order with state-dependent delay. Debbouche and Torres [36] derived the approximate controllability of a class of fractional nonlocal delay semilinear systems. It is well known that many control systems arising from realistic models can be described as partial fractional differential or integrodifferential inclusions (see [23, 26, 37–39] and references therein), so it is natural to extend the concept of approximate controllability to dynamical systems represented by fractional impulsive differential or integrodifferential inclusions. In this paper, we consider the approximate controllability of a class of fractional impulsive partial neutral integrodifferential inclusions with infinite delay in Hilbert spaces of the form

$$
\begin{aligned}
&{}^cD^{\alpha}N(x_t) \in AN(x_t) + \int_0^t Q(t-s)N(x_s)\,ds + Bu(s) + F\left(t, x_t, \int_0^t h(t,s,x_s)\,ds\right),\\
&\qquad t \in J = [0,b], t \neq t_k, k = 1,\dots,m, &(1.1)\\
&x_0 = \varphi \in \mathcal{B}, \quad x'(0) = 0, &(1.2)\\
&\Delta x(t_k) = I_k(x_{t_k}), \quad k = 1,\dots,m, &(1.3)
\end{aligned}
$$

where the unknown $x(\cdot)$ takes values in the Hilbert space H with the norm $\|\cdot\|$, ${}^cD^{\alpha}$ is the Caputo fractional derivative of order $\alpha \in (1,2)$, A, $(Q(t))_{t\geq 0}$, are closed linear operators defined on a common domain which is dense in $(H, \|\cdot\|)$, the control function $u \in L^2(J, U)$, a Hilbert space of admissible control functions. Further, B is a bounded linear operator from U to H and $D_t^\alpha \xi(t)$ represents the Caputo derivative of order $\alpha > 0$ defined by

$$D_t^\alpha \xi(t) = \int_0^t g_{n-\alpha}(t-s)\frac{d^n}{ds^n}\xi(s)\,ds,$$

where n is the smallest integer greater than or equal to α and $g_\beta(t) := \frac{t^{\beta-1}}{\Gamma(\beta)}$, $t > 0$, $\beta \geq 0$. The time history $x_t : (-\infty, 0] \to H$ given by $x_t(\theta) = x(t+\theta)$ belongs to some abstract phase space $\mathcal{B}$ defined axiomatically; and $F : J \times \mathcal{B} \times H \to \mathcal{P}(H)$ is a bounded, closed, convex-valued, multivalued map, $\mathcal{P}(H)$ is the family of all nonempty subsets of H, $G : J \times \mathcal{B} \to H$, $N(\psi) = \psi(0) + G(t, \psi)$, $\psi \in \mathcal{B}$, and $I_k : \mathcal{B} \to H$ $(k = 1, \ldots, m)$, are functions subject to some additional conditions. Moreover, let $0 < t_1 < \cdots < t_m < b$, are prefixed points and the symbol $\Delta x(t_k) = x(t_k^+) - x(t_k^-)$, where $x(t_k^-)$ and $x(t_k^+)$ represent the right and left limits of $x(t)$ at $t = t_k$, respectively.

To the best of our knowledge, there is no work reported on the approximate controllability of the fractional impulsive partial neutral integrodifferential inclusions with infinite delay, which is expressed in the form (1.1)-(1.3), and the aim of this paper is to close the gap. In this paper, motivated by the previously mentioned papers, we will study this interesting problem. Sufficient conditions for the existence are given by means of the nonlinear alternative of Leray-Schauder type for multivalued maps due to O'Regan [40] with the α-resolvent operator combined with approximation techniques. Especially, the well-known results that appeared in [30–36] are generalized to the fractional multivalued settings and the case of infinite delay. Further, the operators I_k $(k = 1, \ldots, m)$ are continuous but without imposing a completely continuous and Lipschitz condition. Therefore, our results have more applications in mathematical physical problems.

The rest of this paper is organized as follows. In Section 2, we introduce some notations and necessary preliminaries. In Section 3 we verify the existence of solutions for the fractional impulsive control system (1.1)-(1.3). In Section 4 we establish the approximate controllability of fractional impulsive control system (1.1)-(1.3). Finally in Section 5, an example is given to illustrate our results.

2 Preliminaries

In this section, we introduce some basic definitions, notations and lemmas which are used throughout this paper.

Let $(H, \|\cdot\|)$ be a Hilbert space. $C(J, H)$ is the Hilbert space of all continuous functions from J into H with the norm $\|x\|_\infty = \sup\{\|x(t)\| : t \in J\}$ and $L(H)$ denotes the Hilbert space of bounded linear operators from H to H. A measurable function $x : J \to H$ is Bochner integrable if and only if $\|x\|$ is Lebesgue integrable. For properties of the Bochner integral see Yosida [41]. $L^1(J, H)$ denotes the Hilbert space of measurable functions $x : J \to H$ which are Bochner integrable normed by $\|x\|_{L^1} = \int_0^b \|x(t)\|\,dt$ for all $x \in L^1(J, H)$. Furthermore, the notation $B_r(x, H)$ stands for the closed ball with center at x and radius $r > 0$ in H.

Let $\mathcal{P}(H)$ denote the class of all nonempty subsets of H. Let $\mathcal{P}_{bd,cl}(H)$, $\mathcal{P}_{cp,cv}(H)$, $\mathcal{P}_{bd,cl,cv}(H)$ and $\mathcal{P}_{cd}(H)$ denote respectively the families of all nonempty bounded-closed, compact-convex, bounded-closed-convex, and compact-acyclic (see [42]) subsets of H. For $x \in H$ and $Y, Z \in \mathcal{P}_{bd,cl}(H)$, we define $D(x, Y) = \inf\{\|x - y\| : y \in Y\}$ and $\tilde{\rho}(Y, Z) =$

$\sup_{a\in Y} D(a,Z)$, and the Hausdorff metric $H_d : \mathcal{P}_{bd,cl}(H) \times \mathcal{P}_{bd,cl}(H) \to \mathbb{R}^+$ by $H_d(\tilde{A},\tilde{B}) = \max\{\tilde{\rho}(\tilde{A},\tilde{B}), \tilde{\rho}(\tilde{B},\tilde{A})\}$.

G is called upper semicontinuous (u.s.c.) on H if, for each $x_0 \in H$, the set $G(x_0)$ is a nonempty, closed subset of H and if, for each open set S of H containing $G(x_0)$, there exists an open neighborhood S of x_0 such that $G(S) \subseteq V$. F is said to be completely continuous if $G(V)$ is relatively compact, for every bounded subset $V \subseteq H$.

If the multivalued map G is completely continuous with nonempty compact values, then G is u.s.c. if and only if F has a closed graph, *i.e.*, $x_n \to x_*$, $y_n \to y_*$, $y_n \in G(x_n)$ imply $y_* \in G(x_*)$.

A multivalued map $G : J \to P_{bd,cl,cv}(H)$ is said to be measurable if for each $x \in H$, the function $t \mapsto D(x, G(t))$ is a measurable function on J.

Definition 2.1 Let $G : H \to P_{bd,cl}(H)$ be a multivalued map. Then G is called a multivalued contraction if there exists a constant $\kappa \in (0,1)$ such that, for each $x, y \in H$, we have

$$H_d\big(G(x) - G(y)\big) \leq \kappa \|x - y\|.$$

The constant κ is called a contraction constant of G.

In this paper, we assume that the phase space $(\mathcal{B}, \|\cdot\|_{\mathcal{B}})$ is a semi-normed linear space of functions mapping $(-\infty, 0]$ into H, and satisfying the following fundamental axioms due to Hale and Kato (see, *e.g.*, [43]).

(A) If $x : (-\infty, \sigma + b] \to H$, $b > 0$, is such that $x|_{[\sigma,\sigma+b]} \in C([\sigma, \sigma + b], H)$ and $x_\sigma \in \mathcal{B}$, then for every $t \in [\sigma, \sigma + b]$ the following conditions hold:
   (i) x_t is in $\mathcal{B}$;
   (ii) $\|x(t)\| \leq \tilde{H}\|x_t\|_{\mathcal{B}}$;
   (iii) $\|x_t\|_{\mathcal{B}} \leq K(t-\sigma)\sup\{\|x(s)\| : \sigma \leq s \leq t\} + M(t-\sigma)\|x_\sigma\|_{\mathcal{B}}$, where $\tilde{H} \geq 0$ is a constant; $K, M : [0,\infty) \to [1,\infty)$, K is continuous and M is locally bounded; $\tilde{H}$, K, M are independent of $x(\cdot)$.

(B) For the function $x(\cdot)$ in (A), the function $t \to x_t$ is continuous from $[\sigma, \sigma + b]$ into $\mathcal{B}$.

(C) The space $\mathcal{B}$ is complete.

Remark 2.1 ([3]) Let $\varphi \in \mathcal{B}$ and $t \leq 0$. The notation φ_t represents the function defined by $\varphi_t = \varphi(t + \theta)$. Consequently, if the function $x(\cdot)$ in axiom (A) is such that $x_0 = \varphi$, then $x_t = \varphi_t$. We observe that φ_t is well defined for $t < 0$ since the domain of φ is $(-\infty, 0]$.

Remark 2.2 In the rest of this paper M_b and K_b are the constants defined by $M_b = \sup_{t\in J} M(t)$ and $K_b = \sup_{t\in J} K(t)$.

To describe appropriately our problems we say that a function $x : [\mu, \tau] \to H$ is a normalized piecewise continuous function on $[\mu, \tau]$ if x is piecewise continuous and continuous on $[\mu, \tau]$. We denote by $\mathcal{PC}([\mu, \tau], H)$ the space formed by the normalized piecewise continuous from $[\mu, \tau]$ into H. In particular, we introduce the space $\mathcal{PC}$ formed by all functions $x : [0, b] \to H$ such that x is continuous at $t \neq t_k$, $x(t_k) = x(t_k^-)$ and $x(t_k^+)$ exist for $k = 1, 2, \ldots, m$. In this paper, we always assume that $\mathcal{PC}$ is endowed with the norm $\|x\|_{\mathcal{PC}} = \sup_{t\in[0,b]} \|x(t)\|$. Then $(\mathcal{PC}, \|\cdot\|_{\mathcal{PC}})$ is a Banach space.

To simplify the notations, we put $t_0 = 0$, $t_{m+1} = b$ and for $x \in \mathcal{PC}$, we denote by $\hat{x}_k \in C([t_k, t_{k+1}]; H)$, $k = 0, 1, \ldots, m$, the function given by

$$\hat{x}_k(t) := \begin{cases} x(t) & \text{for } t \in (t_k, t_{k+1}], \\ x(t_k^+) & \text{for } t = t_k. \end{cases}$$

Moreover, for $B \subseteq \mathcal{PC}$ we denote by $\hat{B}_k$, $k = 0, 1, \ldots, m$, the set $\hat{B}_k = \{\hat{x}_k : x \in B\}$.

Let us recall the following definitions and facts.

Definition 2.2 ([13]) A one-parameter family of bounded linear operators $(\mathcal{R}_\alpha(t))_{t\geq 0}$ on H is called an α-resolvent operator for

$$^cD^\alpha x(t) = Ax(t) + \int_0^t Q(t-s)x(s)\,ds, \tag{2.1}$$

$$x_0 = \varphi \in H, \qquad x'(0) = 0, \tag{2.2}$$

if the following conditions are verified.

(a) The function $\mathcal{R}_\alpha(\cdot) : [0, \infty) \to L(H)$ is strongly continuous and $\mathcal{R}_\alpha(0)x = x$ for all $x \in H$ and $\alpha \in (1, 2)$.

(b) For $x \in D(A)$, $\mathcal{R}_\alpha(\cdot)x \in C([0, \infty), [D(A)]) \cap C^1((0, \infty), H)$, and

$$D_t^\alpha \mathcal{R}_\alpha(t)x = A\mathcal{R}_\alpha(t)x + \int_0^t Q(t-s)\mathcal{R}_\alpha(s)x\,ds,$$

$$D_t^\alpha \mathcal{R}_\alpha(t)x = \mathcal{R}_\alpha(t)Ax + \int_0^t \mathcal{R}_\alpha(t-s)Q(s)x\,ds$$

for every $t \geq 0$.

In this work we will consider the following conditions.

(P1) The operator $A : D(A) \subseteq H \to H$ is a closed linear operator with $[D(A)]$ dense in H. Let $\alpha \in (1, 2)$. For some $\phi_0 \in (0, \frac{\pi}{2}]$, for each $\phi < \phi_0$ there is a positive constant $C_0 = C_0(\phi)$ such that $\lambda \in \rho(A)$ for each

$$\lambda \in \Sigma_{0,\alpha\vartheta} = \left\{\lambda \in \mathbb{C}, \lambda \neq 0, \left|\arg(\lambda)\right| < \alpha\vartheta\right\},$$

where $\vartheta = \phi + \frac{\pi}{2}$ and $\|R(\lambda, A)\| \leq \frac{C_0}{|\lambda|}$ for all $\lambda \in \Sigma_{0,\alpha\vartheta}$.

(P2) For all $t \geq 0$, $Q(t) : D(Q(t)) \subseteq H \to H$ is a closed linear operator, $D(A) \subseteq D(Q(t))$ and $Q(\cdot)x$ is strongly measurable on $(0, \infty)$ for each $x \in D(A)$. There exists $b(\cdot) \in L^1_{\text{loc}}(\mathbb{R}^+)$ such that $\widehat{b}(\lambda)$ exists for $\text{Re}(\lambda) > 0$ and $\|Q(t)x\| \leq b(t)\|x\|_1$ for all $t > 0$ and $x \in D(A)$. Moreover, the operator-valued function $\widehat{Q} : \Sigma_{0,\pi/2} \to L([D(A)], H)$ has an analytical extension (still denoted by $\widehat{Q}$) to $\Sigma_{0,\vartheta}$ such that $\|\widehat{Q}(\lambda)x\| \leq \|\widehat{Q}(\lambda)\| \|x\|_1$ for all $x \in D(A)$, and $\|\widehat{Q}(\lambda)\| = O(\frac{1}{|\lambda|})$, as $|\lambda| \to \infty$.

(P3) There exists a subspace $D \subseteq D(A)$ dense in $[D(A)]$ and a positive constant $\widetilde{C}$ such that $A(D) \subseteq D(A)$, $\widehat{Q}(\lambda)(D) \subseteq D(A)$, and $\|A\widehat{Q}(\lambda)x\| \leq \widetilde{C}\|x\|$ for every $x \in D$ and all $\lambda \in \Sigma_{0,\vartheta}$.

In the sequel, for $r > 0$ and $\theta \in (\frac{\pi}{2}, \vartheta)$,

$$\Sigma_{r,\theta} = \{\lambda \in \mathbb{C}, |\lambda| > r, |\arg(\lambda)| < \theta\},$$

for $\Gamma_{r,\theta}$, $\Gamma^i_{r,\theta}$, $i = 1,2,3$, are the paths

$$\Gamma^1_{r,\theta} = \{te^{i\theta} : t \geq r\},$$
$$\Gamma^2_{r,\theta} = \{te^{i\xi} : |\xi| \leq \theta\},$$
$$\Gamma^3_{r,\theta} = \{te^{-i\theta} : t \geq r\},$$

and $\Gamma_{r,\theta} = \bigcup_{i=1}^3 \Gamma^i_{r,\theta}$ oriented counterclockwise. In addition, $\rho_\alpha(G_\alpha)$ are the sets

$$\rho_\alpha(G_\alpha) = \{\lambda \in \mathbb{C} : G_\alpha(\lambda) := \lambda^{\alpha-1}(\lambda^\alpha I - A - \widehat{Q}(\lambda))^{-1} \in L(H)\}.$$

We now define the operator family $(\mathcal{R}_\alpha(t))_{t\geq 0}$ by

$$\mathcal{R}_\alpha(t) := \begin{cases} \frac{1}{2\pi i}\int_{\Gamma_{r,\theta}} e^{\lambda t} G_\alpha(\lambda)\, d\lambda, & t > 0, \\ I, & t = 0. \end{cases}$$

Lemma 2.1 ([13]) *Assume that conditions* (P1)-(P3) *are fulfilled. Then there exists a unique α-resolvent operator for problem* (2.1)-(2.2).

Lemma 2.2 ([13]) *The function $\mathcal{R}_\alpha : [0,\infty) \to L(H)$ is strongly continuous and $\mathcal{R}_\alpha : (0,\infty) \to L(H)$ is uniformly continuous.*

Definition 2.3 ([13]) Let $\alpha \in (1,2)$, we define the family $(\mathcal{S}_\alpha(t))_{t\geq 0}$ by

$$\mathcal{S}_\alpha(t)x := \int_0^t g_{\alpha-1}(t-s)\mathcal{R}_\alpha(s)\, ds$$

for each $t \geq 0$.

Lemma 2.3 ([13]) *If the function $\mathcal{R}_\alpha(\cdot)$ is exponentially bounded in $L(H)$, then $\mathcal{S}_\alpha(\cdot)$ is exponentially bounded in $L(H)$.*

Lemma 2.4 ([13]) *If the function $\mathcal{R}_\alpha(\cdot)$ is exponentially bounded in $L([D(A)])$, then $\mathcal{S}_\alpha(\cdot)$ is exponentially bounded in $L([D(A)])$.*

Lemma 2.5 ([13]) *If $R(\lambda_0^\alpha, A)$ is compact for some $\lambda_0^\alpha \in \rho(A)$, then $\mathcal{R}_\alpha(t)$ and $\mathcal{S}_\alpha(t)$ are compact for all $t > 0$.*

Let $x_b(x_0; u)$ be the state value of system (1.1)-(1.3) at terminal time b corresponding to the control u and the initial value $x_0 = \varphi \in \mathcal{B}$. Introduce the set

$$\mathcal{B}(b, x_0) = \{x_b(x_0; u)(0) : u(\cdot) \in L^2(J, U)\},$$

which is called the reachable set of system (1.1)-(1.3) at terminal time b, its closure in H is denoted by $\overline{\mathcal{B}(b, x_0)}$.

Definition 2.4 A function $x : (-\infty, b] \to H$ is called a mild solution of problem (1.1)-(1.3) if $x_0 = \varphi \in \mathcal{B}$ for every $s \in J$ and $\Delta x(t_k) = I_k(x_{t_k})$, $k = 1, \dots, m$, the restriction of $x(\cdot)$ to the interval $(t_k, t_{k+1}]$ $(k = 0, 1, \dots, m)$ is continuous, and

$$x(t) \in \begin{cases} \mathcal{R}_\alpha(t)[\varphi(0) - G(0,\varphi)] + G(t,x_t) + \int_0^t \mathcal{S}_\alpha(t-s)Bu(s)\,ds \\ \quad + \int_0^t \mathcal{S}_\alpha(t-s)F(s,x_s,\int_0^s h(s,\tau,x_\tau)\,d\tau)\,ds, & t \in [0,t_1], \\ \mathcal{R}_\alpha(t-t_1)[x(t_1^-) + I_1(x_{t_1}) - G(t_1, x_{t_1^+})] \\ \quad + G(t,x_t) + \int_{t_1}^t \mathcal{S}_\alpha(t-s)Bu(s)\,ds \\ \quad + \int_{t_1}^t \mathcal{S}_\alpha(t-s)F(s,x_s,\int_0^s h(s,\tau,x_\tau)\,d\tau)\,ds, & t \in (t_1,t_2], \\ \vdots \\ \mathcal{R}_\alpha(t-t_m)[x(t_m^-) + I_m(x_{t_m}) - G(t_m, x_{t_m^+})] \\ \quad + G(t,x_t) + \int_{t_m}^t \mathcal{S}_\alpha(t-s)Bu(s)\,ds \\ \quad + \int_{t_m}^t \mathcal{S}_\alpha(t-s)F(s,x_s,\int_0^s h(s,\tau,x_\tau)\,d\tau)\,ds, & t \in (t_m,b]. \end{cases}$$

Definition 2.5 System (1.1)-(1.3) is said to be approximately controllable on the interval $[0,b]$ if $\overline{\mathcal{B}(b,x_0)} = H$ that is, if it is possible to steer the system from the initial point x_0 to within a distance $\varepsilon > 0$ from all the final points in the state space H at time b.

It is convenient at this point to define operators

$$\Gamma_\tau^b = \int_\tau^b \mathcal{S}_\alpha(b-s)BB^*\mathcal{S}_\alpha^*(b-s)\,ds, \quad 0 \le \tau < b,$$

$$\Gamma_{t_{k-1}}^{t_k} = \int_{t_{k-1}}^{t_k} \mathcal{S}_\alpha(t_k-s)BB^*\mathcal{S}_\alpha^*(t_k-s)\,ds, \quad k = 1,2,\dots,m,m+1,$$

$$R(a,\Gamma_{t_{k-1}}^{t_k}) = \left(aI + \Gamma_{t_{k-1}}^{t_k}\right)^{-1} \quad \text{for } a > 0, k = 1,2,\dots,m,m+1,$$

where B^* denotes the adjoint of B and $\mathcal{S}_\alpha^*(t)$ is the adjoint of $\mathcal{S}_\alpha(t)$. It is straightforward that the operator $\Gamma_{t_{k-1}}^{t_k}$ is a linear bounded operator.

Lemma 2.6 *The linear integrodifferential Cauchy problem corresponding to system* (1.1)-(1.3) *is approximately controllable on J if and only if $aR(a,\Gamma_\tau^s) \to 0$, $0 \le \tau < s \le b$, as $a \to 0^+$ in the strong operator topology.*

The proof of this lemma is a straightforward adaptation of the proof of [28], Theorem 2.

Lemma 2.7 *A set $B \subseteq \mathcal{PC}$ is relatively compact in $\mathcal{PC}$ if and only if the set $\hat{B}_k$ is relatively compact in $C([t_k,t_{k+1}];H)$ for every $k = 0,1,\dots,m$.*

Lemma 2.8 ([40] Nonlinear alternative of Leray-Schauder type for multivalued maps due to O'Regan) *Let H be a Hilbert space with V an open, convex subset of H and $y \in H$. Suppose*

(a) $\Phi : \overline{V} \to \mathcal{P}_{cd}(H)$ *has a closed graph, and*

(b) $\Phi : \overline{V} \to \mathcal{P}_{cd}(H)$ *is a condensing map with $\Phi(\overline{V})$ a subset of a bounded set in H hold.*

Then either

(i) Φ *has a fixed point in $\overline{V}$; or*

(ii) *there exist $y \in \partial V$ and $\lambda \in (0,1)$ with $y \in \lambda\Phi(y) + (1-\lambda)\{y_0\}$.*

3 Existence of solutions for fractional impulsive control systems

In this section, we prove the existence of solutions for the fractional impulsive control system (1.1)-(1.3). We make the following hypotheses.

(H1) The operator families $\mathcal{R}_\alpha(t)$ and $\mathcal{S}_\alpha(t)$ are compact for all $t > 0$, and there exist constants M and δ such that $\|\mathcal{R}_\alpha(t)\|_{L(H)} \leq Me^{\delta t}$ and $\|\mathcal{S}_\alpha(t)\|_{L(H)} \leq Me^{\delta t}$ for every $t \in J$.

(H2) The function $G : J \times \mathcal{B} \to H$ is continuous and there exists a $L > 0$ such that

$$\|G(t,\psi_1) - G(t,\psi_2)\| \leq L\big[|t_1 - t_2| + \|\psi_1 - \psi_2\|_{\mathcal{B}}\big], \quad t_1, t_2 \in J, \psi_1, \psi_2 \in \mathcal{B},$$

and

$$\|G(t,\psi)\| \leq L\big(\|\psi\|_{\mathcal{B}} + 1\big), \quad t \in J, \psi \in \mathcal{B}.$$

(H3) (i) For each $(t,s) \in \Lambda$ the function $h(t,s,\cdot) : \mathcal{B} \to H$ is continuous and for each $x \in \mathcal{B}$, the function $h(\cdot,\cdot,x) : \Lambda \to H$ is strongly measurable.

(ii) There exists a continuous function $p : \Lambda \to [0,\infty)$, such that

$$\|h(t,s,\psi)\| \leq p(t,s)\Theta_0\big(\|\psi\|_{\mathcal{B}}\big)$$

for a.e. $t, s \in J$, $\psi \in \mathcal{B}$, where $\Theta_0 : [0,\infty) \to (0,\infty)$ is a continuous nondecreasing function.

(H4) The multivalued map $F : J \times \mathcal{B} \times H \to \mathcal{P}_{bd,cl,cv}(H)$; for each $t \in J$, the function $F(t,\cdot,\cdot) : \mathcal{B} \times H \to \mathcal{P}_{bd,cl,cv}(H)$ is u.s.c. and for each $(\psi,y) \in \mathcal{B} \times H$, the function $F(\cdot,\psi,y)$ is measurable; for each fixed $(\psi,y) \in \mathcal{B} \times H$, the set

$$S_{F,\psi} = \left\{f \in L^1(J,H) : f(t) \in F\left(t,\psi,\int_0^t h(t,s,\psi)\,ds\right) \text{ for a.e. } t \in J\right\}$$

is nonempty.

(H5) There exist a continuous function $m : J \to [0,\infty)$ and a continuous nondecreasing function $\Theta : [0,\infty) \to (0,\infty)$ such that

$$\|F(t,\psi,y)\| = \sup\{\|f\| : f \in F(t,\psi,y)\} \leq m(t)\Theta\big(\|\psi\|_{\mathcal{B}} + \|y\|\big),$$

for a.e. $t \in J$ and each $\psi \in \mathcal{B}$, $y \in H$ with

$$\int_1^\infty \frac{1}{s + \Theta(s) + \Theta_0(s)}\,ds = \infty. \tag{3.1}$$

(H6) The functions $I_k : \mathcal{B} \to H$ are continuous and there exist constants c_k such that

$$\limsup_{\|\psi\|_{\mathcal{B}} \to \infty} \frac{\|I_k(\psi)\|}{\|\psi\|_{\mathcal{B}}} = c_k$$

for every $\psi \in \mathcal{B}$, $k = 1,\ldots,m$.

Lemma 3.1 ([44]) *Let J be a compact interval and H be a Hilbert space. Let F be a multivalued map satisfying* (H4) *and let P be a linear continuous operator from $L^1(J,H)$ to $C(J,H)$. Then the operator*

$$P \circ S_F : C(J,H) \to \mathcal{P}_{cp,cv}(H), \quad x \to (P \circ S_F)(x) := P(S_F, x)$$

is a closed graph in $C(J,H) \times C(J,H)$.

Theorem 3.1 *If assumptions* (H1)-(H6) *are satisfied. Further, suppose that, for all $a > 0$, then system* (1.1)-(1.3) *has at least one mild solution on J, provided that*

$$\max_{1 \le k \le m} \left\{ M_2\left[1 + K_b(Mc_k + ML)\right] + M_3K_bML \right\} < 1, \tag{3.2}$$

where $M_2 = MN_*(1 + \frac{1}{a}M_*^2N_*^2M_1^2b)$, $M_3 = (1 + \frac{1}{a}MM_*N_*M_1^2b)N_*$, $M_* = M\max\{1, e^{\delta b}\}$, $N_* = \max\{1, e^{-\delta b}\}$, $M_1 = \|B\|$.

Proof Consider the space $\mathcal{Y} = \{x : (-\infty, b] \to H; x_0 = 0, x|_J \in \mathcal{PC}(J,H)\}$ endowed with the uniform convergence topology and define the multivalued map $\Phi : \mathcal{Y} \to \mathcal{P}(\mathcal{Y})$ by Φx, the set of $\rho \in \mathcal{Y}$ such that

$$\rho(t) = \begin{cases} 0, & t \in (-\infty, 0], \\ \mathcal{R}_\alpha(t)[\varphi(0) - G(0,\varphi)] + G(t,\bar{x}_t) & \\ \quad + \int_0^t \mathcal{S}_\alpha(t-s)Bu_{\bar{x}}^a(s)\,ds + \int_0^t \mathcal{S}_\alpha(t-s)f(s)\,ds, & t \in [0, t_1], \\ \mathcal{R}_\alpha(t-t_1)[\bar{x}(t_1^-) + I_1(\bar{x}_{t_1}) - G(t_1, \bar{x}_{t_1^+})] + G(t,\bar{x}_t) & \\ \quad + \int_0^t \mathcal{S}_\alpha(t-s)Bu_{\bar{x}}^a(s)\,ds + \int_{t_1}^t \mathcal{S}_\alpha(t-s)f(s)\,ds, & t \in (t_1, t_2], \\ \vdots & \\ \mathcal{R}_\alpha(t-t_m)[\bar{x}(t_m^-) + I_m(\bar{x}_{t_m}) - G(t_m, \bar{x}_{t_m^+})] + G(t,\bar{x}_t) & \\ \quad + \int_0^t \mathcal{S}_\alpha(t-s)Bu_{\bar{x}}^a(s)\,ds + \int_{t_m}^t \mathcal{S}_\alpha(t-s)f(s)\,ds, & t \in (t_m, b], \end{cases}$$

where

$$u_{\bar{x}}^a(s) = \begin{cases} B^*\mathcal{S}_\alpha^*(t_1 - s)R(a, \Gamma_0^{t_1})[x_0 + \frac{x_b - x_0}{m+1} - \mathcal{R}_\alpha(t_1)[\varphi(0) - G(0,\varphi)] & \\ \quad - G(t_1, \bar{x}_{t_1}) - \int_0^{t_1} \mathcal{S}_\alpha(t_1 - \eta)f(\eta)\,d\eta], & s \in [0, t_1], \\ B^*\mathcal{S}_\alpha^*(t_2 - s)R(a, \Gamma_{t_1}^{t_2})[x_0 + \frac{2(x_b - x_0)}{m+1} - \mathcal{R}_\alpha(t_2 - t_1)[\bar{x}(t_1^-) + I_1(\bar{x}_{t_1}) & \\ \quad - G(t_1, \bar{x}_{t_1^+})] - G(t_2, \bar{x}_{t_2}) - \int_{t_1}^{t_2} \mathcal{S}_\alpha(t_2 - \eta)f(\eta)\,d\eta], & s \in (t_1, t_2], \\ \vdots & \\ B^*\mathcal{S}_\alpha^*(b - s)R(a, \Gamma_{t_m}^{b})[x_b - \mathcal{R}_\alpha(b - t_m)[\bar{x}(t_m^-) + I_m(\bar{x}_{t_m}) & \\ \quad - G(t_m, \bar{x}_{t_m^+})] - G(b, \bar{x}_b) - \int_{t_m}^{b} \mathcal{S}_\alpha(b - \eta)f(\eta)\,d\eta], & s \in (t_m, b], \end{cases}$$

and $f \in S_{F,\bar{x}} = \{f \in L^1(J,H) : f(t) \in F(t, \bar{x}_t, \int_0^t h(t,s,\bar{x}_s)\,ds) \text{ a.e. } t \in J\}$, and $\bar{x} : (-\infty, 0] \to H$ is such that $\bar{x}_0 = \varphi$ and $\bar{x} = x$ on J. In what follows, we aim to show that the operator Φ has a fixed point, which is a solution of problem (1.1)-(1.3).

Let $\{\sigma_n : n \in \mathbb{N}\}$ be a decreasing sequence in $(0,t_1) \subset (0,b)$ such that $\lim_{n\to\infty}\sigma_n = 0$. To prove the above theorem, we consider the following problem:

$$^{c}D^{\alpha}\widetilde{N}(x_t) \in A\widetilde{N}(x_t) + \int_0^t Q(t-s)\widetilde{N}(x_s)\,ds + Bu(s) + F\left(t, x_t, \int_0^t h(t,s,x_s)\,ds\right),$$

$$t \in J = [0,b], t \neq t_k, k = 1,\dots,m, \tag{3.3}$$

$$x_0 = \varphi \in \mathcal{B}, \qquad x'(0) = 0, \tag{3.4}$$

$$\Delta x(t_k) = \mathcal{R}_\alpha(\sigma_n)I_k(x_{t_k}), \quad k = 1,\dots,m, \tag{3.5}$$

where $\widetilde{N}(x_t) = \varphi(0) + \mathcal{R}_\alpha(\sigma_n)G(t,x_t)$. We shall show that the problem has at least one mild solution $x_n \in \mathcal{Y}$.

For fixed $n \in \mathbb{N}$, set the multivalued map $\Phi_n : \mathcal{Y} \to \mathcal{P}(\mathcal{Y})$ by $\Phi_n x$ the set of $\rho_n \in \mathcal{Y}$ such that

$$\rho_n(t) = \begin{cases} 0, & t \in (-\infty, 0], \\ \mathcal{R}_\alpha(t)[\varphi(0) - \mathcal{R}_\alpha(\sigma_n)G(0,\varphi)] + \mathcal{R}_\alpha(\sigma_n)G(t,\bar{x}_t) & \\ \quad + \int_0^t \mathcal{S}_\alpha(t-s)Bu_{\bar{x}}^{a}(s)\,ds + \int_0^t \mathcal{S}_\alpha(t-s)f(s)\,ds, & t \in [0,t_1], \\ \mathcal{R}_\alpha(t-t_1)[\bar{x}(t_1^-) + \mathcal{R}_\alpha(\sigma_n)I_1(\bar{x}_{t_1}) - \mathcal{R}_\alpha(\sigma_n)G(t_1,\bar{x}_{t_1^+})] & \\ \quad + \mathcal{R}_\alpha(\sigma_n)G(t,\bar{x}_t) + \int_{t_1}^t \mathcal{S}_\alpha(t-s)Bu_{\bar{x}}^{a}(s)\,ds & \\ \quad + \int_{t_1}^t \mathcal{S}_\alpha(t-s)f(s)\,ds, & t \in (t_1,t_2], \\ \vdots & \\ \mathcal{R}_\alpha(t-t_m)[\bar{x}(t_m^-) + \mathcal{R}_\alpha(\sigma_n)I_m(\bar{x}_{t_m}) - \mathcal{R}_\alpha(\sigma_n)G(t_m,\bar{x}_{t_m^+})] & \\ \quad + \mathcal{R}_\alpha(\sigma_n)G(t,\bar{x}_t) + \int_{t_m}^t \mathcal{S}_\alpha(t-s)Bu_{\bar{x}}^{a}(s)\,ds & \\ \quad + \int_{t_m}^t \mathcal{S}_\alpha(t-s)f(s)\,ds, & t \in (t_m,b], \end{cases}$$

where

$$u_{\bar{x}}^{a}(s) = \begin{cases} B^*\mathcal{S}_\alpha^*(t_1-s)R(a,\Gamma_0^{t_1})[x_0 + \frac{x_b-x_0}{m+1} - \mathcal{R}_\alpha(t_1)[\varphi(0) & \\ \quad - \mathcal{R}_\alpha(\sigma_n)G(0,\varphi)] - \mathcal{R}_\alpha(\sigma_n)G(t_1,\bar{x}_{t_1}) & \\ \quad - \int_0^{t_1} \mathcal{S}_\alpha(t_1-\eta)f(\eta)\,d\eta], & s \in [0,t_1], \\ B^*\mathcal{S}_\alpha^*(t_2-s)R(a,\Gamma_{t_1}^{t_2})[x_0 + \frac{2(x_1-x_0)}{m+1} - \mathcal{R}_\alpha(t_2-t_1)[\bar{x}(t_1^-) & \\ \quad + \mathcal{R}_\alpha(\sigma_n)I_1(\bar{x}_{t_1}) - \mathcal{R}_\alpha(\sigma_n)G(t_1,\bar{x}_{t_1^+})] & \\ \quad - \mathcal{R}_\alpha(\sigma_n)G(t_2,\bar{x}_{t_2}) - \int_{t_1}^{t_2} \mathcal{S}_\alpha(t_2-s)f(\eta)\,d\eta], & s \in (t_1,t_2], \\ \vdots & \\ B^*\mathcal{S}_\alpha^*(b-s)R(a,\Gamma_{t_m}^{b})[x_b - \mathcal{R}_\alpha(b-t_m)[\bar{x}(t_m^-) & \\ \quad + \mathcal{R}_\alpha(\sigma_n)I_m(\bar{x}_{t_m}) - \mathcal{R}_\alpha(\sigma_n)G(t_m,\bar{x}_{t_m^+})] & \\ \quad - \mathcal{R}_\alpha(\sigma_n)G(b,\bar{x}_b) - \int_{t_m}^{b} \mathcal{S}_\alpha(b-s)f(\eta)\,d\eta], & t \in (t_m,b], \end{cases}$$

and $f \in S_{F,\bar{x}}$. It is easy to see that the fixed point of Φ_n is a mild solution of problem (3.3)-(3.5).

Step 1. We shall show there exists an open set $V \subseteq \mathcal{Y}$ with $x \in \lambda\Phi_n x$ for $\lambda \in (0,1)$ and $x \notin \partial V$.

Let $\lambda \in (0,1)$ and let $x \in \lambda \Phi_n x$, then there exists an $f \in S_{F,\bar{x}}$ such that we have

$$x(t) = \begin{cases} \lambda \mathcal{R}_\alpha(t)[\varphi(0) - \mathcal{R}_\alpha(\sigma_n)G(0,\varphi)] + \lambda \mathcal{R}_\alpha(\sigma_n)G(t,\bar{x}_t) \\ \quad + \lambda \int_0^t \mathcal{S}_\alpha(t-s)Bu^a_{\bar{x}}(s)\,ds + \lambda \int_0^t \mathcal{S}_\alpha(t-s)f(s)\,ds, & t \in [0,t_1], \\ \lambda \mathcal{R}_\alpha(t-t_1)[\bar{x}(t_1^-) + \mathcal{R}_\alpha(\sigma_n)I_1(\bar{x}_{t_1}) - \mathcal{R}_\alpha(\sigma_n)G(t_1,\bar{x}_{t_1^+})] \\ \quad + \lambda \mathcal{R}_\alpha(\sigma_n)G(t,\bar{x}_t) + \lambda \int_{t_1}^t \mathcal{S}_\alpha(t-s)Bu^a_{\bar{u}}(s)\,ds \\ \quad + \lambda \int_{t_1}^t \mathcal{S}_\alpha(t-s)f(s)\,ds, & t \in (t_1,t_2], \\ \vdots \\ \lambda \mathcal{R}_\alpha(t-t_m)[\bar{x}(t_m^-) + \mathcal{R}_\alpha(\sigma_n)I_m(\bar{x}_{t_m}) - \mathcal{R}_\alpha(\sigma_n)G(t_m,\bar{x}_{t_m^+})] \\ \quad + \lambda \mathcal{R}_\alpha(\sigma_n)G(t,\bar{x}_t) + \lambda \int_{t_m}^t \mathcal{S}_\alpha(t-s)Bu^a_{\bar{x}}(s)\,ds \\ \quad + \lambda \int_{t_m}^t \mathcal{S}_\alpha(t-s)f(s)\,ds, & t \in (t_m,b], \end{cases}$$

for some $\lambda \in (0,1)$. On the other hand, from condition (H6), we conclude that there exist positive constants ϵ_k $(k = 1,\ldots,m)$, γ_1 such that, for all $\|\psi\|_{\mathcal{B}} > \gamma_1$,

$$\begin{aligned} &\|I_k(\psi)\| \le (c_k + \epsilon_k)\|\psi\|_{\mathcal{B}}, \\ &\max_{1\le k\le m} \{M_2[1 + K_b(M(c_k+\epsilon_k) + ML)] + M_3K_bML\} < 1. \end{aligned} \tag{3.6}$$

Let

$$F_1 = \{\psi : \|\psi\|_{\mathcal{B}} \le \gamma_1\}, \qquad F_2 = \{\psi : \|\psi\|_{\mathcal{B}} > \gamma_1\}, \qquad C_1 = \max\{\|I_k(\psi)\|, x \in F_1\}.$$

Therefore

$$\|I_k(\psi)\| \le C_1 + (c_k+\epsilon_k)\|\psi\|_{\mathcal{B}}. \tag{3.7}$$

Then, by (H1)-(H5) and (3.7), from the above equation, we have for $t \in [0,t_1]$

$$\begin{aligned} \|x(t)\| \le{}& Me^{\delta t}[\tilde{H}\|\varphi\|_{\mathcal{B}} + Me^{\delta\sigma_n}L(\|\varphi\|_{\mathcal{B}}+1)] + Me^{\delta\sigma_n}L(\|\bar{x}_t\|_{\mathcal{B}}+1) \\ &+ Me^{\delta t}\frac{1}{a}Me^{\delta b}N_*^2t_1M_1^2Me^{\delta\sigma_n}L(\|\bar{x}_{t_1}\|_{\mathcal{B}}+1) + Me^{\delta t}\frac{1}{a}Me^{\delta b}N_*^2t_1M_1^2\tilde{N}_0 \\ &+ Me^{\delta t}\int_0^t e^{-\delta s}m(s)\Theta\left(\|\bar{x}_s\|_{\mathcal{B}} + \int_0^s p(s,\tau)\Theta_0(\|\bar{x}_\tau\|_{\mathcal{B}})\,d\tau\right)ds, \end{aligned}$$

where

$$\begin{aligned} \tilde{N}_0 ={}& \|x_0\| + \left\|\frac{x_1-x_0}{m+1}\right\| + M_*[\|\varphi(0)\| + Me^{\delta\sigma_n}L(\|\varphi\|_{\mathcal{B}}+1)] \\ &+ M_*\int_0^{t_1} e^{-\delta\eta}m_f(\eta)\Theta\left(\|\bar{x}_\eta\|_{\mathcal{B}} + \int_0^\eta p(\eta,\vartheta)\Theta_0(\|\bar{x}_\vartheta\|_{\mathcal{B}})\,d\vartheta\right)d\eta, \qquad M_1 = \|B\|. \end{aligned}$$

Similarly, for any $t \in (t_k,t_{k+1}]$, $k = 1,\ldots,m$, we have

$$\begin{aligned} \|x(t)\| \le{}& Me^{\delta(t-t_k)}\{\|\bar{x}(t_k^-)\| + Me^{\delta\sigma_n}[C_1 + (c_k+\epsilon_k)\|\bar{x}_{t_k}\|_{\mathcal{B}}] + Me^{\delta\sigma_n}L(\|\bar{x}_{t_k^+}\|_{\mathcal{B}}+1)\} \\ &+ Me^{\delta\sigma_n}L(\|\bar{x}_t\|_{\mathcal{B}}+1) + Me^{\delta t}\frac{1}{a}Me^{\delta b}N_*^2M_1^2(t_{k+1}-t_k)\{M_*N_*[\|\bar{x}(t_k^-)\| \end{aligned}$$

$$+ Me^{\delta\sigma_n}\big[C_1 + (c_k + \epsilon_k)\|\bar{x}_{t_k}\|_{\mathcal{B}}\big] + Me^{\delta\sigma_n}L\big(\|\bar{x}_{t_k^+}\|_{\mathcal{B}} + 1\big)\big]$$

$$+ Me^{\delta\sigma_n}L\big(\|\bar{x}_{t_{k+1}}\|_{\mathcal{B}} + 1\big)\big\} + Me^{\delta t}\frac{1}{a}Me^{\delta b}N_*^2M_1^2(t_{k+1} - t_k)\tilde{N}_k$$

$$+ Me^{\delta t}\int_{t_k}^{t} e^{-\delta s}m(s)\Theta\bigg(\|\bar{x}_t\|_{\mathcal{B}} + \int_0^s p(s,\tau)\Theta_0\big(\|\bar{x}_\tau\|_{\mathcal{B}}\big)\,d\tau\bigg)\,ds,$$

where

$$\tilde{N}_k = \|x_0\| + \left\|\frac{(k+1)(x_1 - x_0)}{m+1}\right\|$$

$$+ M_*\int_{t_k}^{t_{k+1}} e^{-\delta\eta}m_f(\eta)\Theta\bigg(\|\bar{x}_\eta\|_{\mathcal{B}} + \int_0^\eta p(\eta,\vartheta)\Theta_0\big(\|\bar{x}_\vartheta\|_{\mathcal{B}}\big)\,d\vartheta\bigg)\,d\eta.$$

Then, for all $t \in [0, b]$, we have

$$\|x(t)\| \le \widetilde{M}e^{\delta t} + Me^{\delta t}N_*\bigg(1 + \frac{1}{a}M_*^2N_*^2M_1^2b\bigg)\big[\|\bar{x}(t_k^-)\| + Me^{\delta\sigma_n}(c_k + \epsilon_k)\|\bar{x}_{t_k}\|_{\mathcal{B}}$$

$$+ Me^{\delta\sigma_n}L\|\bar{x}_{t_k^+}\|_{\mathcal{B}}\big] + Me^{\delta t}\frac{1}{a}M_*N_*^2M_1^2bMe^{\delta\sigma_n}L\|\bar{x}_{t_{k+1}}\|_{\mathcal{B}} + Me^{\delta\sigma_n}L\|\bar{x}_t\|_{\mathcal{B}}$$

$$+ Me^{\delta t}\int_0^{t} e^{-\delta s}m(s)\Theta\bigg(\|\bar{x}_s\|_{\mathcal{B}} + \int_0^s p(s,\tau)\Theta_0\big(\|\bar{x}_\tau\|_{\mathcal{B}}\big)\,d\tau\bigg)\,ds,$$

where

$$\widetilde{M} = \max\bigg\{M\big[\tilde{H}\|\varphi\|_{\mathcal{B}} + Me^{\delta\sigma_n}L\big(\|\varphi\|_{\mathcal{B}} + 1\big)\big] + Me^{\delta\sigma_n}L$$

$$+ M\frac{1}{a}M_*N_*^2t_1M_1^2Me^{\delta\sigma_n}L + M\frac{1}{a}M_*N_*^2t_1\tilde{N}_0,$$

$$MN_*\bigg(1 + \frac{1}{a}M_*^2N_*^2M_1^2b\bigg)\big[Me^{\delta\sigma_n}C_1 + Me^{\delta\sigma_n}L\big] + Me^{\delta\sigma_n}L$$

$$+ M\frac{1}{a}M_*N_*^2M_1^2bMe^{\delta\sigma_n}L + M\frac{1}{a}M_*N_*^2M_1^2b\widetilde{N}\bigg\}, \qquad \widetilde{N} = \max_{1\le k\le m}\{\tilde{N}_k\}.$$

Since $\lim_{n\to\infty}\sigma_n = 0$, it follows that

$$\|x(t)\| \le \widetilde{M}e^{\delta t} + Me^{\delta t}N_*\bigg(1 + \frac{1}{a}M_*^2N_*^2M_1^2b\bigg)\big[\|\bar{x}(t_k^-)\| + M(c_k + \epsilon_k)\|\bar{x}_{t_k}\|_{\mathcal{B}}$$

$$+ ML\|\bar{x}_{t_k^+}\|_{\mathcal{B}}\big] + Me^{\delta t}\frac{1}{a}M_*N_*^2M_1^2bML\|\bar{x}_{t_{k+1}}\|_{\mathcal{B}} + ML\|\bar{x}_t\|_{\mathcal{B}}$$

$$+ Me^{\delta t}\int_0^{t} e^{-\delta s}m(s)\Theta\bigg(\|\bar{x}_s\|_{\mathcal{B}} + \int_0^s p(s,\tau)\Theta_0\big(\|\bar{x}_\tau\|_{\mathcal{B}}\big)\,d\tau\bigg)\,ds.$$

It is easy to see that

$$\|\bar{x}_t\|_{\mathcal{B}} \le M_b\|\varphi\|_{\mathcal{B}} + K_b\|x\|_t, \qquad t \in [0, b],$$

where $\|x\|_t = \sup_{0\le s\le t}\|x(s)\|$. If $v(t) = M_b\|\varphi\|_{\mathcal{B}} + K_b\|x\|_t$, $t \in [0, b]$, we obtain

$$\begin{aligned}
v(t) &\le M_b\|\varphi\|_{\mathcal{B}} + \widetilde{M}e^{\delta t} + Me^{\delta t}N_*\left(1 + \frac{1}{a}M_*^2N_*^2M_1^2b\right)v(t) \\
&\quad + K_bMe^{\delta t}N_*\left(1 + \frac{1}{a}M_*^2N_*^2M_1^2b\right)\left[M(c_k + \epsilon_k)v(t) + MLv(t)\right] \\
&\quad + K_bMe^{\delta t}\frac{1}{a}M_*N_*^2M_1^2bMLv(t) + K_bMLv(t) \\
&\quad + K_bMe^{\delta t}\int_0^t e^{-\delta s}m(s)\Theta\left(v(s) + \int_0^s p(s,\tau)\Theta_0\big(v(\tau)\big)\,d\tau\right)ds.
\end{aligned}$$

Since $\widetilde{L} = \max_{1\le k\le m}\{M_2[1 + K_b(M(c_k + \epsilon_k) + ML)] + M_3K_bML\} < 1$, we obtain

$$\begin{aligned}
e^{-\delta t}v(t) &\le \frac{1}{1-\widetilde{L}}\Bigg[N_*M_b\|\varphi\|_{\mathcal{B}} + \widetilde{M} \\
&\quad + K_bM\int_0^t e^{-\delta s}m(s)\Theta\left(v(s) + \int_0^s p(s,\tau)\Theta_0\big(v(\tau)\big)\,d\tau\right)ds\Bigg].
\end{aligned}$$

Denoting by $w(t)$ the right-hand side of the above inequality, we have

$$v(t) \le e^{\delta t}w(t) \quad \text{for all } t \in J,$$

and

$$\begin{aligned}
w(0) &= \frac{1}{1-\widetilde{L}}\left[N_*M_b\|\varphi\|_{\mathcal{B}} + \widetilde{M}\right], \\
w'(t) &= \frac{1}{1-\widetilde{L}}K_bMe^{-\delta t}m(t)\Theta\left(v(t) + \int_0^t p(t,s)\Theta_0\big(v(s)\big)\,ds\right) \\
&\le \frac{1}{1-\widetilde{L}}K_bMe^{-\delta t}m(t)\Theta\left(e^{\delta t}w(t) + \int_0^t p(t,s)\Theta_0\big(e^{\delta s}w(s)\big)\,ds\right), \quad t \in J.
\end{aligned}$$

Let $\xi(t) = e^{\delta t}w(t) + \int_0^t p(t,s)\Theta_0(e^{\delta s}w(s))\,ds$, then $\xi(0) = w(0)$, $e^{\delta t}w(t) \le \xi(t)$, and for each $t \in J$ we have

$$\begin{aligned}
\big(\xi(t)\big)' &= \delta e^{\delta t}w(t) + w'(t)e^{\delta t} + p(t,t)\Theta_0\big(e^{\delta t}w(t)\big) \\
&\le \delta\xi(t) + \frac{1}{1-\widetilde{L}}K_bMm(t)\Theta\big(\xi(t)\big) + p(t,t)\Theta_0\big(\xi(t)\big) \\
&\le \max\left\{\delta, \frac{1}{1-\widetilde{L}}K_bMm(t), p(t,t)\right\}\left[\xi(t) + \Theta\big(\xi(t)\big) + \Theta_0\big(\xi(t)\big)\right], \quad t \in J.
\end{aligned}$$

This implies that

$$\int_{\xi(0)}^{\xi(t)} \frac{d\varsigma}{\varsigma + \Theta(\varsigma) + \Theta_0(\varsigma)} \le \int_0^b \max\left\{\delta, \frac{1}{1-\widetilde{L}}K_bMm(t), p(t,t)\right\}dt < \infty.$$

This inequality shows that there is a constant $\widetilde{K}$ such that $\xi(t) \le \widetilde{K}$, $t \in J$, and hence $\|x\|_{\mathcal{PC}} \le v(t) \le \xi(t) \le \widetilde{K}$, where $\widetilde{K}$ depends only on M, δ, b and on the functions $m(\cdot)$, $p(\cdot,\cdot)$, $\Theta(\cdot)$ and $\Theta_0(\cdot)$. Then there exists r^* such that $\|x\|_{\mathcal{PC}} \ne r^*$. Set

$$V = \left\{x \in \mathcal{Y} : \|x\|_{\mathcal{PC}} < r^*\right\}.$$

From the choice of V, there is no $x \in \partial V$ such that $x \in \lambda\Phi_n x$ for $\lambda \in (0,1)$.

Step 2. Φ_n has a closed graph.

Let $x^{(j)} \to x^*$, $\rho_n^{(j)} \in \Phi_n x^{(j)}$, $x^{(j)} \in \overline{V}$ and $\rho_n^{(j)} \to \rho_n^*$. From axiom (A), it is easy to see that $(\overline{x^{(j)}})_s \to \overline{x^*}_s$ uniformly for $s \in (-\infty, b]$ as $n \to \infty$. We prove that $\rho_n^* \in \Phi_n \overline{x^*}$. Now $\rho_n^{(j)} \in \Phi_n \overline{x^{(j)}}$ means that there exists $f^{(j)} \in S_{F,\overline{x^{(j)}}}$ such that, for each $t \in [0, t_1]$,

$$\begin{aligned}
\rho_n^{(j)}(t) = {} & \mathcal{R}_\alpha(t)\big[\varphi(0) - \mathcal{R}_\alpha(\sigma_n)G(0,\varphi)\big] + \mathcal{R}_\alpha(\sigma_n)G\big(t, (\overline{x^{(j)}})_t\big) \\
& + \int_0^t \mathcal{S}_\alpha(t-s)BB^*\mathcal{S}_\alpha^*(t_1-s)R\big(a, \Gamma_0^{t_1}\big)\bigg[x_0 + \frac{x_b - x_0}{m+1} - \mathcal{R}_\alpha(t_1)\big[\varphi(0) \\
& - \mathcal{R}_\alpha(\sigma_n)G(0,\varphi)\big] - \mathcal{R}_\alpha(\sigma_n)G\big(t_1, (\overline{x^{(j)}})_{t_1}\big) \\
& - \int_0^{t_1} \mathcal{S}_\alpha(t_1-\eta)f^{(j)}(\eta)\,d\eta\bigg]ds + \int_0^t \mathcal{S}_\alpha(t-s)f^{(j)}(s)\,ds, \quad t \in [0, t_1].
\end{aligned}$$

We must prove that there exists $f^* \in S_{F,\overline{x^*}}$ such that, for each $t \in [0, t_1]$,

$$\begin{aligned}
\rho_n^*(t) = {} & \mathcal{R}_\alpha(t)\big[\varphi(0) - \mathcal{R}_\alpha(\sigma_n)G(0,\varphi)\big] + \mathcal{R}_\alpha(\sigma_n)G\big(t, (\overline{x^*})_t\big) \\
& + \int_0^t \mathcal{S}_\alpha(t-s)BB^*\mathcal{S}_\alpha^*(t_1-s)R\big(a, \Gamma_0^{t_1}\big)\bigg[x_0 + \frac{x_b - x_0}{m+1} \\
& - \mathcal{R}_\alpha(t_1)\big[\varphi(0) - \mathcal{R}_\alpha(\sigma_n)G(0,\varphi)\big] - \mathcal{R}_\alpha(\sigma_n)G\big(t_1, (\overline{x^*})_{t_1}\big) \\
& - \int_0^{t_1} \mathcal{S}_\alpha(t_1-\eta)f^{(j)}(\eta)\,d\eta\bigg]ds + \int_0^t \mathcal{S}_\alpha(t-s)f^*(s)\,ds.
\end{aligned}$$

Now, for every $t \in [0, t_1]$, we have

$$\begin{aligned}
\bigg\| \bigg(& \rho_n^{(j)}(t) - \mathcal{R}_\alpha(t)\big[\varphi(0) - \mathcal{R}_\alpha(\sigma_n)G(0,\varphi)\big] - \mathcal{R}_\alpha(\sigma_n)G\big(t, (\overline{x^{(j)}})_t\big) \\
& - \int_0^t \mathcal{S}_\alpha(t-s)BB^*\mathcal{S}_\alpha^*(t_1-s)R\big(a, \Gamma_0^{t_1}\big)\bigg[x_0 + \frac{x_b - x_0}{m+1} - \mathcal{R}_\alpha(t_1)\big[\varphi(0) \\
& - \mathcal{R}_\alpha(\sigma_n)G(0,\varphi)\big] - \mathcal{R}_\alpha(\sigma_n)G\big(t_1, (\overline{x^{(j)}})_{t_1}\big)\bigg]ds\bigg) \\
& - \bigg(\rho_n^*(t) - \mathcal{R}_\alpha(t)\big[\varphi(0) - \mathcal{R}_\alpha(\sigma_n)G(0,\varphi)\big] - \mathcal{R}_\alpha(\sigma_n)G\big(t, (\overline{x^*})_t\big) \\
& - \int_0^t \mathcal{S}_\alpha(t-s)BB^*\mathcal{S}_\alpha^*(t_1-s)R\big(a, \Gamma_0^{t_1}\big)\bigg[x_0 + \frac{x_b - x_0}{m+1} - \mathcal{R}_\alpha(t_1)\big[\varphi(0) \\
& - \mathcal{R}_\alpha(\sigma_n)G(0,\varphi)\big] - \mathcal{R}_\alpha(\sigma_n)G\big(t_1, (\overline{x^*})_{t_1}\big)\bigg]ds\bigg)\bigg\|_{\mathcal{PC}} \to 0 \quad \text{as } j \to \infty.
\end{aligned}$$

Consider the linear continuous operator $\Psi : L^1([0, t_1], H) \to C([0, t_1], H)$,

$$\begin{aligned}
\Psi(f)(t) = {} & \int_0^t \mathcal{S}_\alpha(t-s)BB^*\mathcal{S}_\alpha^*(t_1-s)R\big(a, \Gamma_0^{t_1}\big)\bigg[\int_0^{t_1} \mathcal{S}_\alpha(t_1-\eta)f(\eta)\,d\eta\bigg]ds \\
& + \int_0^t \mathcal{S}_\alpha(t-s)f(s)\,ds.
\end{aligned}$$

From Lemma 3.1, it follows that $\Psi \circ S_F$ is a closed graph operator. Also, from the definition of Ψ, we have, for every $t \in [0, t_1]$,

$$\rho_n^{(j)}(t) - \mathcal{R}_\alpha(t)\big[\varphi(0) - \mathcal{R}_\alpha(\sigma_n)G(0,\varphi)\big] + \mathcal{R}_\alpha(\sigma_n)G\big(t, \big(\overline{x^{(j)}}\big)_t\big)$$
$$- \int_0^t \mathcal{S}_\alpha(t-s)BB^*\mathcal{S}_\alpha^*(t_1 - s)R\big(a, \Gamma_0^{t_1}\big)\bigg[x_0 + \frac{x_b - x_0}{m+1}$$
$$- \mathcal{R}_\alpha(t_1)\big[\varphi(0) - \mathcal{R}_\alpha(\sigma_n)G(0,\varphi)\big] - \mathcal{R}_\alpha(\sigma_n)G\big(t_1, \big(\overline{x^{(j)}}\big)_{t_1}\big)\bigg]\, ds \in \Psi(S_{F,\overline{x^{(j)}}}).$$

Since $\overline{x^{(j)}} \to \overline{x^*}$, for some $f^* \in S_{F,\overline{x^*}}$ it follows that, for every $t \in [0, t_1]$, we have

$$\rho_n^*(t) - \mathcal{R}_\alpha(t)\big[\varphi(0) - \mathcal{R}_\alpha(\sigma_n)G(0,\varphi)\big] + \mathcal{R}_\alpha(\sigma_n)G\big(t, \big(\overline{x^*}\big)_t\big)$$
$$- \int_0^t \mathcal{S}_\alpha(t-s)BB^*\mathcal{S}_\alpha^*(t_1 - s)R\big(a, \Gamma_0^{t_1}\big)\bigg[x_0 + \frac{x_1 - x_0}{m+1}$$
$$- \mathcal{R}_\alpha(t)\big[\varphi(0) - \mathcal{R}_\alpha(\sigma_n)G(0,\varphi)\big] - \mathcal{R}_\alpha(\sigma_n)G\big(t_1, \big(\overline{x^*}\big)_{t_1}\big)\bigg]\, ds$$
$$= \int_0^t \mathcal{S}_\alpha(t-s)BB^*\mathcal{S}_\alpha^*(t_1 - s)R\big(a, \Gamma_0^{t_1}\big)\bigg[\int_0^{t_1} \mathcal{S}_\alpha(t_1 - \eta)f^*(\eta)\, d\eta\bigg]\, ds$$
$$+ \int_0^t \mathcal{S}_\alpha(t-s)f^*(s)\, ds.$$

Similarly, for any $t \in (t_k, t_{k+1}]$, $k = 1, \ldots, m$, we have

$$\rho_n^{(j)}(t) = \mathcal{R}_\alpha(t - t_k)\big[\overline{x^{(j)}}\big(t_k^-\big) + \mathcal{R}_\alpha(\sigma_n)I_k\big(\big(\overline{x^{(j)}}\big)_{t_k}\big) - \mathcal{R}_\alpha(\sigma_n)G\big(t_k, \big(\overline{x^{(j)}}\big)_{t_k^+}\big)\big]$$
$$- \mathcal{R}_\alpha(\sigma_n)G\big(t, \big(\overline{x^{(j)}}\big)_t\big)$$
$$+ \int_{t_k}^t \mathcal{S}_\alpha(t-s)BB^*\mathcal{S}_\alpha^*(t_{k+1} - s)R\big(a, \Gamma_{t_k}^{t_{k+1}}\big)\bigg[x_0 + \frac{(k+1)(x_b - x_0)}{m+1}$$
$$- \mathcal{R}_\alpha(t_{k+1} - t_k)\big[\overline{x^{(j)}}\big(t_k^-\big) + \mathcal{R}_\alpha(\sigma_n)I_k\big(\big(\overline{x^{(j)}}\big)_{t_k}\big) - \mathcal{R}_\alpha(\sigma_n)G\big(t_k, \big(\overline{x^{(j)}}\big)_{t_k^+}\big)\big]$$
$$- \mathcal{R}_\alpha(\sigma_n)G\big(t_{k+1}, \big(\overline{x^{(j)}}\big)_{t_{k+1}}\big) - \int_{t_k}^{t_{k+1}} \mathcal{S}_\alpha(t_{k+1} - \eta)f^{(j)}(\eta)\, d\eta\bigg]\, ds$$
$$+ \int_{t_k}^t \mathcal{S}_\alpha(t-s)f^{(j)}(s)\, ds, \quad t \in (t_k, t_{k+1}].$$

We must prove that there exists $f^* \in S_{F,\overline{x^*}}$ such that, for each $t \in (t_k, t_{k+1}]$,

$$\rho_n^*(t) = \mathcal{R}_\alpha(t - t_k)\big[\overline{x^*}\big(t_k^-\big) + \mathcal{R}_\alpha(\sigma_n)I_k\big(\big(\overline{x^*}\big)_{t_k}\big) - \mathcal{R}_\alpha(\sigma_n)G\big(t_k, \big(\overline{x^*}\big)_{t_k^+}\big)\big]$$
$$- \mathcal{R}_\alpha(\sigma_n)G\big(t, \big(\overline{x^*}\big)_t\big)$$
$$+ \int_{t_k}^t \mathcal{S}_\alpha(t-s)BB^*\mathcal{S}_\alpha^*(t_{k+1} - s)R\big(a, \Gamma_{t_k}^{t_{k+1}}\big)\bigg[x_0 + \frac{(k+1)(x_b - x_0)}{m+1}$$
$$- \mathcal{R}_\alpha(t_{k+1} - t_k)\big[\overline{x^*}\big(t_k^-\big) + \mathcal{R}_\alpha(\sigma_n)I_k\big(\big(\overline{x^*}\big)_{t_k}\big)\big] - \mathcal{R}_\alpha(\sigma_n)G\big(t_k, \big(\overline{x^*}\big)_{t_k^+}\big)$$
$$- \mathcal{R}_\alpha(\sigma_n)G\big(t_{k+1}, \big(\overline{x^*}\big)_{t_{k+1}}\big) - \int_{t_k}^{t_{k+1}} \mathcal{S}_\alpha(t_{k+1} - \eta)f^*(\eta)\, d\eta\bigg]\, ds$$
$$+ \int_{t_k}^t \mathcal{S}_\alpha(t-s)f^*(s)\, ds, \quad t \in (t_k, t_{k+1}].$$

Now, for every $t \in (t_k, t_{k+1}]$, $k = 1, \ldots, m$, we have

$$\Big\| \Big(\rho_n^{(j)}(t) - \mathcal{R}_\alpha(t - t_k)\big[\overline{x^{(j)}}(t_k^-) + \mathcal{R}_\alpha(\sigma_n) I_k\big((\overline{x^{(j)}})_{t_k}\big) - \mathcal{R}_\alpha(\sigma_n) G\big(t_k, (\overline{x^{(j)}})_{t_k^+}\big)\big]$$
$$- \mathcal{R}_\alpha(\sigma_n) G\big(t, (\overline{x^{(j)}})_t\big)$$
$$- \int_{t_k}^t \mathcal{S}_\alpha(t-s) BB^* \mathcal{S}_\alpha^*(t_{k+1} - s) R\big(a, \Gamma_{t_k}^{t_{k+1}}\big)\Big[x_0 + \frac{(k+1)(x_1 - x_0)}{m+1}$$
$$- \mathcal{R}_\alpha(t_{k+1} - t_k)\big[\overline{x^{(j)}}(t_k^-) + \mathcal{R}_\alpha(\sigma_n) I_k\big((\overline{x^{(j)}})_{t_k}\big) - \mathcal{R}_\alpha(\sigma_n) G\big(t_k, (\overline{x^{(j)}})_{t_k^+}\big)\big]$$
$$- \mathcal{R}_\alpha(\sigma_n) G\big(t_{k+1}, (\overline{x^{(j)}})_{t_{k+1}}\big)\Big] ds \Big) - \Big(\rho_n^*(t) - \mathcal{R}_\alpha(t - t_k)\big[\overline{x^*}(t_k^-) + I_k\big((\overline{x^*})_{t_k}\big)$$
$$- \mathcal{R}_\alpha(\sigma_n) G\big(t_k, (\overline{x^*})_{t_k^+}\big)\big] - \mathcal{R}_\alpha(\sigma_n) G\big(t, (\overline{x^*})_t\big)$$
$$- \int_{t_k}^t \mathcal{S}_\alpha(t-s) BB^* \mathcal{S}_\alpha^*(t_{k+1} - s) R\big(a, \Gamma_{t_k}^{t_{k+1}}\big)\Big[x_0 + \frac{(k+1)(x_1 - x_0)}{m+1}$$
$$- \mathcal{R}_\alpha(t_{k+1} - t_k)\big[\overline{x^*}(t_k^-) + \mathcal{R}_\alpha(\sigma_n) I_k\big((\overline{x^*})_{t_k}\big) - \mathcal{R}_\alpha(\sigma_n) G\big(t_k, (\overline{x^*})_{t_k^+}\big)\big]$$
$$- \mathcal{R}_\alpha(\sigma_n) G\big(t_{k+1}, (\overline{x^*})_{t_{k+1}}\big)\Big] ds \Big) \Big\|_{\mathcal{PC}} \to 0 \quad \text{as } j \to \infty.$$

Consider the linear continuous operator $\Psi : L^1((t_k, t_{k+1}], H) \to C((t_k, t_{k+1}], H)$, $k = 1, \ldots, m$,

$$\Psi(f)(t) = \int_{t_k}^t \mathcal{S}_\alpha(t-s) BB^* \mathcal{S}_\alpha^*(t_{k+1} - s) R\big(a, \Gamma_{t_k}^{t_{k+1}}\big)\Big[\int_{t_k}^{t_{k+1}} \mathcal{S}_\alpha(t_1 - \eta) f(\eta)\, d\eta \Big] ds$$
$$+ \int_{t_k}^t \mathcal{S}_\alpha(t-s) f(s)\, ds.$$

From Lemma 3.1, it follows that $\Psi \circ S_F$ is a closed graph operator. Also, from the definition of Ψ, we have, for every $t \in (t_k, t_{k+1}]$, $k = 1, \ldots, m$,

$$\rho_n^{(j)}(t) - \mathcal{R}_\alpha(t - t_k)\big[\overline{x^{(j)}}(t_k^-) + \mathcal{R}_\alpha(\sigma_n) I_k\big((\overline{x^{(j)}})_{t_k}\big) - \mathcal{R}_\alpha(\sigma_n) G\big(t_k, (\overline{x^{(j)}})_{t_k^+}\big)\big]$$
$$- \mathcal{R}_\alpha(\sigma_n) G\big(t, (\overline{x^{(j)}})_t\big)$$
$$- \int_{t_k}^t \mathcal{S}_\alpha(t-s) BB^* \mathcal{S}_\alpha^*(t_{k+1} - s) R\big(a, \Gamma_{t_k}^{t_{k+1}}\big)\Big[x_0 + \frac{(k+1)(x_b - x_0)}{m+1}$$
$$- \mathcal{R}_\alpha(t_{k+1} - t_k)\big[\overline{x^{(j)}}(t_k^-) + \mathcal{R}_\alpha(\sigma_n) I_k\big((\overline{x^{(j)}})_{t_k}\big) - \mathcal{R}_\alpha(\sigma_n) G\big(t_k, (\overline{x^{(j)}})_{t_k^+}\big)\big]$$
$$- \mathcal{R}_\alpha(\sigma_n) G\big(t_{k+1}, (\overline{x^{(j)}})_{t_{k+1}}\big)\Big] ds \in \Psi(S_{F, \overline{x^{(j)}}}).$$

Since $\overline{x^{(j)}} \to \overline{x^*}$, for some $f^* \in S_{F, \overline{x^*}}$ it follows that, for every $t \in (t_k, t_{k+1}]$, we have

$$\rho_n^*(t) - \mathcal{R}_\alpha(t - t_k)\big[\overline{x^*}(t_k^-) + \mathcal{R}_\alpha(\sigma_n) I_k\big((\overline{x^*})_{t_k}\big) - \mathcal{R}_\alpha(\sigma_n) G\big(t_k, (\overline{x^*})_{t_k^+}\big)\big]$$
$$- \mathcal{R}_\alpha(\sigma_n) G\big(t, (\overline{x^*})_t\big)$$
$$- \int_{t_k}^t \mathcal{S}_\alpha(t-s) BB^* \mathcal{S}_\alpha^*(t_{k+1} - s) R\big(a, \Gamma_{t_k}^{t_{k+1}}\big)\Big[x_0 + \frac{(k+1)(x_b - x_0)}{m+1}$$

$$- \mathcal{R}_\alpha(t_{k+1} - t_k)\left[\overline{x^*}(t_k^-) + \mathcal{R}_\alpha(\sigma_n) I_k\left(\left(\overline{x^*}\right)_{t_k}\right) - \mathcal{R}_\alpha(\sigma_n) G\left(t_k, \left(\overline{x^*}\right)_{t_k^+}\right)\right]$$

$$\left. - \mathcal{R}_\alpha(\sigma_n) G\left(t_{k+1}, \left(\overline{x^*}\right)_{t_{k+1}}\right)\right] ds$$

$$= \int_{t_k}^{t} \mathcal{S}_\alpha(t-s) BB^* \mathcal{S}_\alpha^*(t_{k+1} - s) R\left(a, \Gamma_{t_k}^{t_{k+1}}\right)\left[\int_{t_k}^{t_{k+1}} \mathcal{S}_\alpha(t_{k+1} - \eta) f^*(\eta)\, d\eta\right] ds$$

$$+ \int_{t_k}^{t} \mathcal{S}_\alpha(t-s) f^*(s)\, ds.$$

Therefore, Φ_n has a closed graph.

Step 3. We show that the operator Φ_n is condensing.

For this purpose, we decompose Φ_n as $\Lambda_n + \Gamma_n$, where the map $\Lambda_n : \overline{V} \to \mathcal{Y}$ is defined by $\Lambda_n x$, the set $\varpi_n \in \mathcal{Y}$ is such that

$$\varpi_n(t) = \begin{cases} 0, & t \in (-\infty, 0], \\ -\mathcal{R}_\alpha(t)\mathcal{R}_\alpha(\sigma_n) G(0, \varphi) + \mathcal{R}_\alpha(\sigma_n) G(t, \bar{x}_t), & t \in [0, t_1], \\ -\mathcal{R}_\alpha(t - t_1)\mathcal{R}_\alpha(\sigma_n) G(t_1, \bar{x}_{t_1^+}) + \mathcal{R}_\alpha(\sigma_n) G(t, \bar{x}_t), & t \in (t_1, t_2], \\ \vdots & \\ -\mathcal{R}_\alpha(t - t_m)\mathcal{R}_\alpha(\sigma_n) G(t_m, \bar{x}_{t_m^+}) + \mathcal{R}_\alpha(\sigma_n) G(t, \bar{x}_t), & t \in (t_m, b], \end{cases}$$

and the map $\Gamma_n : \overline{V} \to \mathcal{Y}$ is defined by $\Gamma_n x$, the set $\vartheta_n \in \mathcal{Y}$ is such that

$$\vartheta_n(t) = \begin{cases} 0, & t \in (-\infty, 0], \\ \mathcal{R}_\alpha(t)\varphi(0) + \int_0^t \mathcal{S}_\alpha(t-s) B u_{\bar{x}}^a(s)\, ds + \int_0^t \mathcal{S}_\alpha(t-s) f(s)\, ds, & t \in [0, t_1], \\ \mathcal{R}_\alpha(t - t_1)[\bar{x}(t_1^-) + \mathcal{R}_\alpha(\sigma_n) I_1(\bar{x}_{t_1})] & \\ \quad + \int_{t_1}^t \mathcal{S}_\alpha(t-s) B u_{\bar{x}}^a(s)\, ds + \int_{t_1}^t \mathcal{S}_\alpha(t-s) f(s)\, ds, & t \in (t_1, t_2], \\ \vdots & \\ \mathcal{R}_\alpha(t - t_m)[\bar{x}(t_m^-) + \mathcal{R}_\alpha(\sigma_n) I_m(\bar{x}_{t_m})] & \\ \quad + \int_{t_m}^t \mathcal{S}_\alpha(t-s) B u_{\bar{x}}^a(s)\, ds + \int_{t_m}^t \mathcal{S}_\alpha(t-s) f(s)\, ds, & t \in (t_m, b]. \end{cases}$$

We first show that Λ_n is a contraction while Γ_n is a completely continuous operator.

Claim 1. Λ_n is a contraction on $\overline{V}$.

Let $t \in [0, t_1]$ and $x^*, x^{**} \in \mathcal{Y}$. From (H2), we have

$$\begin{aligned} &\left\| \left(\Lambda_n x^*\right)(t) - \left(\Lambda_n x^{**}\right)(t) \right\| \\ &\quad \le \left\| \mathcal{R}_\alpha(\sigma_n)\left[G\left(t, \overline{x^*}_t\right) - G\left(t, \overline{x^{**}}_t\right)\right] \right\| \\ &\quad \le L M e^{\delta \sigma_n} \left\| \overline{x^*}_t - \overline{x^{**}}_t \right\|_{\mathcal{B}} \\ &\quad \le L M e^{\delta \sigma_n} K_b \sup_{s \in [0,b]} \left\| \overline{x^*}(s) - \overline{x^{**}}(s) \right\| \\ &\quad = L M e^{\delta \sigma_n} K_b \sup_{s \in [0,b]} \left\| x^*(s) - x^{**}(s) \right\| \quad (\text{since } \bar{x} = x \text{ on } J) \\ &\quad = L M e^{\delta \sigma_n} K_b \left\| x^* - x^{**} \right\|_{\mathcal{PC}}. \end{aligned}$$

Similarly, for any $t \in (t_k, t_{k+1}]$, $k = 1, \dots, m$, we have

$$\begin{aligned}
&\left\|(\Lambda_n x^*)(t) - (\Lambda_n x^{**})(t)\right\| \\
&\quad \le \left\|\mathcal{R}_\alpha(t-t_k)\mathcal{R}_\alpha(\sigma_n)\left[-G(t_k, \overline{x^*}_{t_k^+}) + G(t_k, \overline{x^{**}}_{t_k^+})\right]\right\| \\
&\qquad + \left\|\mathcal{R}_\alpha(\sigma_n)\left[G(t, \overline{x^*}_t) - G(t, \overline{x^{**}}_t)\right]\right\| \\
&\quad \le Me^{\delta(t-t_k)} LMe^{\delta\sigma_n}\left\|\overline{x^*}_{t_k^+} - \overline{x^{**}}_{t_k^+}\right\|_{\mathcal{B}} + LMe^{\delta\sigma_n}\left\|\overline{x^*}_t - \overline{x^{**}}_t\right\|_{\mathcal{B}} \\
&\quad \le (M_* N_* + 1) LMe^{\delta\sigma_n} K_b \sup_{s\in[0,b]} \left\|\overline{x^*}(s) - \overline{x^{**}}(s)\right\| \\
&\quad = (M_* N_* + 1) LMe^{\delta\sigma_n} K_b \sup_{s\in[0,b]} \left\|x^*(s) - x^{**}(s)\right\| \quad (\text{since } \bar{x} = x \text{ on } J) \\
&\quad = (M_* N_* + 1) LMe^{\delta\sigma_n} K_b \left\|x^* - x^{**}\right\|_{\mathcal{PC}},
\end{aligned}$$

where $M_* = M\max\{1, e^{\delta b}\}$, $N_* = \max\{1, e^{-\delta b}\}$. Thus, for all $t \in [0, b]$, we have

$$\left\|(\Lambda_n x^*)(t) - (\Lambda_n x^{**})(t)\right\| \le (M_* N_* + 1) LMe^{\delta\sigma_n} K_b \left\|x^* - x^{**}\right\|_{\mathcal{PC}}.$$

Since $\lim_{n\to\infty} \sigma_n = 0$, it follows that

$$\left\|(\Lambda_n x^*)(t) - (\Lambda_n x^{**})(t)\right\| \le L_0 \left\|x^* - x^{**}\right\|_{\mathcal{PC}}.$$

Taking the supremum over t,

$$\left\|\Lambda_n x^* - \Lambda_n x^{**}\right\|_{\mathcal{PC}} \le L_0 \left\|x^* - x^{**}\right\|_{\mathcal{PC}},$$

where $L_0 = (M_* N_* + 1) LMK_b$. By (3.2), we see that $L_0 < 1$. Hence, Λ_n is a contraction on $\mathcal{Y}$.

Claim 2. Γ_n is convex for each $x \in \overline{V}$.

In fact, if ϑ_n^1, ϑ_n^2 belong to $\Gamma_n x$, then there exist $f_1, f_2 \in S_{F,\bar{x}}$ such that

$$\begin{aligned}
\vartheta_n^i(t) &= \mathcal{R}_\alpha(t)\varphi(0) + \int_0^t S(t-s)BB^* \mathcal{S}_\alpha^*(t_1 - s) R\left(a, \Gamma_0^{t_1}\right)\left[x_0 + \frac{x_b - x_0}{m+1}\right. \\
&\quad - \mathcal{R}_\alpha(t_1)\left[\varphi(0) - \mathcal{R}_\alpha(\sigma_n) G(0,\varphi)\right] - \mathcal{R}_\alpha(\sigma_n) G(t_1, \bar{x}_{t_1}) \\
&\quad \left. - \int_0^{t_1} \mathcal{S}_\alpha(t_1 - \eta) f_i(\eta)\, d\eta\right] ds + \int_0^t \mathcal{S}_\alpha(t-s) f_i(s)\, ds, \quad t \in [0, t_1], i = 1, 2.
\end{aligned}$$

Let $0 \le \lambda \le 1$. For each $t \in [0, t_1]$ we have

$$\begin{aligned}
&\left(\lambda\vartheta_n^1 + (1-\lambda)\vartheta_n^2\right)(t) \\
&\quad = \mathcal{R}_\alpha(t)\varphi(0) + \int_0^t S(t-s)BB^* \mathcal{S}_\alpha^*(t_1 - s) R\left(a, \Gamma_0^{t_1}\right)\left[x_0 + \frac{x_b - x_0}{m+1}\right. \\
&\qquad - \mathcal{R}_\alpha(t)\left[\varphi(0) - \mathcal{R}_\alpha(\sigma_n) G(0,\varphi)\right] - \mathcal{R}_\alpha(\sigma_n) G(t_1, \bar{x}_{t_1}) \\
&\qquad \left. - \int_0^{t_1} \mathcal{S}_\alpha(t_1 - \eta)\left[\lambda f_1(\eta) + (1-\lambda) f_2(\eta)\right] d\eta\right] ds \\
&\qquad + \int_0^t \mathcal{S}_\alpha(t-s)\left[\lambda f_1(s) + (1-\lambda) f_2(s)\right] ds.
\end{aligned}$$

Similarly, for any $t \in (t_k, t_{k+1}]$, $k = 1, \dots, m$, we have

$$\begin{aligned}
\vartheta_n^i(t) &= \mathcal{R}_\alpha(t-t_k)\big[\bar{x}(t_k^-) + \mathcal{R}_\alpha(\sigma_n)I_k(\bar{x}_{t_k})\big] \\
&\quad + \int_{t_k}^{t} S(t-s)BB^*\mathcal{S}_\alpha^*(t_{k+1}-s)R\big(a,\Gamma_{t_k}^{t_{k+1}}\big)\Bigg[x_0 + \frac{(k+1)(x_b-x_0)}{m+1} \\
&\quad - \mathcal{R}_\alpha(t_{k+1}-t_k)\big[\bar{x}(t_k^-) + \mathcal{R}_\alpha(\sigma_n)I_k(\bar{x}_{t_k}) - \mathcal{R}_\alpha(\sigma_n)G(t_k, \bar{x}_{t_k^+})\big] \\
&\quad - \mathcal{R}_\alpha(\sigma_n)G(t_{k+1}, \bar{x}_{t_{k+1}}) - \int_{t_k}^{t_{k+1}} \mathcal{S}_\alpha(t_{k+1}-\eta)f_i(\eta)\,d\eta\Bigg]\,ds \\
&\quad + \int_{t_k}^{t} \mathcal{S}_\alpha(t-s)f_i(s)\,ds, \quad i = 1, 2.
\end{aligned}$$

Let $0 \le \lambda \le 1$. For each $t \in (t_k, t_{k+1}]$, $k = 1, \dots, m$, we have

$$\begin{aligned}
&\big(\lambda\vartheta_n^1 + (1-\lambda)\vartheta_n^2\big)(t) \\
&\quad = \mathcal{R}_\alpha(t-t_k)\big[\bar{x}(t_k^-) + \mathcal{R}_\alpha(\sigma_n)I_k(\bar{x}_{t_k})\big] \\
&\qquad + \int_{t_k}^{t} S(t-s)BB^*\mathcal{S}_\alpha^*(t_{k+1}-s)R\big(a,\Gamma_{t_k}^{t_{k+1}}\big)\Bigg[x_0 + \frac{(k+1)(x_1-x_0)}{m+1} \\
&\qquad - \mathcal{R}_\alpha(t_{k+1}-t_k)\big[\bar{x}(t_k^-) + \mathcal{R}_\alpha(\sigma_n)I_k(\bar{x}_{t_k}) - \mathcal{R}_\alpha(\sigma_n)G(t_k, \bar{x}_{t_k^+})\big] \\
&\qquad - \mathcal{R}_\alpha(\sigma_n)G(t_k, \bar{x}_{t_{k+1}}) - \int_{t_k}^{t_{k+1}} \mathcal{S}_\alpha(t_{k+1}-\eta)\big[\lambda f_1(\eta) + (1-\lambda)f_2(\eta)\big]\,d\eta\Bigg]\,ds \\
&\qquad + \int_{t_k}^{t} \mathcal{S}_\alpha(t-s)\big[\lambda f_1(s) + (1-\lambda)f_2(s)\big]\,ds.
\end{aligned}$$

Since $S_{F,\bar{x}}$ is convex (because F has convex values) we have $(\lambda\vartheta_n^1 + (1-\lambda)\vartheta_n^2) \in \Gamma_n x$.

Claim 3. $\Gamma_n(\overline{V})$ is completely continuous.

We begin by showing that $\Gamma_n(\overline{V})$ is equicontinuous. If $x \in \overline{V}$, it follows that

$$\|\bar{x}_s\|_{\mathcal{B}} \le M_b\|\varphi\|_{\mathcal{B}} + K_b r^* := r'.$$

Let $0 < \tau_1 < \tau_2 \le t_1$. Then we see for each $x \in \overline{V}$, and $\vartheta_n \in \Gamma_n x$, that there exists $f \in S_{F,\bar{x}}$ such that

$$\vartheta_n(t) = \mathcal{R}_\alpha(t)\varphi(0) + \int_0^t \mathcal{S}_\alpha(t-s)Bu_{\bar{x}}^a(s)\,ds + \int_0^t \mathcal{S}_\alpha(t-s)f(s)\,ds. \tag{3.8}$$

Then

$$\begin{aligned}
&\big\|\vartheta_n(\tau_2) - \vartheta_n(\tau_1)\big\| \\
&\quad \le \big\|\big[\mathcal{R}_\alpha(\tau_2) - \mathcal{R}_\alpha(\tau_1)\big]\varphi(0)\big\| + \left\|\int_0^{\tau_1-\varepsilon}\big[\mathcal{S}_\alpha(\tau_2-s) - \mathcal{S}_\alpha(\tau_1-s)\big]Bu_{\bar{x}}(s)\,ds\right\| \\
&\qquad + \left\|\int_{\tau_1-\varepsilon}^{\tau_1}\big[\mathcal{S}_\alpha(\tau_2-s) - \mathcal{S}_\alpha(\tau_1-s)\big]Bu_{\bar{x}}(s)\,ds\right\| + \left\|\int_{\tau_1}^{\tau_2}\mathcal{S}_\alpha(\tau_2-s)Bu_{\bar{x}}(s)\,ds\right\| \\
&\qquad + \left\|\int_0^{\tau_1-\varepsilon}\big[\mathcal{S}_\alpha(\tau_2-s) - \mathcal{S}_\alpha(\tau_1-s)\big]f(s)\,ds\right\|
\end{aligned}$$

$$
\begin{aligned}
&+\left\|\int_{\tau_1-\varepsilon}^{\tau_1}\big[\mathcal{S}_\alpha(\tau_2-s)-\mathcal{S}_\alpha(\tau_1-s)\big]f(s)\,ds\right\|+\left\|\int_{\tau_1}^{\tau_2}\mathcal{S}_\alpha(\tau_2-s)f(s)\,ds\right\|\\
&\le\big\|\big[\mathcal{R}_\alpha(\tau_2)-\mathcal{R}_\alpha(\tau_1)\big]\varphi(0)\big\|+\int_0^{\tau_1-\varepsilon}\big\|\mathcal{S}_\alpha(\tau_2-s)-\mathcal{S}_\alpha(\tau_1-s)\big\|N_0\,ds\\
&+2M_*\int_{\tau_1-\varepsilon}^{\tau_1}e^{-\delta s}N_0\,ds+Me^{\delta\tau_2}\int_{\tau_1}^{\tau_2}e^{-\delta s}N_0\,ds\\
&+\Theta(r')\int_0^{\tau_1-\varepsilon}\big\|\mathcal{S}_\alpha(\tau_2-s)-\mathcal{S}_\alpha(\tau_1-s)\big\|m(s)\,ds\\
&+2M_*\Theta(r')\int_{\tau_1-\varepsilon}^{\tau_1}e^{-\delta s}m(s)\,ds+Me^{\delta\tau_2}\Theta(r')\int_{\tau_1}^{\tau_2}e^{-\delta s}m(s)\,ds.
\end{aligned}
$$

Similarly, for any $\tau_1,\tau_2\in(t_k,t_{k+1}]$, $\tau_1<\tau_2$, $k=1,\dots,m$, we have

$$
\begin{aligned}
\vartheta_n(t)=\mathcal{R}_\alpha(t-t_k)\big[\bar{x}(t_k^-)+\mathcal{R}_\alpha(\sigma_n)I_k(\bar{x}_{t_k})\big]\\
+\int_{t_k}^{t}\mathcal{S}_\alpha(t-s)Bu_{\bar{x}}^a(s)\,ds+\int_{t_k}^{t}\mathcal{S}_\alpha(t-s)f(s)\,ds.
\end{aligned}
\tag{3.9}
$$

Then

$$
\begin{aligned}
&\big\|\vartheta_n(\tau_2)-\vartheta_n(\tau_1)\big\|\\
&\le\big\|\big[\mathcal{R}_\alpha(\tau_2)-\mathcal{R}_\alpha(\tau_1)\big]\big[\bar{x}(t_k^-)+\mathcal{R}_\alpha(\sigma_n)I_k(\bar{x}_{t_k})\big]\big\|\\
&+\left\|\int_{t_k}^{\tau_1-\varepsilon}\big[\mathcal{S}_\alpha(\tau_2-s)-\mathcal{S}_\alpha(\tau_1-s)\big]Bu_{\bar{x}}^a(s)\,ds\right\|\\
&+\left\|\int_{\tau_1-\varepsilon}^{\tau_1}\big[\mathcal{S}_\alpha(\tau_2-s)-\mathcal{S}_\alpha(\tau_1-s)\big]Bu_{\bar{x}}^a(s)\,ds\right\|+\left\|\int_{\tau_1}^{\tau_2}\mathcal{S}_\alpha(\tau_2-s)Bu_{\bar{x}}^a(s)\,ds\right\|\\
&+\left\|\int_{t_k}^{\tau_1-\varepsilon}\big[\mathcal{S}_\alpha(\tau_2-s)-\mathcal{S}_\alpha(\tau_1-s)\big]f(s)\,ds\right\|\\
&+\left\|\int_{\tau_1-\varepsilon}^{\tau_1}\big[\mathcal{S}_\alpha(\tau_2-s)-\mathcal{S}_\alpha(\tau_1-s)\big]f(s)\,ds\right\|+\left\|\int_{\tau_1}^{\tau_2}\mathcal{S}_\alpha(\tau_2-s)f(s)\,ds\right\|\\
&\le\big\|\big[\mathcal{R}_\alpha(\tau_2)-\mathcal{R}_\alpha(\tau_1)\big]\big[\bar{x}(t_k^-)+\mathcal{R}_\alpha(\sigma_n)I_k(\bar{x}_{t_k})\big]\big\|\\
&+\int_{t_k}^{\tau_1-\varepsilon}\big\|\mathcal{S}_\alpha(\tau_2-s)-\mathcal{S}_\alpha(\tau_1-s)\big\|N_k\,ds\\
&+2M_*\int_{\tau_1-\varepsilon}^{\tau_1}e^{-\delta s}N_k\,ds+Me^{\delta\tau_2}\int_{\tau_1}^{\tau_2}e^{-\delta s}N_k\,ds\\
&+\Theta(r')\int_{t_k}^{\tau_1-\varepsilon}\big\|\mathcal{S}_\alpha(\tau_2-s)-\mathcal{S}_\alpha(\tau_1-s)\big\|m(s)\,ds\\
&+2M_*\Theta(r')\int_{\tau_1-\varepsilon}^{\tau_1}e^{-\delta s}m(s)\,ds+Me^{\delta\tau_2}\Theta(r')\int_{\tau_1}^{\tau_2}e^{-\delta s}m(s)\,ds.
\end{aligned}
$$

The right-hand side of the above inequality is independent of $x\in\overline{V}$ and tends to zero as $\tau_2\to\tau_1$, with ε is sufficiently small, since the compactness of $\mathcal{R}_\alpha(t)$, $\mathcal{S}_\alpha(t)$ for $t>0$ implies imply the continuity in the uniform operator topology and the set $\{\mathcal{R}_\alpha(\sigma_n)I_k(\bar{x}_{t_k}):x\in\overline{V},k=1,\dots,m\}$ is relatively compact in H. It remains to prove that the functions $\Gamma_n x$,

$x \in \overline{V}$, are equicontinuous at $t = 0$. Indeed, this is true since $\mathcal{R}_\alpha(\sigma_n)$ is a compact operator. Thus, the set $\{\Gamma_n x : x \in \overline{V}\}$ is equicontinuous.

Next, we prove that $\Gamma_n(\overline{V})(t) = \{\vartheta_n(t) : \vartheta_n(t) \in \Gamma_n(\overline{V})\}$ is relatively compact for every $t \in [0, b]$.

To this end, we decompose Γ_n by $\Gamma_n(\overline{V}) = \Gamma_n^1(\overline{V}) + \Gamma_n^2(\overline{V})$, where the map Γ_n^1 is defined by $\Gamma_n^1 x$, $x \in \overline{V}$ the set ϑ_n^1 such that

$$\vartheta_n^1(t) = \begin{cases} \int_0^t \mathcal{S}_\alpha(t-s)Bu_{\bar{x}}^a(s)\,ds + \int_0^t \mathcal{S}_\alpha(t-s)f(s)\,ds, & t \in [0, t_1], \\ \int_{t_1}^t \mathcal{S}_\alpha(t-s)Bu_{\bar{x}}^a(s)\,ds + \int_{t_1}^t \mathcal{S}_\alpha(t-s)f(s)\,ds, & t \in (t_1, t_2], \\ \vdots \\ \int_{t_m}^t \mathcal{S}_\alpha(t-s)Bu_{\bar{x}}^a(s)\,ds + \int_{t_m}^t \mathcal{S}_\alpha(t-s)f(s)\,ds, & t \in (t_m, b], \end{cases}$$

and the map Γ_n^2 is defined by $\Gamma_n^2 x$, $x \in \overline{V}$ the set ϑ_n^2 such that

$$\vartheta_n^2(t) = \begin{cases} \mathcal{R}_\alpha(t)\varphi(0), & t \in [0, t_1], \\ \mathcal{R}_\alpha(t-t_1)[\bar{x}(t_1^-) + \mathcal{R}_\alpha(\sigma_n)I_1(\bar{x}_{t_1})], & t \in (t_1, t_2], \\ \vdots \\ \mathcal{R}_\alpha(t-t_m)[\bar{x}(t_m^-) + \mathcal{R}_\alpha(\sigma_n)I_m(\bar{x}_{t_m})], & t \in (t_m, b]. \end{cases}$$

We now prove that $\Gamma_n^1(\overline{V})(t) = \{\vartheta_n^1(t) : \vartheta_n^1(t) \in \Gamma_n^1(\overline{V})\}$ is relatively compact for every $t \in [0, b]$. Let $0 < t \le s \le t_1$ be fixed and let ε be a real number satisfying $0 < \varepsilon < t$. For $x \in \overline{V}$, we define

$$\vartheta_n^{1,\varepsilon}(t) = \int_0^{t-\varepsilon} \mathcal{S}_\alpha(t-s)Bu_{\bar{x}}^a(s)\,ds + \int_0^{t-\varepsilon} \mathcal{S}_\alpha(t-s)f(s)\,ds,$$

where $f \in S_{F,\bar{x}}$. Using the compactness of $\mathcal{S}_\alpha(t)$ for $t > 0$, we deduce that the set $U_\varepsilon(t) = \{\vartheta_n^{1,\varepsilon}(t) : x \in \overline{V}\}$ is relatively compact in H for every ε, $0 < \varepsilon < t$. Moreover, for every $x \in \overline{V}$ we have

$$\begin{aligned} \left\|\vartheta_n^1(t) - \vartheta_n^{1,\varepsilon}(t)\right\| &\le \left\|\int_{t-\varepsilon}^t \mathcal{S}_\alpha(t-s)Bu_{\bar{x}}^a(s)\,ds\right\| + \left\|\int_{t-\varepsilon}^t \mathcal{S}_\alpha(t-s)f(s)\,ds\right\| \\ &\le M_* \int_{t-\varepsilon}^t e^{-\delta s}N_0\,ds + M_*\Theta(r')\int_{t-\varepsilon}^t e^{-\delta s}m(s)\,ds. \end{aligned}$$

Similarly, for any $t \in (t_k, t_{k+1}]$ with $k = 1, \ldots, m$. Let $t_k < t \le s \le t_{k+1}$ be fixed and let ε be a real number satisfying $0 < \varepsilon < t$. For $x \in \overline{V}$, we define

$$\vartheta_n^{1,\varepsilon}(t) = \int_{t_k}^{t-\varepsilon} \mathcal{S}_\alpha(t-s)Bu_{\bar{x}}^a(s)\,ds + \int_{t_k}^{t-\varepsilon} \mathcal{S}_\alpha(t-s)f(s)\,ds,$$

where $f \in S_{F,\bar{x}}$. Using the compactness of $\mathcal{S}_\alpha(t)$ for $t > 0$, we deduce that the set $U_\varepsilon(t) = \{\vartheta_n^{1,\varepsilon}(t) : x \in \overline{V}\}$ is relatively compact in H for every ε, $0 < \varepsilon < t$. Moreover, for every $x \in \overline{V}$ we have

$$\begin{aligned} \left\|\vartheta_n^1(t) - \vartheta_n^{1,\varepsilon}(t)\right\| &\le \left\|\int_{t-\varepsilon}^t \mathcal{S}_\alpha(t-s)Bu_{\bar{x}}(s)\,ds\right\| + \left\|\int_{t-\varepsilon}^t \mathcal{S}_\alpha(t-s)f(s)\,ds\right\| \\ &\le M_* \int_{t-\varepsilon}^t e^{-\delta s}N_k\,ds + M_*\Theta(r')\int_{t-\varepsilon}^t e^{-\delta s}m(s)\,ds. \end{aligned}$$

The right-hand side of the above inequality tends to zero as $\varepsilon \to 0$. Since there are relatively compact sets arbitrarily close to the set $U(t) = \{\vartheta_n^1(t) : x \in \overline{V}\}$. Hence, the Arzelá-Ascoli theorem shows that Γ_n^1 is a compact multivalued map.

Next, we show that $\Gamma_n^2(\overline{V})(t) = \{\vartheta_n^2(t) : \vartheta_n^2(t) \in \Gamma_2(\overline{V})\}$ is relatively compact for every $t \in [0,b]$. For all $t \in [0,t_1]$, since $\vartheta_n^2(t) = \mathcal{R}_\alpha(t)\varphi(0)$, by (H1), it follows that $\{\vartheta_n^2(t) : t \in [0,t_1], x \in \overline{V}\}$ is a compact subset of H. On the other hand, for $t \in (t_k, t_{k+1}]$, $k = 1,\dots,m$, and $x \in \overline{V}$, there exists $r'' > 0$ such that

$$\left[\widehat{\vartheta_n^2}\right]_k(t) \in \begin{cases} \mathcal{R}_\alpha(t-t_k)[\bar{x}(t_k^-) + \mathcal{R}_\alpha(\sigma_n)I_k(\bar{x}_{t_k})], & t \in (t_k,t_{k+1}), x \in B_{r''}(0,\mathcal{Y}),\\ \mathcal{R}_\alpha(t_{k+1}-t_k)[\bar{x}(t_k^-) + \mathcal{R}_\alpha(\sigma_n)I_k(\bar{x}_{t_k})], & t = t_{k+1}, x \in B_{r''}(0,\mathcal{Y}),\\ \bar{x}(t_k^-) + \mathcal{R}_\alpha(\sigma_n)I_k(\bar{x}_{t_k}), & t = t_k, x \in B_{r''}(0,\mathcal{Y}),\end{cases}$$

where $B_{r''}(0,\mathcal{Y})$ is a closed ball of radius r''. From (3.7) we obtain $I_k(\bar{x}_{t_k})$ bounded in H. By the compactness of $(\mathcal{R}_\alpha(t))_{t>0}$, we see $\{\mathcal{R}_\alpha(\sigma_n)I_k(\bar{x}_{t_k}) : x \in B_{r''}(0,\mathcal{Y}), k = 1,2,\dots,m\}$ are relatively compact in H. Also, it follows that $[\widehat{\vartheta_n^2}]_k(t)$ is relatively compact in H, for all $t \in [t_k, t_{k+1}]$, $k = 1,\dots,m$. By Lemma 2.7, we infer that $\Gamma_n^2(\overline{V})$ is relatively compact. Moreover, using the continuity of the operator $\mathcal{R}_\alpha(t)$, for all $t \in [0,b]$, we conclude that operator Γ_n^2 is also a compact multivalued map.

As a consequence of the above Steps 1-3, we conclude that $\Phi_n = \Lambda_n + \Gamma_n$ is a condensing map. All of the conditions of Lemma 2.8 are satisfied, we deduce that Φ_n has a fixed point $x_n \in \mathcal{Y}$, which is in turn a mild solution of problem (3.3)-(3.5). Then we have

$$x_n(t) = \begin{cases} \mathcal{R}_\alpha(t)[\varphi(0) - \mathcal{R}_\alpha(\sigma_n)G(0,\varphi,0)] + \mathcal{R}_\alpha(\sigma_n)G(t,\bar{x}_{n,t}) \\ \quad + \int_0^t \mathcal{S}_\alpha(t-s)Bu_{\bar{x}_n}^a(s)\,ds + \int_0^t \mathcal{S}_\alpha(t-s)f_n(s)\,ds, & t \in [0,t_1],\\ \mathcal{R}_\alpha(t-t_1)[\bar{x}_n(t_1^-) + \mathcal{R}_\alpha(\sigma_n)I_1(\bar{x}_{n,t_1}) - \mathcal{R}_\alpha(\sigma_n)G(t_1,\bar{x}_{n,t_1^+})] \\ \quad + \mathcal{R}_\alpha(\sigma_n)G(t,\bar{x}_{n,t}) + \int_{t_1}^t \mathcal{S}_\alpha(t-s)Bu_{\bar{x}_n}^a(s)\,ds \\ \quad + \int_{t_1}^t \mathcal{S}_\alpha(t-s)f_n(s)\,ds, & t \in (t_1,t_2],\\ \vdots \\ \mathcal{R}_\alpha(t-t_m)[\bar{x}_n(t_m^-) + \mathcal{R}_\alpha(\sigma_n)I_m(\bar{x}_{n,t_m}) - \mathcal{R}_\alpha(\sigma_n)G(t_m,\bar{x}_{n,t_m^+})] \\ \quad + \mathcal{R}_\alpha(\sigma_n)G(t,\bar{x}_{n,t}) + \int_{t_m}^t \mathcal{S}_\alpha(t-s)Bu_{\bar{x}_n}^a(s)\,ds \\ \quad + \int_{t_m}^t \mathcal{S}_\alpha(t-s)f_n(s)\,ds, & t \in (t_m,b],\end{cases} \tag{3.10}$$

for $t \in [0,b]$, and some $f_n \in S_{F,\bar{x}_n}$.

Next we will show that the set $\{x_n : n \in \mathbb{N}\}$ is relatively compact in $\mathcal{Y}$. We consider the decomposition $x_n = x_n^1 + x_n^2$ where

$$x_n^1(t) = \begin{cases} -\mathcal{R}_\alpha(t)\mathcal{R}_\alpha(\sigma_n)G(0,\varphi) + \mathcal{R}_\alpha(\sigma_n)G(t,\bar{x}_{n,t}) \\ \quad + \int_0^t \mathcal{S}_\alpha(t-s)Bu_{\bar{x}_n}^a(s)\,ds + \int_0^t \mathcal{S}_\alpha(t-s)f_n(s)\,ds, & t \in [0,t_1],\\ -\mathcal{R}_\alpha(t-t_1)\mathcal{R}_\alpha(\sigma_n)G(t_1,\bar{x}_{n,t_1^+}) + \mathcal{R}_\alpha(\sigma_n)G(t,\bar{x}_{n,t}) \\ \quad + \int_{t_1}^t \mathcal{S}_\alpha(t-s)Bu_{\bar{x}_n}^a(s)\,ds + \int_{t_1}^t \mathcal{S}_\alpha(t-s)f_n(s)\,ds, & t \in (t_1,t_2],\\ \vdots \\ -\mathcal{R}_\alpha(t-t_m)\mathcal{R}_\alpha(\sigma_n)G(t_m,\bar{x}_{n,t_m^+}) + \mathcal{R}_\alpha(\sigma_n)G(t,\bar{x}_{n,t}) \\ \quad + \int_{t_m}^t \mathcal{S}_\alpha(t-s)Bu_{\bar{x}_n}^a(s)\,ds + \int_{t_m}^t \mathcal{S}_\alpha(t-s)f_n(s)\,ds, & t \in (t_m,b],\end{cases}$$

for some $f_n \in S_{F,\bar{x}_n}$, and

$$x_n^2(t) = \begin{cases} \mathcal{R}_\alpha(t)\varphi(0), & t \in [0, t_1], \\ \mathcal{R}_\alpha(t - t_1)[\bar{x}_n(t_1^-) + \mathcal{R}_\alpha(\sigma_n)I_1(\bar{x}_{n,t_1})], & t \in (t_1, t_2], \\ \vdots \\ \mathcal{R}_\alpha(t - t_m)[\bar{x}_n(t_m^-) + \mathcal{R}_\alpha(\sigma_n)I_m(\bar{x}_{n,t_m})], & t \in (t_m, b]. \end{cases}$$

Step 4. $\{x_n^1(t) : n \in \mathbb{N}\}$ is relatively compact in $\mathcal{Y}$.

Claim 1. $\{x_n^1 : n \in \mathbb{N}\}$ is equicontinuous on J.

For $\varepsilon > 0$, $x_n \in \overline{V}$, there exists a constant $\hat{\eta} > 0$ such that, for all $t \in (0, t_1]$ and $\xi \in (0, \hat{\eta})$ with $t + \xi \leq t_1$, we have

$$\begin{aligned} &\|x_n^1(t + \xi) - x_n^1(t)\| \\ &\quad \leq \|[\mathcal{R}_\alpha(t + \xi) - \mathcal{R}_\alpha(t)]\mathcal{R}_\alpha(\sigma_n)G(0, \varphi)\| + \|\mathcal{R}_\alpha(\sigma_n)[G(t + \xi, \bar{x}_{n,t+\xi},) - G(t, \bar{x}_{n,t})]\| \\ &\quad + \left\|\int_t^{t+\xi} \mathcal{S}_\alpha(t + \xi - s)Bu_{\bar{x}_n}(s)\,ds\right\| + \left\|\int_0^t [\mathcal{S}_\alpha(t + \xi - s) - \mathcal{S}_\alpha(t - s)]Bu_{\bar{x}_n}(s)\,ds\right\| \\ &\quad + \left\|\int_t^{t+\xi} \mathcal{S}_\alpha(t + \xi - s)f_n(s)\,ds\right\| + \left\|\int_0^t [\mathcal{S}_\alpha(t + \xi - s) - \mathcal{S}_\alpha(t - s)]f_n(s)\,ds\right\| \\ &\quad \leq \|[\mathcal{R}_\alpha(t + \xi) - \mathcal{R}_\alpha(t)]\mathcal{R}_\alpha(\sigma_n)G(0, \varphi, 0)\| + Me^{\delta\sigma_n}L[\xi + \|\bar{x}_{n,t+\xi} - \bar{x}_{n,t}\|_\mathcal{B}] \\ &\quad + M_* \int_t^{t+\xi} e^{-\delta s}N_0\,ds + \int_0^t \|\mathcal{S}_\alpha(t + \xi - s) - \mathcal{S}_\alpha(t - s)\|N_0\,ds \\ &\quad + M_*\Theta(r') \int_t^{t+\xi} e^{-\sigma s}m(s)\,ds + \Theta(r') \int_0^t \|\mathcal{S}_\alpha(t + \xi - s) - \mathcal{S}_\alpha(t - s)\|m(s)\,ds. \end{aligned}$$

Similarly, for any $t \in (t_k, t_{k+1}]$, $k = 1, \ldots, m$, we have

$$\begin{aligned} &\|x_n^1(t + \xi) - x_n^1(t)\| \\ &\quad \leq \|[\mathcal{R}_\alpha(t + \xi) - \mathcal{R}_\alpha(t)]\mathcal{R}_\alpha(\delta_n)G(t_k, \bar{x}_{n,t_k^+})\| + Me^{\delta\sigma_n}L[\xi + \|\bar{x}_{n,t+\xi} - \bar{x}_{n,t}\|_\mathcal{B}] \\ &\quad + M_* \int_t^{t+\xi} e^{-\sigma s}N_k\,ds + \int_{t_k}^t \|\mathcal{S}_\alpha(t + \xi - s) - \mathcal{S}_\alpha(t - s)\|N_k\,ds \\ &\quad + M_*\Theta(r') \int_t^{t+\xi} e^{-\delta s}m(s)\,ds + \Theta(r') \int_{t_k}^t \|\mathcal{S}_\alpha(t + \xi - s) - \mathcal{S}_\alpha(t - s)\|m(s)\,ds. \end{aligned}$$

Then, for all $t \in (0, b]$, using the compact operator property, we get

$$\|[\mathcal{R}_\alpha(t + \xi) - \mathcal{R}_\alpha(t)]\mathcal{R}_\alpha(\delta_n)G(0, \varphi)\| < \frac{\varepsilon}{6} \tag{3.11}$$

or

$$\|[\mathcal{R}_\alpha(t + \xi) - \mathcal{R}_\alpha(t)]\mathcal{R}_\alpha(\delta_n)G(t_k, \bar{x}_{n,t_k^+})\| < \frac{\varepsilon}{6}, \tag{3.12}$$

and

$$Me^{\delta\sigma_n}L[\xi + \|\bar{x}_{n,t+\xi} - \bar{x}_{n,t}\|_\mathcal{B}] < \frac{\varepsilon}{6}, \tag{3.13}$$

$$M_* \int_t^{t+\xi} e^{-\delta s} N\, ds < \frac{\varepsilon}{6}, \tag{3.14}$$

$$\int_0^t \|\mathcal{S}_\alpha(t+\xi-s) - \mathcal{S}_\alpha(t-s)\| N\, ds < \frac{\varepsilon}{6}, \tag{3.15}$$

$$M_* \Theta(r') \int_t^{t+\xi} e^{-\delta s} m(s)\, ds < \frac{\varepsilon}{6}, \tag{3.16}$$

$$\Theta(r') \int_0^t \|\mathcal{S}_\alpha(t+\xi-s) - \mathcal{S}_\alpha(t-s)\| m(s)\, ds < \frac{\varepsilon}{6}, \tag{3.17}$$

where $N = \max_{0 \le k \le m}\{N_k\}$. By (3.11)-(3.17) one has

$$\|x_n^1(t+\xi) - x_n^1(t)\| < \varepsilon.$$

Therefore, $\{x_n^1(t) : n \in \mathbb{N}\}$ is equicontinuous for $t \in (0, b]$. Clearly $\{x_n^1(0) : n \in \mathbb{N}\}$ is equicontinuous.

Claim 2. $\{x_n^1(t) : n \in \mathbb{N}\}$ is relatively compact in H.

Let $t \in (0, t_1]$, $\varepsilon > 0$, $x_n \in \overline{V}$, there exists $\xi \in (0, t)$ such that

$$\begin{aligned} \|x_n^1(t) - x_n^\xi(t)\| &\le \int_{t-\xi}^t \|\mathcal{S}(t-s) B u_{\bar{x}_n}^a(s)\|\, ds + \int_{t-\xi}^t \|\mathcal{S}(t-s) f_n(s)\|\, ds \\ &\le M_* \int_{t-\xi}^t e^{-\delta s} N_0\, ds + M_* \Theta(r') \int_{t-\xi}^t e^{-\delta s} m(s)\, ds < \varepsilon, \end{aligned}$$

where $x_n^\xi(t) = -\mathcal{R}_\alpha(t)\mathcal{R}_\alpha(\sigma_n) G(0, \varphi) + \mathcal{R}_\alpha(\sigma_n) G(t, \bar{x}_{n,t}) + \int_0^{t-\xi} \mathcal{S}_\alpha(t-s) f_n(s)\, ds$ for some $f_n \in S_{F, \bar{x}_n}$.

Similarly, for any $t \in (t_k, t_{k+1}]$ with $k = 1, \ldots, m$. $\varepsilon > 0$, $x_n \in \overline{V}$, there exists $\xi \in (0, t)$ such that

$$\begin{aligned} \|x_n^1(t) - x_n^\xi(t)\| &\le \int_{t-\xi}^t \|\mathcal{S}(t-s) B u_{\bar{x}_n}^a(s)\|\, ds + \int_{t-\xi}^t \|\mathcal{S}(t-s) f_n(s)\|\, ds \\ &\le M_* \int_{t-\xi}^t e^{-\delta s} N_k\, ds + M_* \Theta(r') \int_{t-\xi}^t e^{-\delta s} m(s)\, ds < \varepsilon, \end{aligned}$$

where $x_n^\xi(t) = -\mathcal{R}_\alpha(t - t_k)\mathcal{R}_\alpha(\sigma_n) G(t_k, \bar{x}_{n, t_k^+}) + \mathcal{R}_\alpha(\sigma_n) G(t, \bar{x}_{n,t}) + \int_{t_k}^{t-\xi} \mathcal{S}_\alpha(t-s) f_n(s)\, ds$ for some $f_n \in S_{F, \bar{x}_n}$. From (H5), we obtain $G(t_k, \bar{x}_{n, t_k^+})$, $G(t, \bar{x}_{n,t})$ are bounded in H. By the compactness of $\mathcal{R}_\alpha(t)$, $\mathcal{S}_\alpha(t)$ for $t > 0$, we see that the set $\{x_n^\xi(t) : n \in \mathbb{N}\}$ is relatively compact in H. Combining the above inequality, one has $\{x_n^1(t) : n \in \mathbb{N}\}$ is relatively compact in H.

Step 5. $\{x_n^2(t) : n \in \mathbb{N}\}$ is relatively compact in $\mathcal{Y}$.

Claim 1. $\{x_n^2 : n \in \mathbb{N}\}$ is equicontinuous on J.

For any $\varepsilon > 0$ and $0 < t < t_1$. Since $\mathcal{R}_\alpha(\sigma_n)$ is a compact operator, we find that the set $W_1 = \{\mathcal{R}_\alpha(\sigma_n) G(0, \varphi)\}$ is relatively compact in H. From the strong continuity of $(\mathcal{R}_\alpha(t))_{t \ge 0}$, for $\varepsilon > 0$, we can choose $0 < \hat{\eta} < b - t$ such that

$$\|(\mathcal{R}_\alpha(t+\xi) - \mathcal{R}_\alpha(t)) v\| < \varepsilon, \quad v \in W_1,$$

when $|\xi| < \hat{\eta}$. For each $x_n \in \overline{V}$, $t \in (0, t_1)$, such that

$$\|x_n^2(t+\xi) - x_n^2(t)\| \le \|[\mathcal{R}_\alpha(t+\xi) - \mathcal{R}_\alpha(t)] \mathcal{R}_\alpha(\delta_n) G(0, \varphi)\| < \varepsilon.$$

Similarly, for any $t \in (t_k, t_{k+1}]$, $k = 1, \dots, m$, and $\varepsilon > 0$. Since $\mathcal{R}_\alpha(\sigma_n)$ is a compact operator, we find that the set $W_2 = \{\mathcal{R}_\alpha(\sigma_n) I_k(\bar{x}_{n,t_k}) : x_n \in \overline{V}\}$ is relatively compact in H. From the strong continuity of $(\mathcal{R}_\alpha(t))_{t \geq 0}$, for $\varepsilon > 0$, we can choose $0 < \hat{\eta} < b - t$ such that

$$\left\| \left(\mathcal{R}_\alpha(t + \xi) - \mathcal{R}_\alpha(t)\right) v \right\| < \frac{\varepsilon}{2(r+1)}, \quad v \in W_2,$$

when $|\xi| < \hat{\eta}$. For each $x_n \in \overline{V}$, $t \in (t_k, t_{k+1}]$, $k = 1, \dots, m$, such that

$$\begin{aligned} \left\| x_n^2(t + \xi) - x_n^2(t) \right\| &\leq \left\| \mathcal{R}_\alpha(t + \xi - t_k) - \mathcal{R}_\alpha(t - t_k) \right\| r' \\ &\quad + \left\| \left[\mathcal{R}_\alpha(t + \xi - t_k) - \mathcal{R}_\alpha(t - t_k)\right] \mathcal{S}_\alpha(\sigma_n) I_k(\bar{x}_{n,t_k}) \right\| < \varepsilon. \end{aligned}$$

As $\xi \to 0$ and ε sufficiently small, the right-hand side of the above inequality tends to zero independently of x_n, so $[\widehat{x_n^2}]_k$, $k = 1, 2, \dots, m$, are equicontinuous.

Claim 2. $\{x_n^2(t) : n \in \mathbb{N}\}$ is relatively compact in H.

Let $t \in (0, t_1]$, $x_n \in \overline{V}$, by (H1), it follows that $\{x_n^2(t) : t \in [0, t_1], x_n \in \overline{V}\}$ is a compact subset of X. Using similar arguments to Step 2, for $t \in (t_k, t_{k+1}]$, $k = 1, \dots, m$, and $x_n \in \overline{V}$, such that

$$\left[\widehat{x_n^2}\right]_k(t) \in \begin{cases} \mathcal{R}_\alpha(t - t_k)[\bar{x}_n(t_k^-) + \mathcal{R}_\alpha(\sigma_n) I_k(\bar{x}_{n,t_k})], & t \in (t_k, t_{k+1}), x_n \in B_{r''}(0, \mathcal{Y}), \\ \mathcal{R}_\alpha(t_{k+1} - t_k)[\bar{x}_n(t_k^-) + \mathcal{R}_\alpha(\sigma_n) I_k(\bar{x}_{n,t_k})], & t = t_{k+1}, x_n \in B_{r''}(0, \mathcal{Y}), \\ \bar{x}_n(t_k^-) + \mathcal{R}_\alpha(\sigma_n) I_k(\bar{x}_{n,t_k}), & t = t_k, x_n \in B_{r''}(0, \mathcal{Y}), \end{cases}$$

where $B_{r''}(0, \mathcal{Y})$ is a closed ball of radius r''. One sees that $[\widehat{x_n^2}]_k(t)$, $k = 1, 2, \dots, m$, is relatively compact for every $t \in [t_k, t_{k+1}]$, and $\{x_n^2(t) : n \in \mathbb{N}\}$ is relatively compact in H.

Thus, we obtain the result that the set $\{x_n : n \in \mathbb{N}\}$ is relatively compact in $\mathcal{Y}$. We may suppose that

$$x_n \to x_* \in \mathcal{Y} \quad \text{as } n \to \infty.$$

Obviously, $x_* \in \mathcal{Y}$, taking the limits in (3.10) one has

$$x_*(t) = \begin{cases} \mathcal{R}_\alpha(t)[\varphi(0) - G(0, \varphi)] + G(t, \bar{x}_{*,t}) \\ \quad + \int_0^t \mathcal{S}_\alpha(t-s) B u_{\bar{x}_*}(s)\, ds + \int_0^t \mathcal{S}_\alpha(t-s) f_*(s)\, ds, & t \in [0, t_1], \\ \mathcal{R}_\alpha(t - t_1)[\bar{x}_*(t_1^-) + I_1(\bar{x}_{*,t_1}) - G(t_1, \bar{x}_{*,t_1^+})] \\ \quad + G(t, \bar{x}_{*,t}) + \int_{t_1}^t \mathcal{S}_\alpha(t-s) B u_{\bar{x}_*}(s)\, ds \\ \quad + \int_{t_1}^t \mathcal{S}_\alpha(t-s) f_*(s)\, ds, & t \in (t_1, t_2], \\ \vdots \\ \mathcal{R}_\alpha(t - t_m)[\bar{x}_n(t_m^-) + I_m(\bar{x}_{*,t_m}) - G(t_m, \bar{x}_{*,t_m^+})] \\ \quad + \mathcal{R}_\alpha(\sigma_n) G(t, \bar{x}_{*,t}) + \int_{t_m}^t \mathcal{S}_\alpha(t-s) B u_{\bar{x}_*}(s)\, ds \\ \quad + \int_{t_m}^t \mathcal{S}_\alpha(t-s) f_*(s)\, ds, & t \in (t_m, b], \end{cases}$$

for $t \in [0, b]$, and some $f_* \in S_{F, \bar{x}_*}$. This implies that x_* is a mild solution of problem (1.1)-(1.3) and the proof of Theorem 3.1 is complete. □

4 Approximate controllability of fractional impulsive control systems

In this section, we present our main result on the approximate controllability of system (1.1)-(1.3). To do this, we also need the following assumption.

(B1) There exists a constant $\widetilde{C} > 0$ such that

$$\|F(t,\psi,y)\| = \sup\{\|f\| : f \in F(t,\psi,y)\} \leq \widetilde{C}, \quad (t,\psi,y) \in J \times \mathcal{B} \times H.$$

Theorem 4.1 *Assume that assumptions of Theorem* 3.1 *hold and, in addition, assumption* (B1) *is satisfied and the linear system corresponding to system* (1.1)-(1.3) *is approximately controllable on* J. *Then system* (1.1)-(1.3) *is approximately controllable on* J.

Proof Let $x^a(\cdot)$ be a fixed point of Φ in $\mathcal{Y}$. By Theorem 3.1, any fixed point of Φ is a mild solution of system (1.1)-(1.3). This means that there is $x^a \in \Phi(x^a)$, that is, there is $f \in S_{F,\bar{x}^a}$ such that

$$x^a(t) = \begin{cases} \mathcal{R}_\alpha(t)[\varphi(0) - G(0,\varphi)] + G(t,\bar{x}_t^a) \\ \quad + \int_0^t \mathcal{S}_\alpha(t-s)Bu_{\bar{x}}^a(s)\,ds + \int_0^t \mathcal{S}_\alpha(t-s)f(s)\,ds, & t \in [0,t_1], \\ \mathcal{R}_\alpha(t-t_1)[\bar{x}^a(t_1^-) + I_1(\bar{x}_{t_1}^a) - G(t_1,\bar{x}_{t_1^+}^a)] \\ \quad + G(t,\bar{x}_t^a) + \int_0^t \mathcal{S}_\alpha(t-s)Bu_{\bar{x}}^a(s)\,ds + \int_{t_1}^t \mathcal{S}_\alpha(t-s)f(s)\,ds, & t \in (t_1,t_2], \\ \vdots \\ \mathcal{R}_\alpha(t-t_m)[\bar{x}^a(t_m^-) + I_m(\bar{x}_{t_m}^a) - G(t_m,\bar{x}_{t_m^+}^a)] \\ \quad + G(t,\bar{x}_t^a) + \int_0^t \mathcal{S}_\alpha(t-s)Bu_{\bar{x}}^a(s)\,ds + \int_{t_m}^t \mathcal{S}_\alpha(t-s)f(s)\,ds, & t \in (t_m,b], \end{cases}$$

where

$$u_{\bar{x}}^a(s) = \begin{cases} B^*\mathcal{S}_\alpha^*(t_1-s)R(a,\Gamma_0^{t_1})[x_0 + \frac{x_b-x_0}{m+1} - \mathcal{R}_\alpha(t_1)[\varphi(0) - G(0,\varphi)] \\ \quad - G(t_1,\bar{x}_{t_1}^a) - \int_0^{t_1} \mathcal{S}_\alpha(t_1-\eta)f(\eta)\,d\eta], & s \in [0,t_1], \\ B^*\mathcal{S}_\alpha^*(t_2-s)R(a,\Gamma_{t_1}^{t_2})[x_0 + \frac{2(x_b-x_0)}{m+1} - \mathcal{R}_\alpha(t-t_1)[\bar{x}^a(t_1^-) \\ \quad + I_1(\bar{x}_{t_1}^a) - G(t_1,\bar{x}_{t_1^+}^a)] - G(t_2,\bar{x}_{t_2}^a) - \int_{t_1}^{t_2} \mathcal{S}_\alpha(t_2-\eta)f(\eta)\,d\eta], & s \in (t_1,t_2], \\ \vdots \\ B^*\mathcal{S}_\alpha^*(b-s)R(a,\Gamma_{t_m}^b)[x_b - \mathcal{R}_\alpha(b-t_m)[\bar{x}^a(t_m^-) \\ \quad + I_m(\bar{x}_{t_m}^a) - G(t_m,\bar{x}_{t_m^+}^a)] - G(b,\bar{x}_b^a) - \int_{t_m}^b \mathcal{S}_\alpha(b-\eta)f(\eta)\,d\eta], & s \in (t_m,b], \end{cases}$$

and it satisfies

$$\begin{cases} x^a(t_1) = x_0 + \frac{x_b-x_0}{m+1} + aR(a,\Gamma_0^{t_1})[x_0 + \frac{x_b-x_0}{m+1} - \mathcal{R}_\alpha(t_1)[\varphi(0) - G(0,\varphi)] \\ \quad - G(t_1,\bar{x}_{t_1}^a) - \int_0^{t_1} \mathcal{S}_\alpha(t_1-\eta)f(\eta)\,d\eta], \\ x^a(t_2) = x_0 + \frac{2(x_b-x_0)}{m+1} + aR(a,\Gamma_{t_1}^{t_2})[x_0 + \frac{2(x_b-x_0)}{m+1} \\ \quad - \mathcal{R}_\alpha(t_2-t_1)[\bar{x}^a(t_1^-) + I_1(\bar{x}_{t_1}^a) - G(t_1,\bar{x}_{t_1^+}^a)] \\ \quad - G(t_2,\bar{x}_{t_2}^a) - \int_{t_1}^{t_2} \mathcal{S}_\alpha(t_2-\eta)f(\eta)\,d\eta], \\ \vdots \\ x^a(b) = x_b + aR(a,\Gamma_{t_m}^b)[x_b - \mathcal{R}_\alpha(b-t_m)[\bar{x}^a(t_m^-) + I_m(\bar{x}_{t_m}^a) \\ \quad - G(t_m,\bar{x}_{t_m^+}^a)] - G(b,\bar{x}_b^a) - \int_{t_m}^b \mathcal{S}_\alpha(b-\eta)f(\eta)\,d\eta]. \end{cases}$$

By condition (B1), we see that

$$\int_0^b \|f(s)\|^2\,ds \leq \widetilde{C}^2 b.$$

Consequently, the sequence $\{f(\eta)\}$ is bounded in $L^2([0,b],H)$. Thus there are subsequences, still denoted by $\{f(\eta)\}$, that converge weakly to, say, $f^{**}(\eta)$ in $L^2([0,b],H)$. The operator

$$l(t) \to \int_0^t \mathcal{S}_\alpha(t-\eta)l(\eta)\,d\eta$$

is also compact on $L^2([0,b],H)$, so one has

$$\int_0^t \mathcal{S}_\alpha(t-\eta)\big[f(\eta)-f^{**}(\eta)\big]\,d\eta \to 0 \quad \text{as } a \to 0^+$$

for all $t \in [0,b]$.

Define

$$p\big(x^a(\cdot)\big) = \begin{cases} x_0 + \frac{x_b - x_0}{m+1} - \mathcal{R}_\alpha(t_1)[\varphi(0) - G(0,\varphi)] \\ \quad - G(t_1, \bar{x}^a_{t_1}) - \int_0^{t_1} \mathcal{S}_\alpha(t_1-\eta)f(\eta)\,d\eta, & t \in [0,t_1], \\ x_0 + \frac{2(x_b - x_0)}{m+1} - \mathcal{R}_\alpha(t_2-t_1)[\bar{x}^a(t_1^-) + I_1(\bar{x}^a_{t_1}) - G(t_1, \bar{x}^a_{t_1^+})] \\ \quad - G(t_2, \bar{x}^a_{t_2}) - \int_{t_1}^{t_2} \mathcal{S}_\alpha(t_2-\eta)f(\eta)\,d\eta, & t \in [t_1,t_2], \\ \vdots \\ x_b - \mathcal{R}_\alpha(b-t_m)[\bar{x}^a(t_m^-) + I_m(\bar{x}^a_{t_m}) - G(t_m, \bar{x}^a_{t_m^+})] \\ \quad - G(b, \bar{x}^a_b) - \int_{t_m}^{b} \mathcal{S}_\alpha(b-\eta)f(\eta)\,d\eta, & t \in [t_m,b], \end{cases}$$

$$q = \begin{cases} x_0 + \frac{x_b - x_0}{m+1} - \mathcal{R}_\alpha(t_1)[\varphi(0) - G(0,\varphi)] \\ \quad - G(t_1, \bar{x}^a_{t_1}) - \int_0^{t_1} \mathcal{S}_\alpha(t_1-\eta)f^{**}(\eta)\,d\eta, & t \in [0,t_1], \\ x_0 + \frac{2(x_b - x_0)}{m+1} - \mathcal{R}_\alpha(t_2-t_1)[\bar{x}^a(t_1^-) + I_1(\bar{x}^a_{t_1}) - G(t_1, \bar{x}^a_{t_1^+})] \\ \quad - G(t_2, \bar{x}^a_{t_2}) - \int_{t_1}^{t_2} \mathcal{S}_\alpha(t_2-\eta)f^{**}(\eta)\,d\eta, & t \in [t_1,t_2], \\ \vdots \\ x_b - \mathcal{R}_\alpha(b-t_m)[\bar{x}^a(t_m^-) + I_m(\bar{x}^a_{t_m}) - G(t_m, \bar{x}^a_{t_m^+})] \\ \quad - G(b, \bar{x}^a_b) - \int_{t_m}^{b} \mathcal{S}_\alpha(b-\eta)f^{**}(\eta)\,d\eta, & t \in [t_m,b]. \end{cases}$$

Hence, for all $t \in [0,b]$, we find that $\|p(x^a) - q\| \to 0$ as $a \to 0^+$. Moreover, from Lemma 2.6, we get, for $t \in [0,t_1]$,

$$\begin{aligned} \big\|x^a(t_1) - x_{t_1}\big\| &= \big\|aR\big(a,\Gamma_0^{t_1}\big)\big(x^a\big)\big\| \\ &\le \big\|aR\big(a,\Gamma_0^{t_1}\big)q\big\| + \big\|aR\big(a,\Gamma_0^{t_1}\big)\big[p\big(x^a\big) - q\big]\big\| \\ &\le \big\|aR\big(a,\Gamma_0^{t_1}\big)q\big\| + \big\|p\big(x^a\big) - q\big\| \to 0 \quad \text{as } a \to 0^+. \end{aligned}$$

Similarly, for any $t \in (t_k, t_{k+1}]$, $k = 1,\ldots,m$, we have

$$\begin{aligned} \big\|x^a(t_{k+1}) - x_{t_{k+1}}\big\| &= \big\|aR\big(a,\Gamma_{t_k}^{t_{k+1}}\big)\big(x^a\big)\big\| \\ &\le \big\|aR\big(a,\Gamma_{t_k}^{t_{k+1}}\big)q\big\| + \big\|aR\big(a,\Gamma_{t_k}^{t_{k+1}}\big)\big[p\big(x^a\big) - q\big]\big\| \\ &\le \big\|aR\big(a,\Gamma_{t_k}^{t_{k+1}}\big)q\big\| + \big\|p\big(x^a\big) - q\big\| \to 0 \quad \text{as } a \to 0^+. \end{aligned}$$

Thus, for all $t \in [0,b]$, we get $\|x^a(b) - x_b\| \to 0$ as $a \to 0^+$. This proves the approximate controllability of system (1.1)-(1.3). □

5 Application

Consider a fractional impulsive fractional partial neutral functional integrodifferential inclusions of the form

$$\begin{aligned} D_t^\alpha N(z_t)(x) &\in \frac{\partial^2}{\partial x^2}N(z_t)(x) + \int_0^t (t-s)^\sigma e^{-\mu(t-s)}\frac{\partial^2}{\partial x^2}N(z_s)(x)\,ds + \tilde{u}(t,x) \\ &\quad + \int_{-\infty}^t b_2\big(t, s-t, x, z(s,x)\big)\,ds \\ &\quad - \int_0^t \int_{-\infty}^s b_3(t) b_4\big(s, \tau - s, x, z(\tau,x)\big)\,d\tau\,ds, \quad 0 \le t \le b, 0 \le x \le \pi, \end{aligned} \tag{5.1}$$

$$z(t,0) = z(t,\pi) = 0, \quad 0 \le t \le b, \tag{5.2}$$

$$z_t(0,x) = 0, \quad 0 \le x \le \pi, \tag{5.3}$$

$$z(\tau,x) = \varphi(\tau,x), \quad \tau \le 0, 0 \le x \le \pi, \tag{5.4}$$

$$\Delta z(t_k,x) = \int_{-\infty}^{t_k} \eta_k(s-t_k) z(s,x)\,ds, \quad k = 1,2,\dots,m, \tag{5.5}$$

where D_t^α is a Caputo fractional partial derivative of order $\alpha \in (1,2)$, σ, and μ are positive numbers, and $\tilde{u}(\cdot)$ is a real function of bounded variation on $[0,b]$. In this system,

$$N(z_t)(x) = z(t,x) - \int_{-\infty}^t b_1(s-t) z(s,x)\,ds.$$

Let $H = L^2([0,\pi])$ with the norm $\|\cdot\|$ and define the operators $A : D(A) \subseteq H \to H$ by $A\omega = \omega''$ with the domain

$$D(A) := \big\{\omega \in H : \omega, \omega' \text{ are absolutely continuous}, \omega'' \in H, \omega(0) = \omega(\pi) = 0\big\}.$$

Then

$$A\omega = \sum_{n=1}^\infty n^2 \langle \omega, \omega_n \rangle \omega_n, \quad \omega \in D(A),$$

where $\omega_n(x) = \sqrt{\frac{2}{\pi}}\sin(nx)$, $n = 1,2,\dots$ is the orthogonal set of eigenvectors of A. It is well known that A generates a strongly continuous semigroup $T(t)$, $t \ge 0$ which is compact, analytic and self-adjoint in X and A is sectorial of type and (P1) is satisfied. The operator $Q(t) : D(A) \subseteq H \to H$, $t \ge 0$, $Q(t)x = t^\sigma e^{-\mu t}x''$ for $x \in D(A)$. Moreover, it is easy to see that conditions (P2) and (P3) in Section 2 are satisfied with $b(t) = t^\sigma e^{-\mu t}$ and $D = C_0^\infty([0,\pi])$, where $C_0^\infty([0,\pi])$ is the space of infinitely differentiable functions that vanish at $x = 0$ and $x = \pi$.

Let $r \ge 0$, $1 \le p < 1$ and let $h : (-\infty, -r] \to \mathbb{R}$ be a nonnegative measurable function which satisfies the conditions (h-5), (h-6) in the terminology of Hino *et al.* [45]. Briefly, this means that h is locally integrable and there is a nonnegative, locally bounded function γ on $(-\infty, 0]$ such that $h(\xi + \tau) \le \gamma(\xi)h(\tau)$ for all $\xi \le 0$ and $\xi \in (-\infty, -r) \setminus N_\xi$, where $N_\xi \subseteq (-\infty, -r)$ is a set whose Lebesgue measure zero. We denote by $\mathcal{PC}_r \times L^p(h,H)$ the set consisting of all classes of functions $\varphi : (-\infty, 0] \to H$ such that $\varphi_{|[-r,0]} \in \mathcal{PC}([-r,0],H)$,

$\varphi(\cdot)$ is Lebesgue measurable on $(-\infty,-r)$, and $h\|\varphi\|^p$ is Lebesgue integrable on $(-\infty,-r)$. The semi-norm is given by

$$\|\varphi\|_{\mathcal{B}} = \sup_{-r\leq\tau\leq 0} \left\|\varphi(\tau)\right\| + \left(\int_{-\infty}^{-r} h(\tau)\left\|\varphi(\tau)\right\|^p d\tau\right)^{1/p}.$$

The space $\mathcal{B} = \mathcal{PC}_r \times L^p(h,H)$ satisfies axioms (A)-(C). Moreover, when $r = 0$ and $p = 2$, we can take $\tilde{H} = 1$, $M(t) = \gamma(-t)^{1/2}$ and $K(t) = 1 + (\int_{-t}^{0} h(\tau)\,d\tau)^{1/2}$ for $t \geq 0$ (see [45], Theorem 1.3.8, for details).

Additionally, we will assume that

(i) The functions $b_1, b_3 : \mathbb{R} \to \mathbb{R}$ are continuous, and $L_G = (\int_{-\infty}^{0} \frac{(b_1(s))^2}{h(s)}\,ds)^{\frac{1}{2}} < \infty$.

(ii) The functions $b_2, b_4 : \mathbb{R}^4 \to \mathbb{R}$ are continuous and there exist continuous functions $a_1, a_2, a_3, a_4 : \mathbb{R} \to \mathbb{R}$ such that

$$\left|b_2(t,s,x,y)\right| \leq a_1(t)a_2(s)|y|, \quad (t,s,x,y) \in \mathbb{R}^4,$$

$$\left|b_4(t,s,x,y)\right| \leq a_3(t)a_4(s)|y|, \quad (t,s,x,y) \in \mathbb{R}^4,$$

with $\widetilde{L}_1 = (\int_{-\infty}^{0} \frac{(a_2(s))^2}{h(s)}\,ds)^{\frac{1}{2}} < \infty$, $\widetilde{L}_2 = (\int_{-\infty}^{0} \frac{(a_4(s))^2}{h(s)}\,ds)^{\frac{1}{2}} < \infty$.

(iii) The functions $\eta_k : \mathbb{R} \to \mathbb{R}$, $k = 1,2,\ldots,m$, are continuous, and $L_k = (\int_{-\infty}^{0} \frac{(\eta_k(s))^2}{h(s)}\,ds)^{\frac{1}{2}} < \infty$ for every $k = 1,2,\ldots,m$.

Take $\varphi \in \mathcal{B} = \mathcal{PC}_0 \times L^2(h,H)$ with $\varphi(s)(\psi) = \varphi(s,\psi)$. Let $N, G : \mathcal{B} \to H$, $F : [0,b] \times \mathcal{B} \times H \to \mathcal{P}_{bd,cl,cv}(H)$ be the operators defined by

$$N(\psi)(x) = \psi(0,x) + G(\psi)(x),$$

$$G(\psi)(x) = \int_{-\infty}^{0} b_1(s)\psi(s,x)\,ds,$$

$$F(t,\psi,\tilde{B}\psi)(x) = \int_{-\infty}^{0} b_2\big(t,s,x,\psi(s,x)\big)\,ds + \tilde{B}\psi(x),$$

where

$$\tilde{B}\psi(x) = \int_0^t \int_{-\infty}^{0} b_3(t)b_4\big(s,\tau-s,\psi(\tau,x)\big)\,d\tau\,ds.$$

Also define the maps I_k and B by

$$I_k(\psi)(x) = \int_{-\infty}^{0} \eta_k(s)\psi(s,x)\,ds,$$

$$(Bu)(t)(x) = \tilde{u}(t,x).$$

Using these definitions, we can represent system (5.1)-(5.5) in the abstract form (1.1)-(1.3). Moreover, G, I_k are bounded linear operators on $\mathcal{B}$ with $\|G\| \leq L_G$ and $\|I_k\| \leq L_k$, $k = 1,2,\ldots,m$. Using (ii), we see that F is continuous and $\|F\|_{L(\mathcal{B},H)} \leq L_F$, where $L_F = \widetilde{L}_1\|a_1\|_\infty + b\widetilde{L}_2\|b_3\|_\infty\|a_3\|_\infty$. Further, we can impose some suitable conditions on the above-defined functions to verify the assumptions on Theorem 3.1. Therefore, assumptions (H1)-(H6) and (B1) all hold; the associated linear system of (5.1)-(5.5) is not exactly controllable but it is approximately controllable. Hence by Theorems 3.1 and 4.1, system (5.1)-(5.5) is approximately controllable on $[0, b]$.

Competing interests
The authors declare that they have no competing interests.

Authors' contributions
The authors completed the paper, and read and approved the final manuscript.

Acknowledgements
The authors would like to thank the editor and the reviewers for their constructive comments and suggestions. This work is supported by the National Natural Science Foundation of China (Grant No. 11461019) and is supported by the President Found of Scientific Research Innovation and Application of Hexi University (Grant No. xz2013-10).

References
1. Benchohra, M, Henderson, J, Ntouyas, SK: Impulsive Differential Equations and Inclusions. Hindawi Publishing Corporation, New York (2006)
2. Haddad, WM, Chellabonia, V, Nersesov, SG, Sergey, G: Impulsive and Hybrid Dynamical Systems: Stability, Dissipativity, and Control. Princeton University Press, Princeton (2006)
3. Hernández, E, Pierri, M, Goncalves, G: Existence results for an impulsive abstract partial differential equation with state-dependent delay. Comput. Math. Appl. **52**, 411-420 (2006)
4. Kilbas, AA, Srivastava, HM, Trujillo, JJ: Theory and Applications of Fractional Differential Equations. North-Holland Mathematics Studies, vol. 204. Elsevier, Amsterdam (2006)
5. Lakshmikantham, V: Theory of fractional differential equations. Nonlinear Anal. **60**, 3337-3343 (2008)
6. Lakshmikantham, V, Vatsala, AS: Basic theory of fractional differential equations. Nonlinear Anal. **69**, 2677-2682 (2008)
7. El-Borai, MM: Some probability densities and fundamental solutions of fractional evolution equations. Chaos Solitons Fractals **14**, 433-440 (2002)
8. El-Borai, MM: Semigroups and some nonlinear fractional differential equations. Appl. Math. Comput. **149**, 823-831 (2004)
9. Yan, Z: Existence results for fractional functional integrodifferential equations with nonlocal conditions in Banach spaces. Ann. Pol. Math. **97**, 285-299 (2010)
10. Zhou, Y, Jiao, F: Existence of mild solutions for fractional neutral evolution equations. Comput. Math. Appl. **59**, 1063-1077 (2010)
11. Debbouche, A, Nieto, JJ: Sobolev type fractional abstract evolution equations with nonlocal conditions and optimal multi-controls. Appl. Math. Comput. **245**, 74-85 (2014)
12. Debbouche, A, Torres, DFM: Sobolev type fractional dynamic equations and optimal multi-integral controls with fractional nonlocal conditions. Fract. Calc. Appl. Anal. **18**, 95-121 (2015)
13. Santos, JPC, Arjunan, MM, Cuevas, C: Existence results for fractional neutral integro-differential equations with state-dependent delay. Comput. Math. Appl. **62**, 1275-1283 (2011)
14. Agarwal, RP, de Andrade, B, Siracusa, G: On fractional integro-differential equations with state-dependent delay. Comput. Math. Appl. **62**, 1143-1149 (2011)
15. de Andrade, B, Santos, JPC: Existence of solutions for a fractional neutral integro-differential equation with unbounded delay. Electron. J. Differ. Equ. **2012**, 90 (2012)
16. Mophou, GM: Existence and uniqueness of mild solutions to impulsive fractional differential equations. Nonlinear Anal. **72**, 1604-1615 (2010)
17. Shu, X-B, Lai, Y, Chen, Y: The existence of mild solutions for impulsive fractional partial differential equations. Nonlinear Anal. **74**, 2003-2011 (2011)
18. Dabas, J, Chauhan, A: Existence and uniqueness of mild solution for an impulsive neutral fractional integro-differential equation with infinite delay. Math. Comput. Model. **57**, 754-763 (2013)
19. Balachandran, K, Kiruthika, S, Trujillo, JJ: Existence results for fractional impulsive integrodifferential equations in Banach spaces. Commun. Nonlinear Sci. Numer. Simul. **16**, 1970-1977 (2011)
20. Balachandran, K, Park, JY: Controllability of fractional integrodifferential systems in Banach spaces. Nonlinear Anal. Hybrid Syst. **3**, 363-367 (2009)
21. Wang, J, Zhou, Y: Complete controllability of fractional evolution systems. Commun. Nonlinear Sci. Numer. Simul. **17**, 4346-4355 (2012)
22. Sakthivel, R, Mahmudov, NI, Nieto, JJ: Controllability for a class of fractional-order neutral evolution control systems. Appl. Math. Comput. **218**, 10334-10340 (2012)
23. Yan, Z: Controllability of fractional-order partial neutral functional integrodifferential inclusions with infinite delay. J. Franklin Inst. **348**, 2156-2173 (2011)
24. Tai, Z, Wang, X: Controllability of fractional-order impulsive neutral functional infinite delay integrodifferential systems in Banach spaces. Appl. Math. Lett. **22**, 1760-1765 (2009)
25. Debbouchea, A, Baleanu, D: Controllability of fractional evolution nonlocal impulsive quasilinear delay integro-differential systems. Comput. Math. Appl. **62**, 1442-1450 (2011)
26. Liu, Z, Li, X: On the controllability of impulsive fractional evolution inclusions in Banach spaces. J. Optim. Theory Appl. **156**, 167-182 (2013)
27. Triggiani, R: A note on the lack of exact controllability for mild solutions in Banach spaces. SIAM J. Control Optim. **15**, 407-411 (1977)
28. Mahmudov, NI, Denker, A: On controllability of linear stochastic systems. Int. J. Control **73**, 144-151 (2000)

29. Dauer, JP, Mahmudov, NI: Approximate controllability of semilinear functional equations in Hilbert spaces. J. Math. Anal. Appl. **273**, 310-327 (2002)
30. Sakthivel, R, Ren, Y, Mahmudov, NI: On the approximate controllability of semilinear fractional differential systems. Comput. Math. Appl. **62**, 1451-1459 (2011)
31. Sakthivel, R, Ren, Y: Approximate controllability of fractional differential equations with state-dependent delay. Results Math. **63**, 949-963 (2013)
32. Sakthivel, R, Suganya, S, Anthoni, SM: Approximate controllability of fractional stochastic evolution equations. Comput. Math. Appl. **63**, 660-668 (2012)
33. Kumar, S, Sukavanam, N: Approximate controllability of fractional order semilinear systems with bounded delay. J. Differ. Equ. **252**, 6163-6174 (2012)
34. Sukavanam, N, Kumar, S: Approximate controllability of fractional order semilinear delay systems. J. Optim. Theory Appl. **151**, 373-384 (2011)
35. Yan, Z: Approximate controllability of partial neutral functional differential systems of fractional order with state-dependent delay. Int. J. Control **85**, 1051-1062 (2012)
36. Debbouche, A, Torres, DFM: Approximate controllability of fractional nonlocal delay semilinear systems in Hilbert spaces. Int. J. Control **86**, 1577-1585 (2013)
37. Debbouche, A, Torres, DFM: Approximate controllability of fractional delay dynamic inclusions with nonlocal control conditions. Appl. Math. Comput. **243**, 161-175 (2014)
38. Benchohra, M, Litimein, S, N'Guérékata, G: On fractional integro-differential inclusions with state-dependent delay in Banach spaces. Appl. Anal. **92**, 33-35 (2013)
39. Yan, Z: On a nonlocal problem for fractional integrodifferential inclusions in Banach spaces. Ann. Pol. Math. **101**, 87-103 (2011)
40. O'Regan, D: Nonlinear alternatives for multivalued maps with applications to operator inclusions in abstract spaces. Proc. Am. Math. Soc. **127**(12), 3557-3564 (1999)
41. Yosida, K: Functional Analysis, 6th edn. Springer, Berlin (1980)
42. Fitzpatrick, PM, Petryshyn, WV: Fixed point theorems for multivalued noncompact acyclic mappings. Pac. J. Math. **54**, 17-23 (1974)
43. Hale, JK, Kato, J: Phase spaces for retarded equations with infinite delay. Funkc. Ekvacioj **21**, 11-41 (1978)
44. Lasota, A, Opial, Z: An application of the Kakutani-Ky Fan theorem in the theory of ordinary differential equations. Bull. Acad. Pol. Sci., Sér. Sci. Math. Astron. Phys. **13**, 781-786 (1965)
45. Hino, Y, Murakami, S, Naito, T: Functional-Differential Equations with Infinite Delay. Lecture Notes in Mathematics, vol. 1473. Springer, Berlin (1991)

4

Mean square containment control problems of multi-agent systems under Markov switching topologies

Guoying Miao* and Tao Li

*Correspondence:
mgyss66@163.com
School of Information and Control, Nanjing University of Information Science and Technology, Nanjing 210044, P.R. China

Abstract

The paper investigates containment control for multi-agent systems under Markov switching topologies. By using graph theory and the tools of stochastic analysis, sufficient conditions of mean square containment control problems are derived for the second-order multi-agent systems. Then the obtained results are further extended to high-order multi-agent systems.

Keywords: multi-agent systems; containment control; mean square consensus

1 Introduction

Recently, cooperative control for multi-agent systems has attracted increasing attention, due to its many applications in different fields. As is well known, consensus is the basic problem of cooperative control for multi-agent systems, which means that every agent tends to the same value in a team. Moreover, the topic of consensus problems with multiple leaders is interesting, which is called containment control for multi-agent systems.

Containment control for multi-agent systems means that all followers are driven into the convex hull generated by the leaders. For example, when some robots are used to carry poisonous materials, since they do not pollute other places, a group of robots are needed to drive them in the designed route, which is called leaders. At present, many results about containment control have been obtained [1–12]. In [1], dynamic containment algorithms based on observers for the high-order continuous-time multi-agent systems were proposed under the fixed topology. Then the results of continuous-time multi-agent systems were extended to that of the discrete-time multi-agent systems. In [2], the containment control problem of double-integrator dynamics with multiple leaders was discussed, where velocities of leaders are unavailable. In [3], the output feedback algorithm and state feedback algorithm were proposed for solving the containment control problem of the discrete-time multi-agent systems, in which time delay is considered in communication networks of multi-agent systems. By using the Lyapunov functional method and the tools of the linear matrix inequality, containment control problems of second-order multi-agent with time delays were investigated, where the stationary leaders and dynamic leaders were considered in [4], respectively. In [5], containment algorithms based on the sampled data for the second-order multi-agent systems were given, where the necessary and sufficient conditions of multi-agent systems were derived.

There are deterministic topologies in the above literature. In fact, communication topologies of networks usually change randomly. By using the graph theory and knowledge of stochastic analysis, mean square consensus of the discrete-time multi-agent systems was discussed under Markov switching topologies in [6]. Then the authors in [7] extended the results in [6] to leader-following consensus of discrete-time multi-agent systems under Markov switching topologies. Under randomly switching topologies, consensus conditions of continuous-time and discrete-time high-order multi-agent systems were given, respectively, where the random link failures between agents were discussed in [8]. In [9], the convergence speed of the first-order discrete-time multi-agent systems was studied. Then the authors in [10] extended the results in [9] to that of the second-order and high-order multi-agent system, respectively. Moreover, under random switching topologies, target containment control for the second-order multi-agent systems was discussed in [11], where switching topologies were driven by a Markov process. In addition, mean square containment control problems of the first-order and second-order multi-agent systems with communication noises was investigated in [12].

Inspired by the results in [1–12], this paper further investigates the containment control for multi-agent systems under Markov switching topologies. In this paper, containment algorithms for continuous-time and discrete-time multi-agent systems are given, respectively. By using the graph theory and theory of stochastic analysis, sufficient conditions of mean square containment control for multi-agent systems are derived. Then we extend the results of the second-order multi-agent systems to high-order multi-agent systems.

2 Mean square containment control for discrete-time multi-agent systems

Before we give the main results, basic graph theory is introduced. Suppose that there are N agents in the topology. $\mathcal{G} = \{\mathcal{V}, \mathcal{A}, \mathcal{E}\}$ denotes the graph corresponding to the communication topology, where $\mathcal{V} = \{1, \ldots, N\}$ is the set of nodes, $\mathcal{A} = [a_{ij}]_{N\times N}$ is the adjacency matrix. If $(i, j) \in \mathcal{E}$ holds, then $a_{ij} = 1$ and otherwise $a_{ij} = 0$. The Laplacian matrix is defined as $L = [l_{ij}]_{N\times N}$, where $l_{ij} = -a_{ij}$ with $i \neq j$ and $l_{ii} = \sum_{j=1}^{N} a_{ij}$ with $i = j$.

Consider the ith follower's dynamic of the second-order multi-agent, described as follows:

$$\begin{aligned} x_i(k+1) &= x_i(k) + v_i(k)T, \\ v_i(k+1) &= v_i(k) + u_i(k)T, \quad i = 1, \ldots, M, \end{aligned} \tag{1}$$

where $x_i(k)$, $v_i(k)$, $u_i(k)$ represent position, velocity, and input control of the ith agent, respectively.

The leader's dynamic is denoted as follows:

$$\begin{aligned} x_i(k+1) &= x_i(k) + v_i(k)T, \\ v_i(k+1) &= v_i(k), \quad i = M+1, \ldots, N. \end{aligned} \tag{2}$$

The containment algorithm is proposed as follows:

$$u_i(k) = \sum_{j\in F\cup L} a_{ij}\big(x_j(k) - x_i(k)\big), \quad i = 1, \ldots, M, \tag{3}$$

where $F = \{1, \ldots, M\}$ denotes the set of followers and $L = \{M+1, \ldots, N\}$ is the set of leaders.

Definition ([6, 12]) Under Markov switching topologies, the mean square containment control problem of the multi-agent system is solved for any initial distribution, if the followers are driven into the convex hull generated by the leader's states, $\lim_{k\to\infty} E(x_i(k) - x_i^*)^2 = 0$ for $i = 1,\dots,M$, where $x_i^* \in co_L = \{\sum_{j=M+1}^{N} \alpha_j x_j(k), j = M + 1,\dots,N, \alpha_j \geq 0, \sum_{j=M+1}^{N} \alpha_j = 1\}$.

$D = [d_{ij}]_{N\times N}$ is a row stochastic matrix, where $d_{ij} > 0$ if $(i,j) \in \mathcal{E}$ and otherwise $d_{ij} = 0$. Then D is divided into the following form:

$$D = \begin{bmatrix} D_1 & D_2 \\ 0 & I_{N-M} \end{bmatrix}, \tag{4}$$

where $D_1 \in R^{M\times M}$, $D_2 \in R^{M\times(N-M)}$, I_{N-M} is an identity matrix with dimension $N - M$. Set $\eta_i(k) = [x_i^T(k)\ v_i^T(k)]^T$, $\xi_i(k) = \sum_{j=1}^{M} a_{ij}(\eta_i(k) - \eta_j(k))$ for $i = 1,\dots,M$, $\eta_F(k) = [\eta_1^T(k)\cdots \eta_M^T(k)]^T$, $\eta_L(k) = [\eta_{M+1}^T(k)\cdots \eta_N^T(k)]^T$, $\xi(k) = [\xi_1^T(k)\cdots \xi_M^T(k)]^T$. Then we have

$$\xi(k) = \big((I_M - D_1)\otimes I_2\big)\eta_F(k) - (D_2\otimes I_2)\eta_L(k). \tag{5}$$

Following the transform methods in [1, 3], we have

$$\begin{aligned}\eta_F(k+1) = {}& \left(I_M \otimes \begin{bmatrix} 1 & T \\ 0 & 1 \end{bmatrix}\right)\eta_F(k) - (I_M - D_1)\otimes \begin{bmatrix} 0 & 0 \\ T & T \end{bmatrix}\eta_F(k) \\ & + D_2 \otimes \begin{bmatrix} 0 & 0 \\ T & T \end{bmatrix}\eta_L(k). \end{aligned}\tag{6}$$

Then (2) can be rewritten as follows:

$$\eta_L(k+1) = \left(I_{N-M}\otimes \begin{bmatrix} 1 & T \\ 0 & 1 \end{bmatrix}\right)\eta_L(k). \tag{7}$$

Noting (5), we have

$$\eta_F(k) = \big((I_M - D_1)^{-1}\otimes I_2\big)\xi(k) + \big((I_M - D_1)^{-1}D_2\otimes I_2\big)\eta_L(k). \tag{8}$$

Taking together (6) with (8), we have

$$\xi(k+1) = \Omega\xi(k), \tag{9}$$

where $\Omega = I_M \otimes \left[\begin{smallmatrix} 1 & T \\ 0 & 1 \end{smallmatrix}\right] - (I_M - D_1)\otimes \left[\begin{smallmatrix} 0 & 0 \\ T & T \end{smallmatrix}\right]$.

Observe

$$(I_M - D_1)\otimes \begin{bmatrix} 0 & 0 \\ T & T \end{bmatrix} = \begin{bmatrix} 0 & 0 & \cdots & 0 & 0 \\ 1-d_{11}T & 1-d_{11}T & \cdots & -d_{1M}T & -d_{1M}T \\ \vdots & \vdots & \ddots & \vdots & \vdots \\ 0 & 0 & \cdots & 0 & 0 \\ -d_{M1}T & -d_{M1}T & \cdots & 1-d_{MM}T & 1-d_{MM}T \end{bmatrix},$$

which implies

$$\Omega = \bar{A} + \bar{B}T,$$

where $\bar{A} = I_M \otimes I_2$,

$$\bar{B} = \begin{bmatrix} 0 & 1 & \cdots & 0 & 0 \\ -(1-d_{11}) & -(1-d_{11}) & \cdots & -d_{1M} & -d_{1M} \\ \vdots & \vdots & \ddots & \vdots & \vdots \\ 0 & 0 & \cdots & 0 & 1 \\ -d_{M1} & -d_{M1} & \cdots & -(1-d_{MM}) & -(1-d_{MM}) \end{bmatrix}.$$

Then there exists an orthogonal matrix W, such that $W^T\bar{B}W = \tilde{B}$, where

$$\tilde{B} = \begin{bmatrix} -(I_M - D_1) & -(I_M - D_1) \\ I_M & 0 \end{bmatrix}. \tag{10}$$

Setting $\xi(k) = W\tilde{\xi}(k)$, (9) turns into the form

$$\tilde{\xi}(k+1) = \tilde{\Omega}\tilde{\xi}(k), \tag{11}$$

where $\tilde{\Omega} = \bar{A} + \tilde{B}T$.

Assumption 1 Assume that there exists a path from the leaders to each follower.

Theorem 1 *Suppose that Assumption* 1 *holds. In the fixed directed topology, under the algorithm* (3), *the containment control problem for system* (1) *can be solved, that is, follower* (1) *is driven into the leaders' sets* (2).

Proof Since there exists a path from the leaders to each follower, based on Lemma 2 in [1], we know that all eigenvalues of D_1 are in the open unit disk. We know the eigenvalues of $\bar{A}$ are equal to 1 with geometric multiplicity 2M. The characteristic polynomial of $\tilde{B}$ is denoted as follows:

$$\lambda_i^2 + \mu_i\lambda + \mu_i = 0, \quad i = 1, \ldots, M, \tag{12}$$

where λ_i is the eigenvalue of $\tilde{B}$, μ_i is the eigenvalue of $I_M - D_1$. From Lemma 2 in [1], we know that μ_i has the positive real part. Then we see that λ_i is the negative real part. Following the methods in [6], we see that the eigenvalues of $\bar{A}$ are perturbed by small negative real parts of $\lambda_i T$. For small T, $\rho(\Omega) < 1$. $\lim_{k\to\infty}\tilde{\xi}(k) = 0$ is derived, which implies $\lim_{k\to\infty}\eta_F(k) = -\lim_{k\to\infty}((I_M - D_1)^{-1}D_2 \otimes I_2)\eta_L(k)$. Then the containment control problem for system (1) can be solved. □

In the following section, we consider topologies driven by a Markov process. Under Markov switching topologies, (11) can be rewritten as follows:

$$\tilde{\xi}(k+1) = \tilde{\Omega}_{\sigma(k)}\tilde{\xi}(k), \tag{13}$$

where $\sigma(k) = [1, 2, \ldots, s]$ is the Markov switching process.

Assumption 2 Under Markov switching topologies, for each follower, there exists at least one path form one leader to that follower in the union of the topologies $\{\mathcal{G}_1, \ldots, \mathcal{G}_s\}$.

Lemma 1 *Under Assumption* 2, *all the eigenvalues of* $I_M - D_1$ *have positive real parts.*

Proof According to Lemma 2 in [1], we obtain Lemma 1 immediately. □

Theorem 2 *Assume that Assumption* 2 *holds. Under the containment algorithm* (3), *system* (1) *can solve the containment control problem in the mean square sense.*

Proof Containment control problem of system (1) can be converted to the mean square consensus of the systems (13). According to Theorem 3 in [6], system (13) reaches mean square consensus if and only if $\rho(\Lambda) < 1$, where $\Lambda = (\pi^T \otimes I_{4M^2})\,\mathrm{diag}(\tilde{\Omega}_i \otimes \tilde{\Omega}_i)$ with $\pi = [\pi_{ij}]_{M\times M}$ being the transition probability matrix. Then we have

$$\begin{aligned}\Lambda &= \left(\pi^T \otimes I_{4M^2}\right)\mathrm{diag}(\tilde{\Omega}_i \otimes \tilde{\Omega}_i)\\ &= \left(\pi^T \otimes I_{4M^2}\right)\mathrm{diag}\left((\bar{A} + \tilde{B}T) \otimes (\bar{A} + \tilde{B}T)\right)\\ &= \left(\pi^T \otimes I_{4M^2}\right) + T\left(\pi^T \otimes I_{4M^2}\right)\mathrm{diag}(\tilde{B}_i \otimes I_{2M} + I_{2M} \otimes \tilde{B}_i)\\ &\quad + T^2\left(\pi^T \otimes I_{4M^2}\right)\mathrm{diag}(\tilde{B}_i \otimes \tilde{B}_i).\end{aligned} \tag{14}$$

For small sampling time T, the last term can be neglected compared to former ones. Since the eigenvalues $\tilde{B}_i$ have negative real parts and Assumption 2 holds, the eigenvalues of $T(\pi^T \otimes I_{4M^2})\,\mathrm{diag}(\tilde{B}_i \otimes I_{2M} + I_{2M} \otimes \tilde{B}_i)$ have negative real parts. Thus, following Theorem 4 in [6], the unit eigenvalues of $\pi^T \otimes I_{4M^2}$ are perturbed by the negative parts of $T(\pi^T \otimes I_{4M^2})\,\mathrm{diag}(\tilde{B}_i \otimes I_{2M} + I_{2M} \otimes \tilde{B}_i)$. Thus, the eigenvalues of Λ are less than 1. Then $\rho(\Lambda) < 1$ is obtained. System (13) solves the mean square consensus problem. Then $\lim_{k\to\infty} E(\tilde{\xi}(k))^2 = 0$ is obtained. Therefore, we have $\lim_{k\to\infty} E(((I_M - D_1) \otimes I_2)\eta_F(k) - (D_2 \otimes I_2)\eta_L(k))^2 = 0$, which implies $\lim_{k\to\infty} E(\eta_F(k) - ((I_M - D_1)^{-1}D_2 \otimes I_2)\eta_L(k))^2 = 0$. The mean square containment control of system (1) can solve the containment control problem in the mean square sense. The proof is completed. □

Remark 1 We extend the results in [1] to the case under Markov switching topologies. The mean square consensus of multi-agent systems was investigated in [6], while the mean square containment control problems are discussed in this paper.

Remark 2 We can extend the results of Theorem 2 to the mean square containment control problem of high-order discrete-time multi-agent systems. In order to save space, we omit it here.

3 Mean square containment control for continuous-time multi-agent systems

In this section, we discuss containment control for the continuous-time multi-agent systems under Markov switching topologies. Each follower's dynamic is denoted

$$\begin{aligned}\dot{x}_i(t) &= v_i(t),\\ \dot{v}_i(t) &= u_i(t), \quad i = 1, \ldots, M,\end{aligned} \tag{15}$$

where $x_i(t)$, $v_i(t)$, $u_i(t)$ denote position, velocity, input control of the ith follower, respectively. The leader's dynamic can be described as follows:

$$\begin{aligned}\dot{x}_i(t) &= v_i(t),\\ \dot{v}_i(t) &= 0, \quad i = M+1,\ldots,N,\end{aligned} \tag{16}$$

where $x_i(t)$, $v_i(t)$ denote the ith leader's position and velocity, respectively. The containment algorithm is proposed as follows:

$$u_i(t) = \sum_{j\in F\cup L} a_{ij}\left(x_j(t) - x_i(t)\right) + \sum_{j\in F\cup L} a_{ij}\left(v_j(t) - v_i(t)\right), \tag{17}$$

where F and L are the same as those in Theorem 1. $L = [a_{ij}]_{N\times N}$ is the Laplacian matrix, which can be divided into four parts:

$$L = \begin{bmatrix} L_1 & L_2 \\ 0 & 0 \end{bmatrix}, \tag{18}$$

where $L_1 \in R^{M\times M}$, $L_2 \in R^{M\otimes(N-M)}$. Set $\zeta_i(t) = [x_i^T(t)\ v_i^T(t)]^T$, $\zeta_F(t) = [\zeta_1^T(t)\cdots\zeta_M^T(t)]^T$, $\zeta_L(t) = [\zeta_{M+1}^T(t)\cdots\zeta_N^T(t)]^T$, $\xi_i(t) = \sum_{j\in F\cup L} a_{ij}(\zeta_i(t) - \zeta_j(t))$, $\xi(t) = [\xi_1^T(t)\cdots\xi_M^T(t)]^T$. Then we obtain $\xi(t) = (L_1\otimes I_2)\zeta_F + (L_2\otimes I_2)\zeta_L$. Combining (15)-(17), we have

$$\begin{aligned}\dot{\zeta}_F &= \left(I_M\otimes\begin{bmatrix}0 & 1\\ 0 & 0\end{bmatrix}\right)\zeta_F(t) - \left(L_1\otimes\begin{bmatrix}0 & 0\\ 1 & 1\end{bmatrix}\right)\zeta_F(t)\\ &\quad - \left(L_2\otimes\begin{bmatrix}0 & 0\\ 1 & 1\end{bmatrix}\right)\zeta_L(t),\\ \dot{\zeta}_L(t) &= \left(I_{N-M}\otimes\begin{bmatrix}0 & 1\\ 0 & 0\end{bmatrix}\right)\zeta_L(t).\end{aligned} \tag{19}$$

Taking the derivation of $\xi(t)$, we have

$$\dot{\xi}(t) = (L_1\otimes I_2)\dot{\zeta}_F + (L_2\otimes I_2)\dot{\zeta}_L. \tag{20}$$

In view of $\zeta_F(t) = (L_1^{-1}\otimes I_2)\xi(t) - (L_1^{-1}L_2\otimes I_2)\zeta_L(t)$, (20) can be written as follows:

$$\dot{\xi}(t) = \Psi\xi(t), \tag{21}$$

where $\Psi = \left(I_M\otimes\left[\begin{smallmatrix}0 & 1\\ 0 & 0\end{smallmatrix}\right]\right) - \left(L_1\otimes\left[\begin{smallmatrix}0 & 0\\ 1 & 1\end{smallmatrix}\right]\right)$. There exists an orthogonal matrix W, such that

$$\begin{aligned}\tilde{\Psi} &= W^T\psi W\\ &= \begin{bmatrix}-L_1 & -L_1\\ I_M & 0\end{bmatrix}.\end{aligned} \tag{22}$$

Set $\xi(t) = W\tilde{\xi}(t)$. Then we have

$$\dot{\tilde{\xi}}(t) = \tilde{\Psi}\tilde{\xi}(t). \tag{23}$$

Lemma 2 *Under Assumption* 2, *all eigenvalues of* L_1 *have positive real parts, and* $L = \begin{bmatrix} L_1 & L_2 \\ 0 & 0 \end{bmatrix}$ *is the Laplacian matrix corresponding to the union set of* $\{\mathcal{G}_1, \ldots, \mathcal{G}_s\}$.

Proof From Lemma 1 in [1], we obtain the results. □

From Lemma 2, we know all eigenvalues of L_1 have positive real parts. For the characteristic polynomial of the matrix $\tilde{\Psi}$, we have

$$\lambda_i^2 + \mu_i \lambda_i + \mu_i = 0, \tag{24}$$

where λ_i is the eigenvalue of $\tilde{\Psi}$, μ_i is the eigenvalue of L_1, $i = 1, \ldots, M$. From Lemma 2, we know the μ_i have a positive real part. Then we find that the λ_i have a negative real part. Then all eigenvalues of $\tilde{\Psi}$ have negative real parts.

Under Markov switching topologies, systems (23) is changed into the following form:

$$\dot{\tilde{\xi}}(t) = \tilde{\Psi}_{\sigma(t)} \tilde{\xi}(t), \tag{25}$$

where $\sigma(t)$ is a Markov switching process and takes values from the sets $S = \{1, \ldots, s\}$. We have

$$\Pr\{\sigma(t+h) = j \mid \sigma(t) = i\} = \pi_{ij} h + o(h), \quad i \neq j, h > 0, \tag{26}$$

where π_{ij} is the transition rate from i to j and $\pi_{ii} = -\sum_{j \neq i} \pi_{ij}$ for $i = j$.

Theorem 3 *Suppose Assumption* 2 *is satisfied. Under Markov switching topologies and the containment algorithm* (17), *system* (15) *can solve the mean square containment control problem.*

Proof Since the union communication topology satisfies Assumption 2, we know all eigenvalues of $\tilde{\Psi}_i$ have negative real parts. Then there exists a positive definite matrix P, such that $\tilde{\Psi}_i^T P + P \tilde{\Psi}_i < 0$ holds. Choose the Lyapunov function as follows:

$$V_i(t) = E\left[\tilde{\xi}^T(t) P \tilde{\xi}(t) \mathbf{1}_{\sigma(t)=i}\right], \quad i \in S. \tag{27}$$

Then, calculating the derivative of $V_i(t)$, we have

$$\begin{aligned} dV_i(t) &= E\left[\left(d\tilde{\xi}(t)\right)^T P \tilde{\xi}(t) \mathbf{1}_{\sigma(t)=i} + \tilde{\xi}^T(t) P\, d\tilde{\xi}(t) \mathbf{1}_{\sigma(t)=i}\right] + \sum_{j=1}^{s} \pi_{ij} V_j \\ &= E\left[\tilde{\xi}^T(t)\left(\tilde{\Psi}_i^T(t) P + P \tilde{\Psi}_i(t)\right)\tilde{\xi}(t)\right] \\ &< 0. \end{aligned} \tag{28}$$

Then, following the method in [13], we obtain $\lim_{t\to\infty} E(\tilde{\xi}(t))^2 = 0$. Thus, we have $\lim_{t\to\infty} E(\zeta_F(t) + (L_1^{-1} L_2 \otimes I_2)\zeta_L(t))^2 = 0$. Then system (15) can solve the mean square containment control problem. The proof is completed. □

Remark 3 In [11], containment control for second-order multi-agent systems was discussed under random switching topologies. However, in this paper, we have given another method to solve mean square containment control problem, which is different from the one in [11].

Next, we extend Theorem 3 to the case of a high-order multi-agent system. Consider the follower's behavior of the high-order multi-agent system as follows:

$$\dot{x}_i(t) = Ax_i(t) + Bu_i(t), \quad i = 1, \ldots, M, \tag{29}$$

where $x_i(t)$, $u_i(t)$ are position and input control of the ith agent, respectively, A and B are constant matrices.

The behavior of the leader's dynamic is described as follows:

$$\dot{x}_i(t) = Ax_i(t), \quad i = M+1, \ldots, N, \tag{30}$$

where $x_i(t)$ is the ith leader's position, A is a constant matrix. The containment protocol is given

$$u_i(t) = K \sum_{j \in F \cup L} a_{ij}\big(x_j(t) - x_i(t)\big), \tag{31}$$

where K is the gain matrix, F, L is the same as the ones in Theorem 1. Set $\xi_i(t) = \sum_{j \in F \cup L} a_{ij}(x_i(t) - x_j(t))$, $\xi(t) = [\xi_1^T(t) \cdots \xi_M^T(t)]^T$, $X_F(t) = [x_1^T(t) \cdots x_M^T(t)]^T$, $X_L(t) = [x_{M+1}(t) \cdots x_N(t)]$. By using similar methods in Theorem 3, we have

$$\begin{aligned} \dot{X}_F(t) &= \big((I_M \otimes A) - (L_1 \otimes BK)\big)X_F(t) - (L_2 \otimes BK)X_L(t) \\ \dot{X}_L(t) &= (I_M \otimes A)X_L(t). \end{aligned} \tag{32}$$

In view of $\xi_i(t) = \sum_{j \in F \cup L} a_{ij}(x_i(t) - x_j(t))$, we have

$$\dot{\xi}(t) = \big((I_M \otimes A) - (L_1 \otimes BK)\big)\xi(t). \tag{33}$$

Since switching topologies are driven by a Markov process, (33) can be rewritten as follows:

$$\dot{\xi}(t) = F_{\sigma(t)}\xi(t), \tag{34}$$

where $F = (I_M \otimes A) - (L_1 \otimes BK)$, $\sigma(t)$, π and π_{ij} are the same as the ones in Theorem 3.

Theorem 4 *Assume that* (A, B) *is stabilizable and Assumption* 2 *holds. Under the containment control protocol* (31) *with* $K = \theta B^T P$, *system* (29) *can be achieved by the mean square containment control under the Markov switching topologies, where* $\theta \geq \frac{1}{\min\{\bar{\pi}_i\}\epsilon}$, ϵ *is the minimum eigenvalue of the matrix* $L_1 + L_1^T$, P *is defined in* (35).

Proof Since (A, B) is stabilizable, we have

$$A^T P + PA - 2PBB^T P < 0. \tag{35}$$

Then we choose the Lyapunov function as follows:

$$V_i(t) = E\big[\xi^T(t)(I_N \otimes P)\xi(t)\mathbf{1}_{\sigma=i}\big], \quad i \in \{1,\dots,s\}. \tag{36}$$

Taking the derivative of $V_i(t)$, we have

$$dV_i(t) = E\big[2\big(d\xi(t)\big)^T(I_M \otimes P)\xi(t)\mathbf{1}_{\sigma=i}\big] + \sum_{j=1}^{s}\pi_{ij}V_j(t)\,dt + o(dt). \tag{37}$$

Assume that $\sigma(t)$ begins in the invariant distribution $\bar{\pi} = [\bar{\pi}_1,\dots,\bar{\pi}_s]^T$, we have

$$\frac{dV(t)}{dt} \le E\left[\xi^T(t)\left(I_M \otimes \big(A^TP + PA\big) - \frac{L_1 + L_1^T}{\lambda_{\min}(L_1 + L_1^T)} \otimes PBB^TP\right)\xi(t)\right],$$

where $\lambda_{\min}(L_1 + L_1^T)$ is the minimum eigenvalue of $L + L^T$. Then we have

$$\begin{aligned}
&\xi^T(t)\left(I_M \otimes \big(A^TP + PA\big) - \frac{L_1 + L_1^T}{\lambda_{\min}(L_1 + L_1^T)} \otimes PBB^TP\right)\xi(t)\\
&= \sum_{j=1}^{M}\xi_j^T(t)\left(A^TP + PA - \frac{2\lambda_j(L_1 + L_1^T)}{\lambda_{\min}(L_1 + L_1^T)}PBB^TP\right)\xi_j(t)\\
&\le \sum_{j=1}^{M}\xi_j^T(t)\big(A^TP + PA - 2PBB^TP\big)\xi_j(t)\\
&< 0.
\end{aligned} \tag{38}$$

Then, following the method in Theorem 3, we obtain $\lim_{t\to\infty}E(\xi(t))^2 = 0$. Then $\lim_{t\to\infty}E(X_F(t) + L_1^{-1}L_2X_L(t))^2 = 0$. Therefore, mean square containment control problem for system (29) can be achieved. The proof is completed. □

Remark 4 In [8], the mean square consensus problem of the high-order multi-agent systems was investigated, while the mean square containment control of high-order multi-agent systems has been discussed in this paper.

4 Conclusions

In the paper, we have investigated the mean square containment control for discrete- and continuous-time second-order multi-agent systems under Markov switching topologies, respectively. By using graph theory and the tools of stochastic analysis, sufficient conditions for mean square containment control are derived. In addition, we extend the results of the second-order multi-agent systems to that of the high-order multi-agent system. Moreover, the topic of containment control problems of multi-agent systems with communication noise is interesting, which is our future work.

Competing interests
The authors declare that they have no competing interests.

Authors' contributions
GM has carried out the main part of this manuscript. TL has checked and corrected the language and theoretical results. All authors have read and approved the final manuscript.

Acknowledgements

This work was supported by Natural Science Fundamental Research Project of Jiangsu Colleges and Universities under Grants 14KJB120006, the Natural Science Foundation of Jiangsu Province under Grant BK20140045, the Scientific Research Foundation of Nanjing University of Information Science and Technology under Grant 2013X047, C-MEIC in School of Information and Control of Nanjing University of Information Science and Technology, Jiangsu Collaborative Innovation Center on Atmospheric Environment and Equipment Technology of Nanjing University of Information Science and Technology.

References

1. Li, ZK, Ren, W, Liu, XD, Fu, MY: Distributed containment control of multi-agent systems with general linear dynamics in the presence of multiple leaders. Int. J. Robust Nonlinear Control **23**, 534-547 (2013)
2. Li, JZ, Ren, W, Xu, SY: Distributed containment control with multiple dynamic leaders for double-integrator dynamics using only position measurements. IEEE Trans. Autom. Control **57**, 1553-1559 (2012)
3. Ma, Q, Lewis, FL, Xu, SY: Cooperative containment of discrete-time linear multi-agent systems. Int. J. Robust Nonlinear Control **25**, 1007-1018 (2015)
4. Liu, K, Xie, G, Wang, L: Containment control for second-order multi-agent systems with time-varying delays. Syst. Control Lett. **67**, 24-31 (2014)
5. Liu, HY, Xie, GM, Wang, L: Necessary and sufficient conditions for containment control of networked multi-agent systems. Automatica **48**, 1415-1422 (2012)
6. Zhang, Y, Tian, YP: Consentability and protocol design of multi-agent systems with stochastic switching topology. Automatica **45**, 1195-1201 (2009)
7. Zhao, HY, Ren, W, Yuan, DM, Chen, J: Distributed discrete-time coordinated tracking with Markovian switching topologies. Syst. Control Lett. **61**, 766-772 (2012)
8. You, KY, Li, ZK, Xie, LH: Consensus conditions for linear multi-agent systems over randomly switching topologies. Automatica **49**, 3125-3132 (2013)
9. Zhou, J, Wang, Q: Convergence speed in distributed consensus over dynamically switching random networks. Automatica **45**, 1455-1461 (2009)
10. Sun, FL, Guan, ZH, Zhan, XS: Consensus of second-order and high-order discrete-time multi-agent systems with random networks. Nonlinear Anal., Real World Appl. **13**, 1979-1990 (2012)
11. Lou, Y, Hong, Y: Target containment control of multi-agent systems with random switching interconnection topologies. Automatica **48**, 879-885 (2012)
12. Wang, YP, Cheng, L, Hou, ZG, Tian, M: Containment control of multi-agent systems in a noisy communication environment. Automatica **50**, 1922-1928 (2014)
13. Zhu, Q, Yang, X, Wang, H: Stochastically asymptotic stability of delayed recurrent neural networks with both Markovian jump parameters and nonlinear disturbances. J. Franklin Inst. **347**, 1489-1510 (2010)

5

The almost sure stability of coupled system of stochastic delay differential equations on networks

Rui Wang[1], Wenxue Li[2] and Xiaoyue Li[1*]

*Correspondence:
lixy209@nenu.edu.cn
[1]School of Mathematics and Statistics, Northeast Normal University, Changchun, 130024, P.R. China
Full list of author information is available at the end of the article

Abstract

This paper investigates the coupled systems of stochastic differential equations with variable delays (CSDDEs) on networks. We analyze the existence and uniqueness of solution by combining the method of graph theory with the Lyapunov function analysis. Furthermore, we utilize the graph theory technique and the nonnegative semimartingale convergence theorem to obtain the almost sure stability of sample solutions and the sufficient principles to locate their limit sets, which correlate closely with the topology property of CSDDEs. Finally we illustrate our main results by examples from population dynamics and vibration systems.

MSC: 60H10; 34K50; 34K20

Keywords: nonlinear models and systems; neural networks; almost sure stability; graph theory; nonnegative semimartingale convergence theorem

1 Introduction

Coupled systems of nonlinear differential equations on networks have been applied widely, especially in the mechanical, electronic, and biological fields [1–9]. A network is always described by a directed graph consisting of vertices and directed arcs connecting them. At each vertex, the local dynamics is given by a system of differential equations called a vertex system [2, 5]. The study of mathematical questions on coupled systems (including synchronization, clustering and transitions) has been introduced by [10]. Among various dynamical properties of the coupled systems on networks overall stability based on given vertex systems is very important and interesting from the viewpoint of controlling complex dynamical systems. However, the stability analysis for the coupled systems on networks is generally a complex and formidable task. It is inspiring that for a coupled deterministic system Li *et al.* [2, 5] gave a systematic method to construct an appropriate Lyapunov function making use of the graph-theoretic technique, and then applied this method to epidemic models [2, 3, 5, 6], oscillator models [5], and ecological models [5], obtaining the global stability.

On the other hand, stochastic differential equations (SDEs) have become a powerful tool in the modeling of realistic systems due to various disturbances; refer to [11]. As a matter of fact, the theory of SDEs has been developed very quickly since Itô introduced his stochastic calculus. The stability theory of stochastic delay differential equations (SDDEs) has been studied extensively; see *e.g.* [12–15]. However, so far as we know there are few papers to deal with almost sure asymptotic stability of the sample paths and their limit sets for stochastic systems except for [16–20]. It is worthy to mention that Mao *et al.* [16–19] made an important development by extending the LaSalle-type theorem from ordinary differential equations (ODEs) to stochastic versions after LaSalle discovered the internal relationship between Birkoff's positive limit set and the Lyapunov function [21].

Motivated by the previous works, the purpose of this paper is to investigate the coupled systems of SDEs with variable delays (CSDDEs) on networks described by

$$\begin{aligned} dx_i(t) = \Bigg[& f_i\big(x_i(t), x_i(t - r_i(t)), t\big) \\ & + \sum_{j=1}^{l} P_{ij}\big(x_i(t), x_i(t - r_i(t)), x_j(t), x_j(t - r_j(t)), t\big)\Bigg] dt \\ & + g_i\big(x_i(t), x_i(t - r_i(t)), t\big)\, dB_i(t), \quad t \geq 0, i = 1, 2, \ldots, l, \end{aligned} \tag{1.1}$$

with the initial data

$$x_i(\theta) = \xi_i(\theta) \in \mathbb{R}^{n_i}, \quad -r_i(0) \leq \theta \leq 0, \tag{1.2}$$

where $\xi(t) = (\xi_1(t), \xi_2(t), \ldots, \xi_l(t)) \in \mathbb{R}^{n_1} \times \mathbb{R}^{n_2} \times \cdots \times \mathbb{R}^{n_l} =: \mathbb{R}^N$, $x(t) = (x_1(t), x_2(t), \ldots, x_l(t)) \in \mathbb{R}^N$, and $B(t) = (B_1(t), B_2(t), \ldots, B_l(t))$ is a $\mathbb{R}^{m_1} \times \mathbb{R}^{m_2} \times \cdots \times \mathbb{R}^{m_l} =: \mathbb{R}^M$-valued Brownian motion.

We analyze the existence and uniqueness of a solution to (1.1) by the combined method of graph theory and Lyapunov function analysis. Furthermore, we utilize the graph theory technique and the nonnegative semimartingale convergence theorem to investigate the almost sure stability of sample solutions and give the sufficient principles to locate their limit sets, which is helpful to understand the dynamical behaviors.

The paper is organized as follows: in Section 2, we prepare some notations and lemmas to be used. In Section 3 we discuss the existence of solution and its uniqueness. In Section 4, we give the sufficient conditions for the almost sure asymptotic stability which relate to the topology structure of the network closely. Moreover, we obtain the limit sets of sample paths with probability 1. Finally, we illustrate our main results through some examples.

2 Preliminaries

For convenience we first state some notations. Throughout the paper, let $(\Omega, \mathcal{F}, \{\mathcal{F}_t\}_{t\geq 0}, P)$ be a complete probability space with a filtration $\{\mathcal{F}_t\}_{t\geq 0}$ satisfying the usual conditions. Let $B(t)$, $t \geq 0$, be a standard $\mathbb{R}^M$-valued Brownian motion defined on this probability space. For any constant sequence $\{c_i\}$ $(i = 1, 2, \ldots, l)$, define $\check{c} = \max_{1\leq i\leq l} c_i$, $\hat{c} = \min_{1\leq i\leq l} c_i$. For each $x = (x_1, x_2, \ldots, x_l) \in \mathbb{R}^N$, define $\|x\| = [\sum_{i=1}^{l} (x^i)^T \cdot (x^i)]^{\frac{1}{2}}$. For any positive constant r, let $C([-r, 0]; \mathbb{R}^N)$ be the family of all continuous $\mathbb{R}^N$-valued functions ρ on $[-r, 0]$ with a norm $|\rho| = \sup_{-r\leq\theta\leq 0} \|\rho(\theta)\|$. For $h \in C(\mathbb{R}^N; \mathbb{R})$, define $\mathrm{Ker}(h) = \{x \in \mathbb{R}^N | h(x) = 0\}$. Let $C^b_{\mathcal{F}_0}([-r, 0]; \mathbb{R}^N)$ be the family of all $\mathcal{F}_0$-measurable bounded $C([-r, 0]; \mathbb{R}^N)$-valued random variables ξ. For each $V_i \in C^{2,1}(\mathbb{R}^{n_i} \times \mathbb{R}_+; \mathbb{R}_+)$, define a function $\mathcal{L}V_i$ from $\mathbb{R}^N \times \mathbb{R}^N \times$

$\mathbb{R}_+$ to $\mathbb{R}$ by

$$\mathcal{L}V_i(x,y,t) = \frac{\partial V_i(x_i,t)}{\partial t} + \frac{\partial V_i(x_i,t)}{\partial x_i} f_i(x_i,y_i,t)$$
$$+ \sum_{j=1}^{l} \frac{\partial V_i(x_i,t)}{\partial x_i} P_{ij}(x_i,y_i,x_j,y_j,t)$$
$$+ \frac{1}{2}\,\mathrm{trace}\left[g_i^T(x_i,y_i,t)\frac{\partial^2 V_i(x_i,t)}{\partial x_i^2} g_i(x_i,y_i,t)\right],$$

where, for $x_i = (x_i^{(1)}, x_i^{(2)}, \ldots, x_i^{(n_i)})$,

$$\frac{\partial V_i(x_i,t)}{\partial x_i} = \left(\frac{\partial V_i(x_i,t)}{\partial x_i^{(1)}}, \frac{\partial V_i(x_i,t)}{\partial x_i^{(2)}}, \ldots, \frac{\partial V_i(x_i,t)}{\partial x_i^{(n_i)}}\right)$$

and

$$\frac{\partial^2 V_i(x_i,t)}{\partial x_i^2} = \left(\frac{\partial^2 V_i(x_i,t)}{\partial x_i^{(j)}\partial x_i^{(k)}}\right)_{n_i\times n_i}.$$

A function $V(x,t): \mathbb{R}^N \times \mathbb{R}_+ \to \mathbb{R}_+$ is said to be radially unbounded, if

$$\lim_{\|x\|\to\infty} \inf_{0\le t<\infty} V(x,t) = \infty.$$

We denote by $\Psi(\mathbb{R}_+;\mathbb{R}_+)$ the family of all continuous functions $\psi: \mathbb{R}_+ \to \mathbb{R}_+$ with the property that for any $\delta > 0$ we have $\liminf_{t\to\infty} \int_t^{t+\delta} \psi(s)\,ds > 0$; see [16] for details.

For self-completeness we cite the following concepts and theorems on graph theory given by [5, 22, 23]. A directed digraph $\mathcal{G} = (\mathcal{V},\mathcal{E})$ contains a vertex set denoted by $\mathcal{V} = \{1,2,\ldots,l\}$ and an arc set denoted by $\mathcal{E}$. Denote by (j,i) an arc leading from initial vertex j to terminal vertex i. A directed digraph $\mathcal{G}$ is weighted if each arc (j,i) is assigned a positive weight a_{ij}. The weight $w(\mathcal{G})$ of a subgraph $\mathcal{G}$ is the product of the weights on all its arcs. Given a weighted digraph $\mathcal{G}$ with l vertices, define the weight matrix $A = (a_{ij})$ whose entry a_{ij} equals the weight of arc (j,i) if it exists, and 0 otherwise. A weight matrix A is cogredient to a matrix E if there exists some permutation matrix P such that $PAP^T = E$. A is *reducible* if it is cogredient to $E = \left[\begin{smallmatrix} B & 0 \\ C & D\end{smallmatrix}\right]$, where B and D are square matrices, or if $l = 1$ and $A = 0$. Otherwise, A is *irreducible*. A directed path $\mathcal{H}$ in $\mathcal{G}$ is a subgraph with distinct vertices $\{i_1, i_2, \ldots, i_k\}$ such that its set of arcs is $\{(i_m, i_{m+1}) : m = 1,2,\ldots,k-1\}$. If $i_k = i_1$, we call $\mathcal{H}$ a directed cycle. A connected subgraph $\mathcal{T}$ is a tree if it contains no cycle, directed or undirected. A tree $\mathcal{T}$ is rooted at vertex i, called the root, if i is not a terminal vertex of any arcs, and each of the remaining vertices is a terminal vertex of exactly one arc. A subgraph $\mathcal{Q}$ is unicyclic if it is a disjoint union of rooted trees whose roots form a directed cycle. A digraph $\mathcal{G} = (\mathcal{V},\mathcal{E})$ is strongly connected if, for any pair of distinct vertices, there exists a directed path from one to the other. The Laplacian matrix of $(\mathcal{G}, A)$ is defined as

$$L = \begin{bmatrix} \sum_{i\neq 1} a_{1i} & -a_{12} & \cdots & -a_{1l} \\ -a_{21} & \sum_{i\neq 2} a_{2i} & \cdots & -a_{2l} \\ \vdots & \vdots & \ddots & \vdots \\ -a_{l1} & -a_{l2} & \cdots & \sum_{i\neq l} a_{li} \end{bmatrix}. \tag{2.1}$$

Let c_i denote the cofactor of the ith diagonal element of L. We mention the Kirchhoff's matrix tree theorem.

Lemma 2.1 [5, 23] *Assume $l \geq 2$. If $(\mathcal{G}, A)$ is strongly connected, then each $c_i > 0$, $i \in \mathcal{V}$.*

Lemma 2.2 [23] *A weighted digraph $(\mathcal{G}, A)$ is strongly connected, if and only if A is irreducible.*

Lemma 2.3 [5] *Assume $l \geq 2$. Let c_i be given in Lemma 2.1. Then the following identity holds:*

$$\sum_{i,j=1}^{l} c_i a_{ij} U_{ij}(x_i, x_j) = \sum_{\mathcal{Q} \in \mathbb{Q}} w(\mathcal{Q}) \sum_{(u,v) \in \mathcal{E}(C_{\mathcal{Q}})} U_{vu}(x_v, x_u).$$

Here for any $i, j \in \mathcal{V}$, $U_{ij}(x_i, x_j)$ is an arbitrary function, $\mathbb{Q}$ is the set of all spanning unicyclic graphs of $(\mathcal{G}, A)$, $w(\mathcal{Q})$ is the weight of spanning unicycle graph $\mathcal{Q}$, and $C_{\mathcal{Q}}$ denotes the directed cycle of $\mathcal{Q}$.

Consider the CSDDEs (1.1) on $t \geq 0$ with initial data (1.2) satisfying $\xi_i \in C^b_{\mathcal{F}_0}([-r_i(0), 0]; \mathbb{R}^{n_i})$, $i = 1, 2, \ldots, l$, where, for each $1 \leq i, j \leq l$,

$$f_i : \mathbb{R}^{n_i} \times \mathbb{R}^{n_i} \times \mathbb{R}_+ \to \mathbb{R}^{n_i},$$
$$g_i : \mathbb{R}^{n_i} \times \mathbb{R}^{n_i} \times \mathbb{R}_+ \to \mathbb{R}^{n_i \times m_i},$$
$$P_{ij} : \mathbb{R}^{n_i} \times \mathbb{R}^{n_i} \times \mathbb{R}^{n_j} \times \mathbb{R}^{n_j} \times \mathbb{R}_+ \to \mathbb{R}^{n_i},$$

$f = (f_1, f_2, \ldots, f_l)^T$, $g = (g_1, g_2, \ldots, g_l)^T$, $P_j = (P_{1j}, P_{2j}, \ldots, P_{lj})^T$ are all Borel-measurable functions. For CSDDEs (1.1), we propose the following assumptions.

Assumption 1 Time delays $r_i : \mathbb{R}_+ \to \mathbb{R}_+$, $i = 1, \ldots, l$, are differentiable and their derivatives are less than 1. That is to say, there exist constants $d_i < 1$, $i = 1, \ldots, l$, such that

$$\frac{dr_i(t)}{dt} \leq d_i, \quad i = 1, 2, \ldots, l.$$

For each $1 \leq i \leq l$, Assumption 1 implies $t - r_i(t)$ is an increasing function of t, so

$$-r_i(0) \leq t - r_i(t) \leq t, \quad \forall t \geq 0. \tag{2.2}$$

Assumption 2 For each $1 \leq i, j \leq l$, f_i, g_i, and P_{ij} satisfy the local Lipschitz conditions. That is to say, for each $h > 0$ there are positive constants $C_i(h)$ and $\bar{C}_{ij}(h)$, such that

$$\begin{aligned}
&\left\| f_i(x_i, y_i, t) - f_i(\bar{x}_i, \bar{y}_i, t) \right\| \vee \left\| g_i(x_i, y_i, t) - g_i(\bar{x}_i, \bar{y}_i, t) \right\| \\
&\quad \leq C_i(h)\big(\|x_i - \bar{x}_i\| + \|y_i - \bar{y}_i\|\big), \\
&\left\| P_{ij}(x_i, y_i, x_j, y_j, t) - P_{ij}(\bar{x}_i, \bar{y}_i, \bar{x}_j, \bar{y}_j, t) \right\| \\
&\quad \leq \bar{C}_{ij}(h)\big(\|x_i - \bar{x}_i\| + \|y_i - \bar{y}_i\| + \|x_j - \bar{x}_j\| + \|y_j - \bar{y}_j\|\big)
\end{aligned}$$

for all $t \geq 0$, $x_i, \bar{x}_i, y_i, \bar{y}_i \in \mathbb{R}^{n_i}$, $x_j, \bar{x}_j, y_j, \bar{y}_j \in \mathbb{R}^{n_j}$ with $\|x_i\| \vee \|\bar{x}_i\| \vee \|y_i\| \vee \|\bar{y}_i\| \vee \|x_j\| \vee \|\bar{x}_j\| \vee \|y_j\| \vee \|\bar{y}_j\| \leq h$.

Assumption 3 $\sup_{0\leq t<\infty}(\|f_i(0,0,t)\| \vee \|g_i(0,0,t)\| \vee \|P_{ij}(0,0,0,0,t)\|) < \infty$, $i,j = 1,2,\ldots,l$.

3 Existence and uniqueness of solution

In order to investigate the sophisticated properties of the solution, we need the existence and uniqueness of the global solution to CSDDEs (1.1) firstly, while their coefficients are usually required to satisfy the linear growth condition and local Lipschitz condition (see *e.g.* Friedman [12] and Mao [14]). However, the coefficients of CSDDEs (1.1) may not satisfy the linear growth condition, though they are locally Lipschitz continuous. We therefore wonder what other alternative conditions proposed can avoid the explosion at a finite time. We answer this question in the following theorem.

Theorem 3.1 *Let Assumptions* 1 *and* 2 *hold. Assume for all* $1 \leq i,j \leq l$, *there exist functions* $V_i(x_i,t) \in C^{2,1}(\mathbb{R}^{n_i} \times \mathbb{R};\mathbb{R}_+)$ *(radially unbounded)*, $F_{ij} \in C(\mathbb{R}^N \times \mathbb{R}^N \times \mathbb{R}_+;\mathbb{R})$, $q_i \in C(\mathbb{R}_+;\mathbb{R}_+)$, $p_j \in C(\mathbb{R}^{n_j} \times [-r_j(0),\infty];\mathbb{R}_+)$, *and nonnegative constants* $a_{ij} \geq 0$ $(A = (a_{ij})$ *is irreducible), such that*

$$\mathcal{L}V_i(x,y,t) \leq q_i(t)\left[1 + \sum_{j=1}^{l} V_j(x_j,t) + \sum_{j=1}^{l} V_j\big(y_j, t - r_i(t)\big)\right] - \sum_{j=1}^{l} p_j(x_j,t)$$
$$+ \sum_{j=1}^{l}(1-d_j)p_j\big(y_j, t - r_j(t)\big) + \sum_{j=1}^{l} a_{ij}F_{ij}(x,y,t). \tag{3.1}$$

Assuming that along each directed cycle $\mathcal{C}_\mathcal{Q}$ *of weighted digraph* $(\mathcal{G},A)$, *there exist functions* $h^\mathcal{Q} \in C(\mathbb{R}_+;\mathbb{R}_+)$, $k_j^\mathcal{Q} \in C(\mathbb{R}^{n_j} \times [-r_j(0),\infty];\mathbb{R}_+)$, $j = 1,2,\ldots,l$, *such that*

$$\sum_{(u,v)\in\mathcal{E}(\mathcal{C}_\mathcal{Q})} F_{vu}(x,y,t) \leq h^\mathcal{Q}(t)\left[1 + \sum_{j=1}^{l} V_j(x_j,t) + \sum_{j=1}^{l} V_j\big(y_j, t - r_j(t)\big)\right]$$
$$- \sum_{j=1}^{l} k_j^\mathcal{Q}(x_j,t) + \sum_{j=1}^{l}(1-d_j)k_j^\mathcal{Q}\big(y_j, t - r_j(t)\big). \tag{3.2}$$

Then for any given initial data ξ, *CSDDEs* (1.1)-(1.2) *have a unique global solution on* $[0,\infty)$.

Proof Under Assumptions 1 and 2, CSDDEs (1.1)-(1.2) have a unique maximal local solution $x(t)$ on $t \in [[0,\sigma_\infty[[$ for any given initial data ξ, where σ_∞ is the explosion time. We then need only to show that $\sigma_\infty = \infty$ a.s. Therefore, choose sufficiently large N, such that $N \geq |\xi_i|$ $(i = 1,2,\ldots,l)$, then for any $n > N$, define the stopping time

$$\tau_n = \inf\big\{t \in [0,\sigma_\infty) : \big\|x_i(t)\big\| \geq n \text{ for some } 1 \leq i \leq l\big\},$$

and set $\inf\emptyset = \infty$. Clearly, the τ_n are increasing so we define the limit $\tau_\infty = \lim_{k\to\infty}\tau_n$. Obviously, $\tau_\infty \leq \sigma_\infty$ a.s. Next, we use the stochastic Lyapunov analysis method to prove $\tau_\infty = \infty$, then we will obtain the required equality $\sigma_\infty = \infty$. Let c_i is the cofactor of the ith diagonal element of Laplacian matrix of $(\mathcal{G}, A)$. From Lemma 2.1 and Lemma 2.2, we know each $c_i > 0$. Then we define $V(x,t) = \sum_{i=1}^{l} c_i V_i(x_i, t)$. Utilizing inequalities (3.1), (3.2), and Lemma 2.3, we compute

$$\begin{aligned}
&\mathcal{L}V(x,y,t) \\
&= \sum_{i=1}^{l} c_i \mathcal{L}V_i(x,y,t) \\
&\leq \sum_{i=1}^{l} c_i \Bigg\{ q_i(t) \Bigg[1 + \sum_{j=1}^{l} V_j(x_j,t) + \sum_{j=1}^{l} V_j(y_j, t - r_j(t)) \Bigg] - \sum_{j=1}^{l} p_j(x_j,t) \\
&\quad + \sum_{j=1}^{l} (1-d_j) p_j(y_j, t - r_j(t)) \Bigg\} + \sum_{i,j=1}^{l} c_i a_{ij} F_{ij}(x,y,t) \\
&= \sum_{i=1}^{l} c_i q_i(t) + \left(\sum_{i=1}^{l} c_i \right) \sum_{i=1}^{l} V_i(x_i,t) \\
&\quad + \left(\sum_{i=1}^{l} c_i \right) \sum_{i=1}^{l} V_i(y_i, t - r_i(t)) - \sum_{i,j=1}^{l} c_i p_j(x_j,t) \\
&\quad + \sum_{i,j=1}^{l} c_i (1-d_j) p_j(y_j, t - r_j(t)) + \sum_{\mathcal{Q}\in\mathbb{Q}} W(\mathcal{Q}) \sum_{(u,v)\in\mathcal{E}(\mathcal{C}_{\mathcal{Q}})} F_{vu}(x,y,t) \\
&\leq \left[\sum_{i=1}^{l} c_i q_i(t) + \sum_{\mathcal{Q}\in\mathbb{Q}} W(\mathcal{Q}) h^{\mathcal{Q}}(t) \right] + \left[\sum_{i=1}^{l} c_i + \sum_{\mathcal{Q}\in\mathbb{Q}} W(\mathcal{Q}) h^{\mathcal{Q}}(t) \right] \sum_{i=1}^{l} V_i(x_i,t) \\
&\quad + \left[\sum_{i=1}^{l} c_i + \sum_{\mathcal{Q}\in\mathbb{Q}} W(\mathcal{Q}) h^{\mathcal{Q}}(t) \right] \sum_{i=1}^{l} V_i(y_i, t - r_i(t)) \\
&\quad - \left(\sum_{i=1}^{l} c_i \right) \sum_{i=1}^{l} p_i(x_i,t) - \sum_{i=1}^{l} \sum_{\mathcal{Q}\in\mathbb{Q}} W(\mathcal{Q}) k_i^{\mathcal{Q}}(x_i,t) \\
&\quad + \left(\sum_{i=1}^{l} c_i \right) \sum_{i=1}^{l} (1-d_i) p_i(y_i, t - r_i(t)) + \sum_{i=1}^{l} (1-d_i) \sum_{\mathcal{Q}\in\mathbb{Q}} W(\mathcal{Q}) k_i^{\mathcal{Q}}(y_i, t - r_i(t)) \\
&\leq \eta(t) \left[\hat{c} + \sum_{i=1}^{l} c_i V_i(x_i,t) + \sum_{i=1}^{l} c_i V_i(y_i, t - r_i(t)) \right] \\
&\quad - \sum_{i=1}^{l} \phi_i(x_i,t) + \sum_{i=1}^{l} (1-d_i) \phi_i(y_i, t - r_i(t)),
\end{aligned} \tag{3.3}$$

where

$$\eta(t) = \frac{1}{\hat{c}} \left[\sum_{i=1}^{l} c_i (1 + q_i(t)) + \sum_{\mathcal{Q}\in\mathbb{Q}} W(\mathcal{Q}) h^{\mathcal{Q}}(t) \right],$$

$$\phi_i(x_i,t) = p_i(x_i,t) \sum_{j=1}^{l} c_j + \sum_{\mathcal{Q}\in\mathbb{Q}} W(\mathcal{Q}) k_i^{\mathcal{Q}}(x_i,t).$$

For any $n > N$ and $t \geq 0$, utilizing Itô's formula, we get

$$\begin{aligned}
&\mathbb{E}\big[V\big(x(t\wedge\tau_n), t\wedge\tau_n\big)\big] \\
&= \mathbb{E}\left[\sum_{i=1}^{l} c_i V_i\big(x_i(t\wedge\tau_n), t\wedge\tau_n\big)\right] \\
&= \mathbb{E}\left[\sum_{i=1}^{l} c_i V_i\big(x_i(0), 0\big)\right] + \mathbb{E}\left[\int_0^{t\wedge\tau_n} \sum_{i=1}^{l} c_i \mathcal{L} V_i\big(x_i(s), y_i(s), s\big)\, ds\right].
\end{aligned}$$

From Assumption 1, we know for each $1 \leq i \leq l$, $t - r_i(t)$ is an increasing function and

$$0 < \frac{1}{1 - r_i'(s)} \leq \frac{1}{1 - d_i} \leq \frac{1}{1 - \check{d}}.$$

Then inequality (3.3) implies

$$\begin{aligned}
&\mathbb{E}\big[V\big(x(t\wedge\tau_n), t\wedge\tau_n\big)\big] \\
&\leq \sum_{i=1}^{l} c_i \mathbb{E} V_i\big(x_i(0), 0\big) + \mathbb{E}\int_0^{t\wedge\tau_n} \hat{c}\eta(s)\, ds + \mathbb{E}\int_0^{t\wedge\tau_n} \eta(s) \sum_{i=1}^{l} c_i V_i\big(x_i(s), s\big)\, ds \\
&\quad + \mathbb{E}\int_0^{t\wedge\tau_n} \eta(s) \sum_{i=1}^{l} c_i V_i\big(x_i\big(s - r_i(s)\big), s - r_i(s)\big)\, ds \\
&\quad - \mathbb{E}\int_0^{t\wedge\tau_n} \sum_{i=1}^{l} \phi_i\big(x_i(s), s\big)\, ds + \mathbb{E}\int_0^{t\wedge\tau_n} \sum_{i=1}^{l} (1 - d_i)\phi_i\big(x_i\big(s - r_i(s)\big), s - r_i(s)\big)\, ds \\
&\leq \sum_{i=1}^{l} c_i \mathbb{E} V_i\big(\xi_i(0), 0\big) + \mathbb{E}\int_0^{t\wedge\tau_n} \hat{c}\eta(s)\, ds + \mathbb{E}\sum_{i=1}^{l} \int_0^{t\wedge\tau_n} c_i \eta(s) V_i\big(x_i(s), s\big)\, ds \\
&\quad + \sum_{i=1}^{l} \mathbb{E}\int_{-r_i(0)}^{t\wedge\tau_n - r_i(t\wedge\tau_n)} \frac{c_i}{1 - r_i'(s)} \eta\big(s + r_i(s)\big) V_i\big(x_i(s), s\big)\, ds \\
&\quad - \sum_{i=1}^{l} \mathbb{E}\int_0^{t\wedge\tau_n} \phi_i\big(x_i(s), s\big)\, ds + \sum_{i=1}^{l} \mathbb{E}\int_{-r_i(0)}^{t\wedge\tau_n - r_i(t\wedge\tau_n)} \frac{1 - d_i}{1 - r_i'(s)} \phi_i\big(x_i(s), s\big)\, ds \\
&\leq \sum_{i=1}^{l} c_i \mathbb{E} V_i\big(\xi_i(0), 0\big) + \mathbb{E}\int_0^{t\wedge\tau_n} \hat{c}\eta(s)\, ds \\
&\quad + \sum_{i=1}^{l} \int_{-r_i(0)}^{0} \frac{c_i}{1 - d_i} \mathbb{E}\eta\big(s + r_i(s)\big) V_i\big(\xi_i(s), s\big)\, ds \\
&\quad + \sum_{i=1}^{l} \int_{-r_i(0)}^{0} \mathbb{E}\phi_i\big(\xi_i(s), s\big)\, ds + \mathbb{E}\int_0^{t\wedge\tau_n} \sum_{i=1}^{l} c_i \eta(s) V_i\big(x_i(s), s\big)\, ds \\
&\quad + \frac{1}{1 - \check{d}} \mathbb{E}\int_0^{t\wedge\tau_n} \sum_{i=1}^{l} c_i \eta\big(s + r_i(s)\big) V_i\big(x_i(s), s\big)\, ds \\
&\leq C(t) + \frac{2 - \check{d}}{1 - \check{d}}\Big[\sup_{0\leq r\leq t} \eta(r)\Big] \int_0^t \mathbb{E} V\big(x(s\wedge\tau_n), s\wedge\tau_n\big)\, ds,
\end{aligned} \tag{3.4}$$

where

$$C(t) = \sum_{i=1}^{l} c_i \mathbb{E}V_i(\xi_i(0),0) + \int_0^t \hat{c}\eta(s)\,ds + \sum_{i=1}^{l}\int_{-r_i(0)}^{0} c_i\eta(s+r_i(s))\mathbb{E}V_i(\xi_i(s),s)\,ds$$
$$+ \sum_{i=1}^{l}\int_{-r_i(0)}^{0} \mathbb{E}\phi_i(\xi_i(s),s)\,ds.$$

The Gronwall inequality yields

$$\mathbb{E}\big[V(x(r\wedge\tau_n), r\wedge\tau_n)\big] \le C(t)e^{\frac{2-\check{d}}{1-\check{d}}t(\sup_{0\le r\le t}\eta(r))}, \quad \forall t\ge 0. \tag{3.5}$$

On the other hand, define $\mu : R_+ \to R_+$ by

$$\mu(a) = \inf_{\|x\|\ge a, 0\le t<\infty} V(x,t) \quad \text{for } a\ge 0. \tag{3.6}$$

Clearly, $\mu(\|x(t)\|) \le V(x(t),t)$. Since $V_i(x_i,t)$, $i = 1,2,\ldots,l$ is radially unbounded,

$$\lim_{a\to\infty}\mu(a) = \infty.$$

It therefore follows from (3.5) that

$$C(t)e^{\frac{2-\check{d}}{1-\check{d}}t(\sup_{0\le r\le t}\eta(r))} \ge \mathbb{E}\mu(\|x(t\wedge\tau_n)\|) \ge \mu(n)\mathbb{P}(\tau_n\le t).$$

Letting $n\to\infty$, we have

$$\mathbb{P}(\tau_\infty \le t) = 0, \quad \forall t > 0.$$

Then $t\to\infty$, and we obtain

$$\mathbb{P}(\tau_\infty < \infty) = 0.$$

That is, $\tau_\infty = \infty$ a.s. We, therefore, must have $\sigma_\infty = \infty$ a.s. This completes the proof. □

Corollary 3.1 *Suppose that the assumptions of Theorem* 3.1 *are satisfied except that inequalities* (3.1) *and* (3.2) *are replaced by*

$$\mathcal{L}V_i(x_i,y_i,t) \le \gamma_i(t) + \sum_{j=1}^{l}\alpha_{ij}(t)V_j(x_j,t) + \sum_{j=1}^{l}\beta_{ij}(t)V_j(y_j,t-r_j(t)) + \sum_{j=1}^{l}a_{ij}F_{ij}(x,y,t)$$

and

$$\sum_{(u,v)\in\mathcal{E}(\mathcal{C}_Q)} F_{vu}(x,y,t) \le \delta^{Q}(t) + \sum_{j=1}^{l}\theta_j^{Q}(t)V_j(x_j,t) + \sum_{j=1}^{l}\vartheta_j^{Q}(t)V_j(y_j,t-r_j(t)),$$

where $\gamma_i(t), \alpha_{ij}(t), \beta_{ij}(t), \delta^Q(t), \theta_j^Q(t), \vartheta_j^Q(t) \in C(\mathbb{R}_+;\mathbb{R}_+)$. *Then the result of Theorem* 3.1 *still holds.*

Proof Define $q_i(t) = \max_{1\leq j\leq l}\{\gamma_i(t), \alpha_{ij}(t), \beta_{ij}(t)\}$, $h^{\mathcal{Q}}(t) = \max_{1\leq j\leq l}\{\delta^{\mathcal{Q}}(t), \theta_j^{\mathcal{Q}}(t), \vartheta_j^{\mathcal{Q}}(t)\}$, $p_j(x_j, t) = k_j^{\mathcal{Q}}(x_j, t) \equiv 0$, then the required assertion follows from Theorem 3.1. □

Theorem 3.2 *Suppose that all of the assumptions of Theorem* 3.1 *are satisfied, moreover, for each* $1 \leq i \leq l$, $V_i(x_i, t) \in C^{2,1}(\mathbb{R}_+^{n_i} \times \mathbb{R}; \mathbb{R}_+)$ *(radially unbounded) satisfies*

$$\lim_{x_i^{(j)} \to 0^+} \inf_{0\leq t<\infty} V_i(x_i, t) = \infty, \quad j = 1, 2, \ldots, n_i, \tag{3.7}$$

where $x_i = (x_i^{(1)}, x_i^{(2)}, \ldots, x_i^{(n_i)})$. *Then for any given initial data* ξ $(\xi(t) \in \mathbb{R}_+^N)$, *CSDDEs* (1.1)-(1.2) *have a unique global solution* $x(t)$ *on* $[0, \infty)$ *and* $x(t) \in \mathbb{R}_+^N$ *for all* $t \in [0, \infty)$ *a.s.*

Proof Let $N > 0$ be sufficiently large for

$$\frac{1}{N} < \min_{1\leq i\leq l, 1\leq j\leq n_i, -r_i(0)\leq t\leq 0} \xi_i^{(j)}(t) \leq \max_{1\leq i\leq l, 1\leq j\leq n_i, -r_i(0)\leq t\leq 0} \xi_i^{(j)}(t) < N.$$

For each integer $n \geq N$, define the stopping time

$$\tau_n = \inf\left\{t \in [0, \sigma_\infty) : x_i^{(j)}(t) \in (1/n, n) \text{ for some } 1 \leq i \leq l \text{ and some } 1 \leq j \leq n_i\right\}. \tag{3.8}$$

The left proof is a modification of that of Theorem 3.1 directly, we omit it for avoiding duplication. □

Example 3.1 (n-Dimensional stochastic diffusion population model) Consider the environmentally perturbed n-dimensional diffusion population system given by the SDEs

$$\begin{aligned} dx_i(t) = &\left[x_i(t)\big(a_i(t) - b_i(t)x_i(t - r_i(t))\big) + \sum_{j=1, j\neq i}^{n} d_{ij}\big(x_j^2(t) - x_i^2(t)\big)\right] dt \\ &+ \sum_{j=1}^{l} \sigma_{ij}(t)x_i(t)\, dB_j(t), \quad i = 1, 2, \ldots, n, \end{aligned} \tag{3.9}$$

where $a_i(t), b_i(t), \sigma_{ij}(t) \in C(\mathbb{R}_+; \mathbb{R}_+)$, $d_{ij} \geq 0$ are the diffusion coefficients and $r_i(t)$ satisfy Assumption 1. The above system describes the phenomenon that species always migrate from high concentration regions to low concentration ones, *i.e.*, the movement is a function of species density [24]. Let $d_{ii} = 0$. In order to yield the existence of the global positive solution we define C^2-functions $V_i : R_+ \to R_+$, $i = 1, 2, \ldots, n$, by

$$V_i(x_i) = x_i - 1 - \log x_i.$$

Obviously, V_i is radially unbounded and $\lim_{x_i\to 0^+} \inf_{0\leq t<\infty} V_i(x_i) = \infty$. By Itô's formula, compute

$$\begin{aligned} \mathcal{L}V_i(x, y, t) = &\ x_i(a_i - b_i y_i) + \sum_{j=1}^{n} d_{ij}\big(x_j^2 - x_i^2\big) \\ &- \left[a_i - b_i y_i + \sum_{j=1}^{n} d_{ij}\left(\frac{x_j^2}{x_i} - x_i\right)\right] + \frac{1}{2}\sum_{j=1}^{l} \sigma_{ij}^2 \end{aligned}$$

$$
\begin{aligned}
&= \sum_{j=1}^{n} d_{ij}\left(x_j^2 - \frac{x_j^2}{x_i}\right) - \sum_{j=1}^{n} d_{ij}\left(x_i^2 - x_i\right) + a_i x_i \\
&\quad - \left(a_i - \frac{1}{2}\sum_{j=1}^{l}\sigma_{ij}^2\right) - b_i x_i y_j + b_i y_i \\
&= \sum_{j=1}^{n} d_{ij}\left(x_j^2 - \frac{x_j^2}{x_i} - x_i^2 + \frac{x_i^2}{x_j}\right) - \sum_{j=1}^{n} d_{ij}\frac{x_i^2}{x_j} \\
&\quad + \left(a_i + \sum_{j=1}^{n} d_{ij}\right)x_i - \left(a_i - \frac{1}{2}\sum_{j=1}^{l}\sigma_{ij}^2\right) - b_i x_i y_j + b_i y_i \\
&\le \sum_{j=1}^{n} d_{ij}\left(x_j^2 - \frac{x_j^2}{x_i} - x_i^2 + \frac{x_i^2}{x_j}\right) + \left(a_i + \sum_{j=1}^{n} d_{ij}\right)x_i + \frac{1}{2}\sum_{j=1}^{l}\sigma_{ij}^2 + b_i y_i \\
&=: \sum_{j=1}^{l} u_{ij}F_{ij}(x) + \left(a_i + \sum_{j=1}^{n} d_{ij}\right)x_i + \frac{1}{2}\sum_{j=1}^{l}\sigma_{ij}^2 + b_i y_i, \qquad (3.10)
\end{aligned}
$$

dropping (t) from a, b, σ, where

$$
u_{ij} = \begin{cases} d_{ij}, & j \neq i, \\ \delta_i, & j = i, \end{cases} \qquad F_{ij}(x) = x_j^2 - \frac{x_j^2}{x_i} - x_i^2 + \frac{x_i^2}{x_j},
$$

here δ_i may be any nonnegative constant for $i = 1, 2, \ldots, n$, because $F_{ii}(x) \equiv 0$.

Utilizing the inequality for $x > 0$, $2(x - 1 - \log x) > x$, we have

$$
\left(a_i(t) + \sum_{j=1}^{n} d_{ij}\right)x_i + \frac{1}{2}\sum_{j=1}^{l}\sigma_{ij}^2(t) \le K(t)\big(1 + V_i(x_i)\big), \qquad b_i(t)y_i \le K(t)V_j(y_j),
$$

where $K(t) \in C(\mathbb{R}_+; \mathbb{R}_+)$. Obviously,

$$
\sum_{(u,v)\in\mathcal{E}(\mathcal{C}_Q)} F_{vu}(x) = 0.
$$

Then by Theorem 3.2, we obtain the following result.

Corollary 3.2 *If Assumption* 1 *hold, moreover, there is a series of nonnegative constants* $(\delta_1, \delta_2, \ldots, \delta_n)$ *such that*

$$
G = \begin{pmatrix} \delta_1 & d_{12} & \cdots & d_{1n} \\ d_{21} & \delta_2 & \cdots & d_{2n} \\ \vdots & \vdots & \ddots & \vdots \\ d_{n1} & d_{n2} & \cdots & \delta_n \end{pmatrix} \qquad (3.11)
$$

is irreducible. Then SDE (3.9) *with initial data* (1.2) *have a unique global solution* $x(t)$ *on* $[0, \infty)$ *and* $x(t) \in \mathbb{R}_+^n$ *for all* $t \in [0, \infty)$ *a.s.*

4 The almost sure stability

One of the important issues in the study of coupled systems is the automatic control with consequent emphasis being placed on the analysis of stability. In this paper, we will mainly discuss the almost surely asymptotic stability of the CSDDEs. The main result is as follows.

Theorem 4.1 *Let Assumptions* 1-3 *hold. For all* $1 \leq i,j \leq l$, *there exists a radially unbounded function* $V_i(x_i,t) \in C^{2,1}(\mathbb{R}^{n_i} \times \mathbb{R};\mathbb{R}_+)$, $\beta_i \in L(\mathbb{R}_+;\mathbb{R}_+)$, $w_i \in C(\mathbb{R}^N \times \mathbb{R};\mathbb{R}_+)$, $\bar{w}_{ij} \in C(\mathbb{R}^{n_j} \times [-r_j(0),\infty];\mathbb{R}_+)$, $\alpha_i \in \Psi(\mathbb{R}_+;\mathbb{R}_+)$, $h_i \in C(\mathbb{R}^N;\mathbb{R}_+)$, $F_{ij} \in C(\mathbb{R}^N \times \mathbb{R}^N \times \mathbb{R}_+;\mathbb{R})$ *and constants* $a_{ij} \geq 0$ ($A = (a_{ij})$ *is irreducible*) *such that*

$$\mathcal{L}V_i(x,y,t) \leq \beta_i(t) - w_i(x,t) + \sum_{j=1}^{l}(1-d_j)\bar{w}_{ij}\big(y_j, t - r_j(t)\big) + \sum_{j=1}^{l} a_{ij}F_{ij}(x,y,t), \tag{4.1}$$

where

$$w_i(x,t) - \sum_{j=1}^{l} \bar{w}_{ij}(x_j,t) \geq \alpha_i(t)h_i(x). \tag{4.2}$$

Assume along each directed cycle $\mathcal{C}_{\mathcal{Q}}$ *of a weighted digraph* $(\mathcal{G},A)$ *that there are functions* $\beta^{\mathcal{Q}} \in L(\mathbb{R}_+;\mathbb{R}_+)$, $w^{\mathcal{Q}} \in C(\mathbb{R}^N \times \mathbb{R};\mathbb{R}_+)$, $\bar{w}_j^{\mathcal{Q}} \in C(\mathbb{R}^{n_j} \times [-r_j(0),\infty];\mathbb{R}_+)$, $\alpha^{\mathcal{Q}} \in \Psi(\mathbb{R}_+;\mathbb{R}_+)$, $h^{\mathcal{Q}} \in C(\mathbb{R}^N;\mathbb{R}_+)$ *such that*

$$\sum_{(u,v)\in\mathcal{E}(\mathcal{C}_{\mathcal{Q}})} F_{vu}(x,y,t) \leq \beta^{\mathcal{Q}}(t) - w^{\mathcal{Q}}(x,t) + \sum_{j=1}^{l}(1-d_j)\bar{w}_j^{\mathcal{Q}}\big(y_j, t - r_j(t)\big), \tag{4.3}$$

where

$$w^{\mathcal{Q}}(x,t) - \sum_{j=1}^{l} \bar{w}_j^{\mathcal{Q}}(x_j,t) \geq \alpha^{\mathcal{Q}}(t)h^{\mathcal{Q}}(x). \tag{4.4}$$

Then $S = (\bigcap_{i=1}^{l} \mathrm{Ker}(h_i)) \cap (\bigcap_{\mathcal{Q}\in\mathbb{Q}} \mathrm{Ker}(h^{\mathcal{Q}})) \neq \emptyset$ *and for any initial data* ξ, *the CSDDEs* (1.1)-(1.2) *has a unique global solution on* $[0,\infty)$ *denoted by* $x(t;\xi)$ *satisfying*

$$\lim_{t\to\infty} d\big(x(t,\xi),S\big) = 0 \quad a.s. \tag{4.5}$$

Proof For any given initial data, the global existence of the unique solution on $t \geq 0$ is a direct application of Theorem 3.1. Next, we borrow the technique in [16] and combine it with the graph method from [5] to get the limit set. Because the proof is rather technical, we divide it into three steps.

Step 1. Write the solution $x_i(t;\xi) = x_i(t)$ for simplicity. Let c_i be the cofactor of the ith diagonal element of Laplacian matrix of $(\mathcal{G},A)$, and compute

$$\begin{aligned}&\sum_{i=1}^{l} c_i\mathcal{L}V_i(x,y,t)\\ &\quad\leq \sum_{i=1}^{l} c_i\beta_i(t) + \sum_{\mathcal{Q}\in\mathbb{Q}} W(\mathcal{Q}) \sum_{(u,v)\in\mathcal{E}(\mathcal{C}_{\mathcal{Q}})} F_{vu}(x,y,t)\end{aligned}$$

$$
\begin{aligned}
&- \sum_{i=1}^{l} c_i \left[w_i(x,t) - \sum_{j=1}^{l} (1-d_j)\bar{w}_{ij}\big(y_j, t - r_j(t)\big) \right] \\
&\le \zeta(t) - \sum_{i=1}^{l} c_i \left[w_i(x,t) - \sum_{j=1}^{l} (1-d_j)\bar{w}_{ij}\big(y_j, t - r_j(t)\big) \right] \\
&\quad - \sum_{\mathcal{Q}\in\mathbb{Q}} W(\mathcal{Q}) \left[w^{\mathcal{Q}}(x,t) - \sum_{j=1}^{l} (1-d_j)\bar{w}_j^{\mathcal{Q}}\big(y_j, t - r_j(t)\big) \right],
\end{aligned}
$$

dropping (t) from x and y, where

$$
\zeta(t) = \sum_{i=1}^{l} c_i \beta_i(t) + \sum_{\mathcal{Q}\in\mathbb{Q}} W(\mathcal{Q}) \beta^{\mathcal{Q}}(t) \in L(\mathbb{R}_+; \mathbb{R}_+).
$$

Then

$$
\begin{aligned}
&\sum_{i=1}^{l} c_i V_i\big(x_i(t), t\big) \\
&\quad \le \sum_{i=1}^{l} c_i V_i\big(\xi_i(0), 0\big) + \int_0^t \zeta(s)\, ds + \sum_{i,j=1}^{l} c_i \int_{-r_j(0)}^{0} \bar{w}_{ij}\big(\xi_j(s), s\big)\, ds \\
&\quad + \sum_{j=1}^{l} \sum_{\mathcal{Q}\in\mathbb{Q}} W(\mathcal{Q}) \int_{-r_j(0)}^{0} \bar{w}_j^{\mathcal{Q}}\big(\xi_j(s), s\big)\, ds - \sum_{i=1}^{l} c_i \int_0^t \alpha_i(s) h_i\big(x(s)\big)\, ds \\
&\quad - \sum_{\mathcal{Q}\in\mathbb{Q}} W(\mathcal{Q}) \int_0^t \alpha^{\mathcal{Q}}(s) h^{\mathcal{Q}}\big(x(s)\big)\, ds + \sum_{i=1}^{l} M_i(t),
\end{aligned}
\tag{4.6}
$$

where

$$
M_i(t) = \int_0^t c_i \frac{\partial V_i}{\partial x_i}\big(x_i(s), s\big) g_i\big(x_i(s), x_i(s - r_i(s)), s\big)\, dB_i(s).
$$

Next, we prove $\sum_{i=1}^{l} M_i(t)$ is a local martingale. Choose sufficiently large N, such that $N \ge |\xi_i|$ $(i = 1, 2, \ldots, l)$, then for any $n > N$, define the stopping time

$$
\tau_n = \inf\{t \ge 0 : \|x_i(t)\| \ge n \text{ for some } 1 \le i \le l\}.
$$

So this stopping sequence $\tau_n \uparrow \infty$ a.s. From Assumptions 2 and 3, we know that, for each i, g_i is local bounded. Notice that $\frac{\partial V_i}{\partial x_i}(x_i(s), s)$ is continuous in s and $g_i(x_i(s), x_i(s - r_i(s)), s) I_{\{0 \le s \le \tau_n\}}$ is uniformly bounded in $s \in [0, t]$, where I_A is the indicator function of set A. So $M_i(t \wedge \tau_n) = \int_0^t c_i \frac{\partial V_i}{\partial x_i}(x_i(s), s) g_i(x_i(s), x_i(s - r_i(s)), s) I_{\{0 \le s \le \tau_n\}}\, dB_i(s)$ is a martingale (see [14] for the details of the martingale proof). Thus each $M_i(t)$ is a local martingale. Then the required assertion follows.

Using the nonnegative semimartingale convergence theorem (*cf.* Liptser and Shiryayev [25] or Mao [14]) we get

$$
\limsup_{t\to\infty} \sum_{i=1}^{l} c_i V_i\big(x_i(t), t\big) < \infty \quad \text{a.s.}
\tag{4.7}
$$

On the other hand, the left side of (4.6) is positive, taking the expectations on the right hand and letting $t \to \infty$, we obtain

$$\sum_{i=1}^{l} c_i \mathbb{E} \int_0^{\infty} \alpha_i(s) h_i\big(x(s)\big)\, ds + \sum_{\mathcal{Q} \in \mathbb{Q}} W(\mathcal{Q}) \mathbb{E} \int_0^{\infty} \alpha^{\mathcal{Q}}(s) h^{\mathcal{Q}}\big(x(s)\big)\, ds < \infty. \tag{4.8}$$

This implies

$$\int_0^{\infty} \alpha_i(s) h_i\big(x(s)\big)\, ds < \infty, \quad i = 1, 2, \ldots, l \text{ a.s.}, \tag{4.9}$$

$$\int_0^{t} \alpha^{\mathcal{Q}}(s) h^{\mathcal{Q}}\big(x(s)\big)\, ds < \infty, \quad \mathcal{Q} \in \mathbb{Q} \text{ a.s.} \tag{4.10}$$

Step 2. From (4.8) and $\alpha \in \Psi(\mathbb{R}_+, \mathbb{R}_+)$, it is not difficult to prove that, for each $1 \leq i \leq l$ and each $\mathcal{Q} \in \mathbb{Q}$,

$$\liminf_{t \to \infty} h_i\big(x(t)\big) = 0, \qquad \liminf_{t \to \infty} h^{\mathcal{Q}}\big(x(t)\big) = 0 \quad \text{a.s.} \tag{4.11}$$

We now claim that, for each $1 \leq i \leq l$ and each $\mathcal{Q} \in \mathbb{Q}$,

$$\limsup_{t \to \infty} h_i\big(x(t)\big) = 0, \qquad \limsup_{t \to \infty} h^{\mathcal{Q}}\big(x(t)\big) = 0 \quad \text{a.s.} \tag{4.12}$$

If this is false, then there is some $i_1 \in \mathbb{L}$ such that

$$\mathbb{P}\Big\{\limsup_{t \to \infty} h_{i_1}\big(x(t)\big) > 0\Big\} > 0, \tag{4.13}$$

or there is some $\mathcal{Q}_1 \in \mathbb{Q}$ such that

$$\mathbb{P}\Big\{\limsup_{t \to \infty} h^{\mathcal{Q}_1}\big(x(t)\big) > 0\Big\} > 0. \tag{4.14}$$

Without loss of generality we suppose (4.13) holds. Hence there is a small constant $0 < \epsilon_1 < \frac{1}{3}$ such that

$$\mathbb{P}(A_{i_1}^1) \geq 3\epsilon_1, \tag{4.15}$$

where

$$A_{i_1}^1 = \Big\{\limsup_{t \to \infty} h_{i_1}\big(x(t)\big) > 2\epsilon_1\Big\}.$$

For each $i \in \mathbb{L}$, define $\mu_i : R_+ \to R_+$ by

$$\mu_i(a) = \inf_{\|x_i\| \geq a, 0 \leq t < \infty} V_i(x_i, t) \quad \text{for } a \geq 0. \tag{4.16}$$

So by (4.7) and the continuity of both $x_i(t)$ and $V_i(x_i, t)$, we have

$$\sup_{0 \leq t < \infty} \mu_i\big(\|x_i(t)\|\big) \leq \sup_{0 \leq t < \infty} V_i\big(x_i(t), t\big) < \infty \quad \text{a.s.} \tag{4.17}$$

On the other side, the radial unboundedness of each $V_i(x_i, t)$ implies

$$\sup_{0\le t<\infty} \|x(t)\| < \infty \quad \text{a.s.} \tag{4.18}$$

Recalling the boundedness of the initial data we can then find a positive number χ, which depends on ϵ_1, sufficiently large such that $|\xi| < \chi$, and

$$\mathbb{P}(A^2) \ge 1 - \epsilon_1, \tag{4.19}$$

where

$$A^2 = \left\{ \sup_{0\le t<\infty} \|x(t)\| < \chi \right\}.$$

It is easy to see from (4.15) and (4.19) that

$$\mathbb{P}(A^1_{i_1} \cap A^2) \ge 2\epsilon_1. \tag{4.20}$$

For any fixed number $\eta > 0$, let us now define a sequence of stopping times

$$\begin{aligned}
\tau &= \inf\{t \ge 0 : \|x(t)\| \ge \chi\},\\
\sigma^1_{i_1} &= \inf\{t \ge 0 : h_{i_1}(x(t)) \ge 2\epsilon_1\},\\
\sigma^2_{i_1} &= \inf\{t \ge \sigma^1_{i_1} + \eta : h_{i_1}(x(t)) \le \epsilon_1\},\\
\sigma^{2k+1}_{i_1} &= \inf\{t \ge \sigma^{2k}_{i_1} : h_{i_1}(x(t)) \ge 2\epsilon_1\}, \quad k = 1, 2, \ldots,\\
\sigma^{2k+2}_{i_1} &= \inf\{t \ge \sigma^{2k+1}_{i_1} + \eta : h_{i_1}(x(t)) \le \epsilon_1\}, \quad k = 1, 2, \ldots.
\end{aligned}$$

Throughout this paper we set $\inf \emptyset = \infty$. From (4.11), for each $\omega \in A^1_{i_1} \cap A^2$, we have

$$\tau(\omega) = \infty \quad \text{and} \quad \sigma^k_i(\omega) < \infty, \quad k = 1, 2, \ldots. \tag{4.21}$$

By Hölder's inequality and Doob's martingale inequality (see [14]), from the CSDDEs (1.1), for any $T > 0$, we have

$$\begin{aligned}
&\mathbb{E}\left[I_{A^{3,k}_{i_1}} \sup_{0\le t\le T} \left\| x_{i_1}\left(\tau \wedge (\sigma^{2k-1}_{i_1} + t)\right) - x_{i_1}\left(\tau \wedge \sigma^{2k-1}_{i_1}\right) \right\|^2 \right]\\
&\quad \le 6(T+2)TK_\chi,
\end{aligned} \tag{4.22}$$

where $A^{3,k}_{i_1} = \{\tau \wedge \sigma^{2k-1}_{i_1} < \infty\}$, $K_\chi > 0$ is a constant only dependent on χ. Since $h_{i_1}(\cdot)$ is uniformly continuous in the closed ball $\bar{S}^\chi_{i_1} = \{x \in \mathbb{R}^{n_{i_1}} : \|x\| \le \chi\}$, we can therefore choose $\delta = \delta(\epsilon_3) > 0$ such that, for any pair $x_i, \bar{x}_i \in \bar{S}^\chi_{i_1}$ with $\|x_i - \bar{x}_i\| < \delta$, we have

$$\left\| h_{i_1}(x_i) - h_{i_1}(\bar{x}_i) \right\| < \epsilon_1. \tag{4.23}$$

We furthermore choose a positive constant T_1 ($< \eta$) sufficiently small such that

$$\frac{6(T_1+2)T_1K_\chi}{\delta^2} < \epsilon_1,$$

then from (4.21) and (4.22), we have

$$\mathbb{P}(A_{i_1}^1 \cap A^2 \cap A_{i_1}^{4,k}) \leq \mathbb{P}(A_{i_1}^{3,k} \cap A_{i_1}^{4,k}) < \epsilon_1, \tag{4.24}$$

where

$$A_{i_1}^{4,k} = \Big\{ \sup_{0 \leq t \leq T_1} \big\| x_{i_1}\big(\tau \wedge (\sigma_{i_1}^{2k-1} + t)\big) - x_{i_1}\big(\tau \wedge \sigma_{i_1}^{2k-1}\big) \big\| \geq \delta \Big\}.$$

Recalling (4.20) and (4.24), we further compute

$$\mathbb{P}(A_{i_1}^1 \cap A^2 \cap A_{i_1}^{5,k}) \geq \mathbb{P}(A_{i_1}^1 \cap A^2 \cap (A_{i_1}^{4,k})^c) \geq \epsilon_1, \tag{4.25}$$

where

$$A_{i_1}^{5,k} = \Big\{ \sup_{0 \leq t \leq T_1} \big\| h_{i_1}\big(x_{i_1}(\sigma_{i_1}^{2k-1} + t)\big) - h_{i_1}\big(x_{i_1}(\sigma_{i_1}^{2k-1})\big) \big\| < \epsilon_1 \Big\}.$$

On the other side, it follows from $\alpha_{i_1} \in \Psi(R_+, R_+)$ that for $T_1 > 0$ there exist two constants $\epsilon_2 = \epsilon_2(T_1) > 0$ and $t_3 = t_3(\epsilon_2) > 0$ such that

$$\int_t^{t+T_1} \alpha_{i_1}(s)\, ds \geq \epsilon_2, \quad \text{whenever } t \geq t_3.$$

By the definition of $\sigma_{i_1}^k$, there exists a positive integer N such that, for any $\omega \in A_{i_1}^1 \cap A^2$,

$$\sigma_{i_1}^{2k-1} \geq t_3, \quad \text{whenever } k \geq N+1.$$

Therefore, by (4.8), we compute

$$\begin{aligned} \infty > \mathbb{E} &\int_0^\infty \alpha_{i_1}(s) h_{i_1}(x(s))\, ds \\ &\geq \epsilon_1 \mathbb{E}\left[\sum_{k=1}^\infty I_{A_{i_1}^1 \cap A^2} \int_{\sigma_{i_1}^{2k-1}}^{\sigma_{i_1}^{2k}} \alpha_{i_1}(s)\, ds \right] \\ &\geq \epsilon_1 \mathbb{E}\left[\sum_{k=N+1}^\infty I_{A_{i_1}^1 \cap A^2 \cap A_{i_1}^{5,k}} \int_{\sigma_{i_1}^{2k-1}}^{\sigma_{i_1}^{2k-1}+T_1} \alpha_{i_1}(s)\, ds \right] \\ &\geq \epsilon_1 \epsilon_2 \sum_{k=N+1}^\infty \epsilon_1 = \infty, \end{aligned} \tag{4.26}$$

which is a contradiction. Thus (4.12) must hold, implying for each i and $\mathcal{Q}$,

$$\lim_{t \to \infty} h_i(x(t)) = 0, \qquad \lim_{t \to \infty} h^{\mathcal{Q}}(x(t)) = 0 \quad \text{a.s.} \tag{4.27}$$

Step 3. Observe from (4.27) that there exists an $A^0 \subset \Omega$ with $\mathbb{P}(A^0) = 1$ such that, for each $1 \leq i \leq l$ and each $\mathcal{Q} \in \mathbb{Q}$,

$$\begin{aligned} &\lim_{t \to \infty} h_i(x(t,\omega)) = 0, \qquad \lim_{t \to \infty} h^{\mathcal{Q}}(x(t,\omega)) = 0 \quad \text{and} \\ &\sup_{0 \leq t < \infty} \|x(t,\omega)\| < \infty, \quad \forall \omega \in A^0. \end{aligned} \tag{4.28}$$

Fix an $\omega \in A^0$, then $\{x(t,\omega)\}_{t\geq 0}$ is bounded in $\mathbb{R}^N$. So there must be an increasing sequence $\{t_j\}_{j\geq 1}$ such that $\{x(t_j,\omega)\}_{j\geq 1}$ converges to some $x^* \in \mathbb{R}^N$. Hence

$$h_i(x^*) = \lim_{j\to\infty} h_i\big(x(t_j,\omega)\big) = 0, \quad i = 1,2,\dots,l,$$

$$h^{\mathcal{Q}}(x^*) = \lim_{t\to\infty} h^{\mathcal{Q}}\big(x(t_j,\omega)\big) = 0, \quad \mathcal{Q} \in \mathbb{Q},$$

which implies $x^* \in (\bigcap_{i=1}^{l} \mathrm{Ker}(h_i)) \cap (\bigcap_{\mathcal{Q}\in\mathbb{Q}} \mathrm{Ker}(h^{\mathcal{Q}}))$, so $S \neq \emptyset$. We shall now show that

$$\lim_{t\to\infty} d\big(x(t,\omega),S\big) = 0 \quad \text{for all } \omega \in A^0. \tag{4.29}$$

If this is false, then there is some $\bar{\omega} \in A^0$, such that

$$\lim_{t\to\infty} d\big(x(t,\bar{\omega}),S\big) = 0 \quad \text{for all } \omega \in A^0.$$

Then there exists either an i_2, $1 \leq i_2 \leq l$ such that

$$\limsup_{t\to\infty} d\big(x(t,\bar{\omega}),\mathrm{Ker}(h_{i_2})\big) > 0, \tag{4.30}$$

or a $\mathcal{Q}_2 \in \mathbb{Q}$ such that

$$\limsup_{t\to\infty} d\big(x(t,\bar{\omega}),\mathrm{Ker}(h^{\mathcal{Q}_2})\big) > 0. \tag{4.31}$$

Without loss of generality we suppose inequality (4.30) holds. Then there is a constant $\epsilon_3 > 0$ and a sequence $\{x(t_j,\bar{\omega})\}_{j\geq 1}$ of $\{x(t,\bar{\omega})\}_{t\geq 0}$ such that

$$d\big(x(t_j,\bar{\omega}),\mathrm{Ker}(h_{i_2})\big) \geq \epsilon_3, \quad \forall j \geq 1.$$

Since $\{x(t_j,\bar{\omega})\}_{j\geq 1}$ is bounded, we can find a subsequence $\{x(t_{j_k},\bar{\omega})\}_{k\geq 1}$ which converges to z. Clearly, $z \notin \mathrm{Ker}(h_{i_2})$, so $h_{i_2}(z) > 0$. However, by (4.28),

$$h_{i_2}(z) = \lim_{k\to\infty} h_{i_2}\big(x(t_{j_k},\bar{\omega})\big) = 0,$$

which contradicts $h_{i_2}(z) > 0$. Hence (4.29) must hold and the required assertion (4.5) follows from $\mathbb{P}(A^0) = 1$. The proof is therefore complete. □

Theorem 4.1 implies the solutions to the CSFDEs (1.1) will asymptotically approach the set S almost surely. Then we deduce some useful corollaries.

Corollary 4.1 *Let Assumptions* 1-3 *hold. For all* $1 \leq i,j \leq l$, *there exists a radially unbounded function* $V_i(x_i,t) \in C^{2,1}(\mathbb{R}^{n_i} \times \mathbb{R};\mathbb{R}_+)$, $\beta_i \in L(\mathbb{R}_+;\mathbb{R}_+)$, $\alpha_i \in \Psi(\mathbb{R}_+;\mathbb{R}_+)$, $\bar{\alpha}_{ij} \in C([-r_j(0),\infty),\mathbb{R}_+)$, $h_i \in C(\mathbb{R}^N;\mathbb{R}_+)$, $\bar{h}_{ij} \in C(\mathbb{R}^{n_j};\mathbb{R}_+)$, $F_{ij} \in C(\mathbb{R}^N \times \mathbb{R}^N \times \mathbb{R}_+;\mathbb{R})$ *and constants* $a_{ij} \geq 0$ *(*$A = (a_{ij})$ *is irreducible) such that*

$$\mathcal{L}V_i(x,y,t) \leq \beta_i(t) - \alpha_i(t)h_i(x) + \sum_{j=1}^{l}(1-d_j)\bar{\alpha}_{ij}\big(t-r_j(t)\big)\bar{h}_{ij}(y_j) + \sum_{j=1}^{l} a_{ij}F_{ij}(x,y,t),$$

where

$$\alpha_i(t) \geq \bar{\alpha}_{ij}(t), \qquad h_i(x) \geq \bar{h}_i(x) := \sum_{j=1}^{l} \bar{h}_{ij}(x_j), \quad \forall i \in \mathbb{L}.$$

Assume along each directed cycle $\mathcal{C}_{\mathcal{Q}}$ *of weighted digraph* $(\mathcal{G}, A)$, *there are functions* $\beta^{\mathcal{Q}} \in L(\mathbb{R}_+;\mathbb{R}_+)$, $\alpha^{\mathcal{Q}} \in \Psi(\mathbb{R}_+;\mathbb{R}_+)$, $\bar{\alpha}_j^{\mathcal{Q}} \in C([-r_j(0),\infty),\mathbb{R}_+)$, $h^{\mathcal{Q}} \in C(\mathbb{R}^N;\mathbb{R}_+)$, $\bar{h}_j^{\mathcal{Q}} \in C(\mathbb{R}^{n_j};\mathbb{R}_+)$ *such that*

$$\sum_{(u,v)\in\mathcal{E}(\mathcal{C}_{\mathcal{Q}})} F_{vu}(x,y,t) \leq \beta^{\mathcal{Q}}(t) - \alpha^{\mathcal{Q}}(t)h^{\mathcal{Q}}(x) + \sum_{j=1}^{l}(1-d_j)\bar{\alpha}_j^{\mathcal{Q}}(t-r_j(t))\bar{h}_j^{\mathcal{Q}}(y_j),$$

where

$$\alpha^{\mathcal{Q}}(t) \geq \bar{\alpha}_j^{\mathcal{Q}}(t), \qquad h^{\mathcal{Q}}(x) \geq \bar{h}^{\mathcal{Q}}(x) := \sum_{j=1}^{l} \bar{h}_j^{\mathcal{Q}}(x_j), \quad \forall \mathcal{Q} \in \mathbb{Q}.$$

Then

$$\bar{S} := \left(\bigcap_{i=1}^{l} \mathrm{Ker}(h_i - \bar{h}_i)\right) \cap \left(\bigcap_{\mathcal{Q}\in\mathbb{Q}} \mathrm{Ker}\big(h^{\mathcal{Q}} - \bar{h}^{\mathcal{Q}}\big)\right) \neq \emptyset \tag{4.32}$$

and for any initial data ξ, *the CSDDEs* (1.1)-(1.2) *have a unique global solution on* $[0,\infty)$ *denoted by* $x(t,\xi)$ *satisfying*

$$\lim_{t\to\infty} d\big(x(t,\xi), S\big) = 0 \quad a.s.$$

Proof Define $w_i(x,t) = \alpha_i(t)h_i(x)$, $\bar{w}_{ij}(x_j,t) = \bar{\alpha}_{ij}(t)\bar{h}_{ij}(x_j)$, $w^{\mathcal{Q}}(x,t) = \alpha^{\mathcal{Q}}(t)h^{\mathcal{Q}}(x)$, and $\bar{w}_j^{\mathcal{Q}}(x_j, t) = \bar{\alpha}_j^{\mathcal{Q}}(t)\bar{h}_j^{\mathcal{Q}}(x_j)$, then we have

$$\begin{aligned} & w_i(x,t) - \sum_{j=1}^{l} \bar{w}_{ij}(x_j,t) \\ &= \alpha_i(t)\left(h_i(x) - \sum_{j=1}^{l}\bar{h}_{ij}(x_j)\right) + \sum_{j=1}^{l}\big(\alpha_i(t) - \bar{\alpha}_{ij}(t)\big)\bar{h}_{ij}(x_j) \\ &\geq \alpha_i(t)\left(h_i(x) - \sum_{j=1}^{l}\bar{h}_{ij}(x_j)\right) = \alpha_i(t)\big(h_i(x) - \bar{h}_i(x)\big), \end{aligned} \tag{4.33}$$

$$\begin{aligned} & w^{\mathcal{Q}}(x,t) - \sum_{j=1}^{l} \bar{w}_j^{\mathcal{Q}}(x_j,t) \\ &= \alpha^{\mathcal{Q}}(t)\left(h^{\mathcal{Q}}(x) - \sum_{j=1}^{l}\bar{h}_j^{\mathcal{Q}}(x_j)\right) + \sum_{j=1}^{l}\big(\alpha^{\mathcal{Q}}(t) - \bar{\alpha}_j^{\mathcal{Q}}(t)\big)\bar{h}_j^{\mathcal{Q}}(x_j) \\ &\geq \alpha^{\mathcal{Q}}(t)\left(h^{\mathcal{Q}}(x) - \sum_{j=1}^{l}\bar{h}_j^{\mathcal{Q}}(x_j)\right) = \alpha^{\mathcal{Q}}(t)\big(h^{\mathcal{Q}}(x) - \bar{h}^{\mathcal{Q}}(x)\big). \end{aligned} \tag{4.34}$$

Hence, the result is obtained by Theorem 4.1 directly. □

Notice that if $h_i(x) \equiv \bar{h}_i(x)$ and $h^{\mathcal{Q}}(x) \equiv \bar{h}^{\mathcal{Q}}(x)$, the result of above corollary is useless. Therefore, the stronger conditions are needed to locate the limit set of the CSDDEs (1.1).

Corollary 4.2 *Let all the assumptions of Corollary* 4.1 *hold, moreover, assume there are subsets* $\mathbb{I}, \mathbb{J} \subseteq \mathbb{L}$ *and* $\mathbb{P} \subseteq \mathbb{Q}$ *such that, for each* $i \in \mathbb{I}$, $j \in \mathbb{J}$, $\mathcal{Q} \in \mathbb{P}$, $(\alpha_i(t) - \bar{\alpha}_{ij}(t)), (\alpha^{\mathcal{Q}}(t) - \bar{\alpha}_j^{\mathcal{Q}}(t)) \in \Psi(\mathbb{R}_+; \mathbb{R}_+)$. *Then*

$$\tilde{S} := \bar{S} \cap \left(\bigcap_{i\in\mathbb{I}, j\in\mathbb{J}} \mathrm{Ker}(\bar{h}_{ij}) \right) \cap \left(\bigcap_{j\in\mathbb{J}, \mathcal{Q}\in\mathbb{P}} \mathrm{Ker}(\bar{h}_j^{\mathcal{Q}}) \right) \neq \emptyset, \tag{4.35}$$

where $\bar{S}$ *is defined as equality* (4.32) *and for any initial data* ξ, *the CSDDEs* (1.1)-(1.2) *have a unique global solution on* $[0, \infty)$ *denoted by* $x(t, \xi)$ *satisfying*

$$\lim_{t\to\infty} d\big(x(t,\xi), \tilde{S}\big) = 0 \quad a.s.$$

Proof Obviously, the conditions of Corollary 4.1 are satisfied. So the conclusions of Corollary 4.1 still hold. By inequalities (4.33) and (4.34), we get

$$w_i(x,t) - \sum_{j=1}^{l} \bar{w}_{ij}(x_j, t) \geq \sum_{j=1}^{l} \big(\alpha_i(t) - \bar{\alpha}_{ij}(t)\big)\bar{h}_{ij}(x_j),$$

$$w^{\mathcal{Q}}(x,t) - \sum_{j=1}^{l} \bar{w}_j^{\mathcal{Q}}(x_j, t) \geq \sum_{j=1}^{l} \big(\alpha^{\mathcal{Q}}(t) - \bar{\alpha}_j^{\mathcal{Q}}(t)\big)\bar{h}_j^{\mathcal{Q}}(x_j).$$

By Theorem 4.1, the required assertions are obtained. □

From Theorem 4.1, S is the limit set of solutions to the CSDDEs (1.1). Therefore, the asymptotic properties of the solutions, such as the asymptotic boundedness, can be obtained more precisely from the information of S.

Corollary 4.3 *Let all the assumptions of Theorem* 4.1 *hold. If* S *is bounded, then with the initial data* ξ, *the solution has the property*

$$\lim_{t\to\infty} \|x(t,\xi)\| \leq C \quad a.s.,$$

where $C = \sup_{x\in S} \|x\|$.

If S only contains the origin of $\mathbb{R}^N$, we can obtain the following result on the globally asymptotic stability with probability 1.

Corollary 4.4 *Let all the assumptions of Theorem* 4.1 *hold. If* $h_i(x) = 0$, $h^{\mathcal{Q}}(x) = 0$, $i \in \mathbb{L}$, $\mathcal{Q} \in \mathbb{Q}$ *iff* $x = 0$, *then the solution with the initial data* ξ *has the property*

$$\lim_{t\to\infty} x(t,\xi) = 0 \quad a.s.$$

Furthermore, if the rate of $V_i(x_i)$ tending to infinity is known, the rate of the solutions converging to Ker(h) is obtained. Exponential stability is a simple application of Theorem 4.1, which is arranged in the following corollary.

Corollary 4.5 *Let all the assumptions of Theorem 4.1 hold. Assume for each $i \in \mathbb{L}$, there are pairs of positive constants λ_i and p_i, such that*

$$V_i(x_i, t) \geq e^{\lambda_i t}\|x_i\|^{p_i}. \tag{4.36}$$

Then the solution with the initial data ξ has the property

$$\lim_{t\to\infty} \frac{1}{t} \ln\|x(t,\xi)\| \leq -\frac{\hat{\lambda}}{\hat{p}} \quad a.s.$$

Proof From (4.7) in the proof of Theorem 4.1, for almost all $\omega \in \Omega$, the solution $x(t,\xi)$ satisfies

$$\limsup_{t\to\infty} \sum_{i=1}^{l} c_i V_i\big(x_i(t,\xi), t\big) < \infty \quad \text{a.s.}$$

Equation (4.36) implies

$$Ke^{\hat{\lambda} t}\|x(t,\xi)\|^{\hat{p}} \leq \sum_{i=1}^{l} c_i e^{\lambda_i t}\|x_i(t,\xi)\|^{p_i} \leq \sum_{i=1}^{l} c_i V_i\big(x_i(t,\xi), t\big) \quad \text{a.s.,}$$

where K is a positive constant. Then the required assertion follows directly. □

Example 4.1 (Coupled stochastic oscillators with unbounded delays) Given a network represented by digraph $\mathcal{Q}$ with l vertices, a couple system can be built by assigning the ith vertex ($i = 1, 2, \ldots, l$) its own oscillator satisfying the Itô equation

$$\ddot{z}_i(t) + p_i(t)\dot{z}_i(t) + f_i\big(z_i(t)\big) = -q_i(t)\dot{z}_i(t - \theta_i t)\dot{B}_i(t), \tag{4.37}$$

and then coupling these oscillators based on directed arcs in the digraph. For each i, define $x_i = (x_i^{(1)}, x_i^{(2)})^T = (z_i, \dot{z}_i)^T$, then we obtain the following self-excited coupled system (SCS) on $\mathcal{Q}$:

$$\begin{aligned} dx_i(t) = &\begin{pmatrix} x_i^{(2)}(t) \\ -p_i(t)x_i^{(2)}(t) - f_i(x_i^{(1)}(t)) - \sum_{j=1}^{l} a_{ij}(x_i^{(2)}(t) - x_j^{(2)}(t)) \end{pmatrix} dt \\ &+ \begin{pmatrix} 0 \\ -q_i(t)x_i^{(2)}(t - \theta_i t) \end{pmatrix} dB_i(t), \quad 1 \leq i \leq l, \end{aligned} \tag{4.38}$$

where, for each $1 \leq i, j \leq l$, $\theta_i \in (0,1)$, $p_i(\cdot), q_i(\cdot), f_i(\cdot) \in C(\mathbb{R};\mathbb{R})$, $x_i^{(1)} f_i(x_i^{(1)}) > 0$ for $x_i^{(1)} \neq 0$, $\int_0^\infty f_i(u)\,du = \infty$, $a_{ij} \geq 0$, and $A = (a_{ij})$ is irreducible. Then the total energy function $V_i(x_i, t) = \frac{(x_i^{(2)})^2}{2} + \int_0^{x_i^{(1)}} f_i(u)\,du$ is radially unbounded. Compute

$$\mathcal{L}V_i(x, y, t) =: -w_i(x, t) + \sum_{j=1}^{l} (1 - \theta_j)\bar{w}_{ij}(y_j, t) + \sum_{j=1}^{l} a_{ij}F_{ij}(x, y, t), \tag{4.39}$$

where $w_i(x,t) = p_i(t)(x_i^{(2)})^2$,

$$\bar{w}_{ij}(x_j,t) = \begin{cases} \frac{1}{2(1-\theta_j)}(q_j(t)x_j^{(2)})^2, & j = i, \\ 0, & j \neq i, \end{cases}$$

$$F_{ij}(x,y,t) = \frac{1}{2}\left(x_j^{(2)}\right)^2 - \frac{1}{2}\left(x_i^{(2)}\right)^2 - \frac{1}{2}\left(x_i^{(2)} - x_j^{(2)}\right)^2.$$

Compute

$$w_i(x,t) - \sum_{j=1}^{l}(1-\theta_j)\bar{w}_{i,j}(x_j,t) = \left[p_i(t) - \frac{1}{2(1-\theta_i)}q_i^2(t)\right]\left(x_i^{(2)}\right)^2.$$

Assume along each directed cycle $\mathcal{C}_{\mathcal{Q}}$ of the weighted digraph $(\mathcal{G}, A)$,

$$\sum_{(u,v)\in\mathcal{E}(\mathcal{C}_{\mathcal{Q}})} F_{vu}(x,y,t) \leq -\frac{1}{2}\sum_{(u,v)\in\mathcal{E}(\mathcal{C}_{\mathcal{Q}})}\left(x_u^{(2)} - x_v^{(2)}\right)^2.$$

Suppose that there exists a $1 \leq k \leq l$ such that $p_k(t) - \frac{1}{2(1-\theta_k)}q_k^2(t) \in \Psi(\mathbb{R}_+;\mathbb{R}_+)$, by Theorem 4.1, the point $(x_1,x_2) \in S$ satisfies $x_k^{(2)} = 0$. By the strong connectivity of $(\mathcal{Q}, A)$, each vertex i belongs to a connective circle $\mathcal{C}_{\mathcal{Q}}$ including vertex k. Using Theorem 4.1 again, we know $x_i^{(2)} = 0$, $i = 1,2,\ldots,l$. Denote by $(x_1(t), x_2(t))$ the global solution of SCS (4.38) with an initial value, then

$$\lim_{t\to\infty} x_i^{(2)}(t) = 0 \quad \text{a.s. } i = 1,2,\ldots,l.$$

Furthermore, by the second equation of (4.38), the stochastic oscillator has the property

$$\lim_{t\to\infty} x_i^{(1)}(t) = 0 \quad \text{a.s. } i = 1,2,\ldots,l.$$

For example, see Figure 1 for a computer simulation of the paths of three independent oscillators described by (4.37), where $f_1(x) = x|x|$, $\theta_1 = 0.5$, $p_1 = 1 + 0.3\cos t$, $q_1 = 0.5\sin t$; $f_2(x) = x$, $\theta_2 = 0.6$, $p_2 = 0.05 + 1.25e^{-t}$, $q_2 = 0.2 - e^{-\frac{1}{2}t}$; $f_3(x) = x$, $\theta_3 = 0.25$, $p_3 = 1 + e^{-t}$, $q_3 = 0.2 + e^{-t}$.

Couple these oscillators based on directed arcs in the digraph described by Figure 2, simulate the paths of the coupled oscillators and obtain Figure 3, which supports the main results clearly.

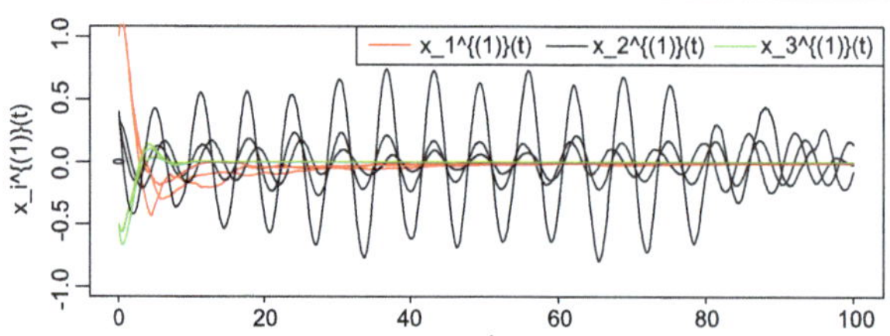

Figure 1 **Computer simulation of the paths of $x_1^{(1)}(t)$, $x_2^{(1)}(t)$ and $x_3^{(1)}(t)$ for the three discrete-time independent oscillators described by (4.37) using the Euler-Maruyama method with step size 0.02 and initial values $x_1^{(1)}(0) = 1$, $x_1^{(2)}(0) = 0.6$, $x_2^{(1)}(0) = 0.4$, $x_2^{(2)}(0) = 0.7$, $x_3^{(1)} = -0.5$, $x_3^{(2)}(0) = -0.5$.**

Figure 2 A strongly connected digraph with three vertices where $a_{13} = 3$, $a_{21} = 2$, $a_{23} = 4$, $a_{32} = 1$, $a_{11} = a_{12} = a_{22} = a_{31} = a_{33} = 0$.

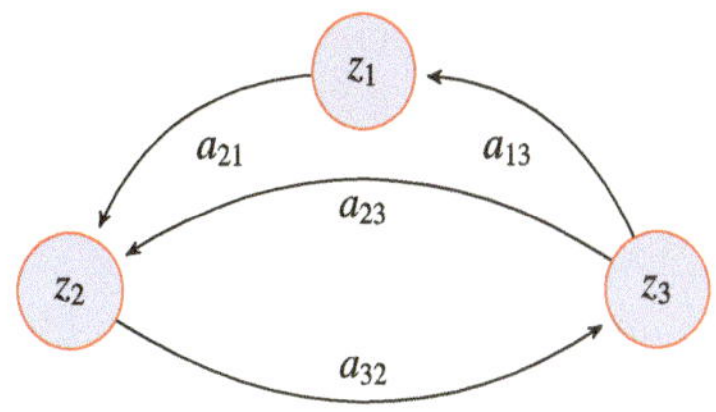

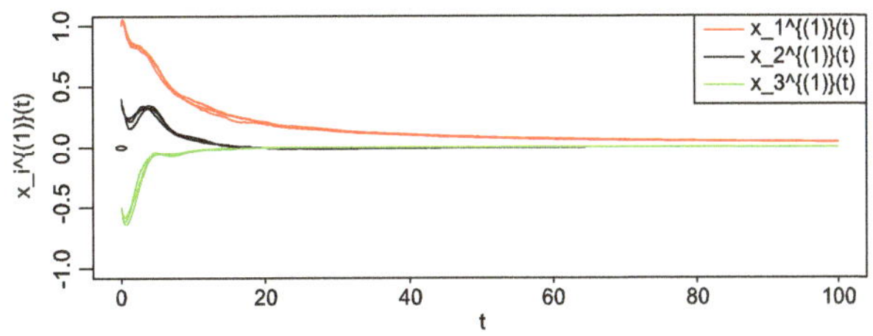

Figure 3 Computer simulation of the paths of $x_1^{(1)}(t)$, $x_2^{(1)}(t)$, and $x_3^{(1)}(t)$ for the discrete-time coupled oscillators described by (4.38) using the Euler-Maruyama method with the same step size, parameters and initial values as those of Figure 1.

Competing interests
The authors declare that they have no competing interests.

Authors' contributions
The authors indicated in parentheses made substantial contributions to the following tasks of research: drafting the manuscript (RW); participating in the design of the study (WL, XL); writing and revising of the paper (XL). All authors read and approved the final manuscript.

Author details
[1]School of Mathematics and Statistics, Northeast Normal University, Changchun, 130024, P.R. China. [2]Department of Mathematics, Harbin Institute of Technology, Weihai, 264209, P.R. China.

Acknowledgements
The authors are grateful to the editor and anonymous reviewers for their helpful comments and valuable suggestions, which have greatly improved our original manuscript. This work is financially supported by the National Natural Science Foundation of China (11171056, 11171081) and Natural Science Foundation of Jilin Province (20101593).

References
1. Chen, H, Sun, J: Global stability for coupled nonlinear systems with time functional on networks. Physica A **391**, 528-534 (2012)
2. Guo, H, Li, MY, Shuai, Z: A graph-theoretic approach to the method of global Lyapunov functions. Proc. Am. Math. Soc. **136**, 2793-2802 (2008)
3. Guo, H, Li, MY, Shuai, Z: Global dynamics of a general class of multistage models for infectious diseases. SIAM J. Appl. Math. **72**, 261-279 (2012)
4. Kao, Y, Wang, C: Global stability analysis for stochastic coupled reaction-diffusion systems on network. Nonlinear Anal., Real World Appl. **14**, 1457-1465 (2013)
5. Li, MY, Shuai, Z: Global-stability problem for coupled systems of differential equations on networks. J. Differ. Equ. **248**, 1-20 (2010)
6. Li, MY, Shuai, Z, Wang, C: Global stability of multi-group epidemic models with distributed delays. J. Math. Anal. Appl. **361**, 38-47 (2010)
7. Li, W, Su, H, Wang, K: Global stability analysis for stochastic coupled systems on network. Automatica **47**, 215-220 (2011)
8. Roy, R, Murphy, TW, Maier, TD, Gills, Z, Hunt, ER: Dynamical control of a chaotic laser: experimental stabilization of a globally coupled system. Phys. Rev. Lett. **68**, 1259-1262 (1992)
9. Shu, H, Fan, D, Wei, J: Global stability of multi-group SEIR epidemic models with distributed delays and nonlinear transmission. Nonlinear Anal., Real World Appl. **13**, 1581-1592 (2012)
10. Hoppensteadt, FC, Izhikevich, EM: Weakly Connected Neural Networks. Springer, New York (1997)
11. Øksendal, B: Stochastic Differential Equations. Springer, New York (2005)
12. Friedman, A: Stochastic Differential Equations and Their Applications. Academic Press, San Diego (1976)
13. Kolmanovskii, VB, Nosov, VR: Stability and Periodic Models of Control Systems with Aftereffect. Nauka, Moscow (1981)
14. Mao, X: Stochastic Differential Equations and Applications. Horwood, Chichester (1997)

15. Shaikhet, LE: Some new aspects of Lyapunov-type theorems for stochastic differential equations of neutral type. SIAM J. Control Optim. **48**, 4481-4499 (2010)
16. Li, X, Mao, X: The improved LaSalle-type theorems for stochastic differential delay equations. Stoch. Anal. Appl. **30**, 568-589 (2012)
17. Li, X, Mao, X: A note on almost sure asymptotic stability of neutral stochastic delay differential equations with Markovian switching. Automatica **48**, 2329-2334 (2012)
18. Mao, X: Stochastic versions of the LaSalle theorem. J. Differ. Equ. **153**, 175-195 (1999)
19. Mao, X: A note on the LaSalle-type theorem for stochastic differential delay equations. J. Math. Anal. Appl. **268**, 125-142 (2002)
20. Rodkina, A, Schurz, H, Shaikhet, L: Almost sure stability of some stochastic dynamical systems with memory. Discrete Contin. Dyn. Syst. **21**, 571-593 (2008)
21. LaSalle, JP: Stability theory of ordinary differential equations. J. Differ. Equ. **4**, 57-65 (1968)
22. Berman, A, Plemmons, RJ: Nonnegative Matrices in the Mathematical Sciences. Academic Press, New York (1979)
23. Moon, JW: Counting Labelled Trees. Canadian Mathematical Congress, Montreal (1970)
24. Lu, Z, Takeuchi, Y: Global asymptotic behavior in single-species discrete diffusion systems. J. Math. Biol. **32**, 66-77 (1993)
25. Liptser, RS, Shiryayev, AN (eds.): Theory of Martingales. Kluwer Academic, Dordrecht (1989)

6

Stability of the zero solution of nonlinear differential equations under the influence of white noise

Irada Dzhalladova[1], Miroslava Růžičková[2*] and Viera Štoudková Růžičková[3]

*Correspondence: miroslava.ruzickova@fpv.uniza.sk
[2]University of Žilina, Žilina, Slovakia
Full list of author information is available at the end of the article

Abstract

The paper deals with a system of nonlinear differential equations under the influence of white noise. This system can be used as a mathematical model of various real problems in finance, mathematical biology, climatology, signal theory and others. Necessary and sufficient conditions for the asymptotic mean square stability of the zero solution of this system are derived in the paper. The paper introduces a new approach to studying such problems - construction of a suitable deterministic system with the use of Lyapunov function.

MSC: 34K50; 60H10; 60H30; 65C30

Keywords: stochastic systems; white noise; mean square stability; Lyapunov function

1 Introduction

We can come across stochastic behavior while examining many important problems of a global character in various fields of research, for example, in the theory of climate change. Detailed understanding of extreme events in climate, of phenomena that are beyond our normal expectations, is a very important topic in climatology, meteorology and related fields. Common methods of studying extreme events, such as the statistical approach, the empirical-physical approach or the numerical modeling approach, have some limitations, and study of them has been largely empirical.

The idea of replacing the whole deterministic system with a stochastic differential equation was introduced by Hasselmann in his work [1] on stochastic climate models that appeared in 1976. There he proposed to improve deterministic models for the 'climate' (slow variables) by incorporating the influence of the 'weather' (fast variables) in the form of random noise. The univariate linear systems that appear in the work have been successful in describing various modes of climate variability. Success of these models has inspired researchers to consider the stochastic forcing as a possible source of more complex dynamics, for example, in [2]. The direction of stochastic parametrizations in which the development of the climate models will be possible in the coming years is formulated, for example, in [3].

Hasselmann's works can be seen as the beginning of describing extreme events in climate by a stochastic system of differential equations in which random weather changes are expressed by a nonlinear stochastic perturbation in the form of white noise. Stochastic characteristics of white noise in the system are either known or are determined by the climate's median value.

Many works, dedicated to the study of evolution models, use white noise to express the amount of ice surface of the Arctic Ocean and then estimate the likelihood that it would spread on the entire surface of the Earth. Part of the mystery of the Earth's periodic ice ages was uncovered in connection with the so-called stochastic resonance that was first discovered while studying the periodic recurrence of Earth's ice ages, see [4, 5]. Since its discovery, the stochastic resonance has been used in various experimental and theoretical studies. For example, it occurs in bi-stable systems, where the input-output relationship is nonlinear, the periodic input signal is weak and there is random, uncorrelated variation added to the signal of interest. In climatic models, the stochastic resonance manifests at its best when regular 'weak' weather changes are in certain proportion to the random weather fluctuations. By the term 'weak' we mean changes, that are not capable of causing climate changes on their own, but the presence of random weather changes causes disturbances in periodicity of the system, which then transforms into a different mode.

Our work deals with a stochastic model with nonlinear member. Some related problems regarding such systems were studied in [6]. In this paper we focus on the determination of conditions that are necessary and sufficient for stable behavior of the above mentioned processes. The present paper contains sufficient conditions for the asymptotic mean square stability of the zero solution of systems with white noise. Further, sufficient conditions for instability of the zero solution of this system and sufficient conditions for a more general system are derived in the paper. The results are followed by several examples. We construct Lyapunov functions as a tool to study stability of a stochastic system that works under the influence of white noise.

Let $(\Omega, \mathcal{F}, F, \mathbb{P})$ be a filtered probability space, or stochastic basis, consisting of a probability space $(\Omega, \mathcal{F}, \mathbb{P})$ and a filtration $F = \{\mathcal{F}_t, \forall t \geq 0\}$ contained in $\mathcal{F}$. On the probability space we consider the stochastic system of nonlinear differential equations of the form

$$dx_i(t) = f_i(x)\,dt + h_i(x)\,dw_i(t), \quad i = 1, 2, \ldots, m, \quad t \geq 0, \tag{1}$$

where the state function $x(t) = (x_1, \ldots, x_m)^T$ (the operation T denotes transposition) is a continuously differentiable m-dimensional column vector-function, $f = (f_1, \ldots, f_m)^T$, $h = (h_1, \ldots, h_m)^T$ are also continuously differentiable m-dimensional column vector-functions such that $f(o) = o$, $h(o) = o$, $o = (0, 0, \ldots, 0)^T \in \mathbb{R}^m$ hold. The function f represents a slow deterministic process, the products $h_i(x)\,dw_i(t)$, $i = 1, 2, \ldots, m$, are the stochastic approximations to a fast phenomenon. The m-dimensional column vector-function $w = (w_1, \ldots, w_m)^T$ indicates a standard Wiener process. The m-dimensional Wiener process is said to be standard Wiener process if

$$\begin{aligned} &w(0) = o, \\ &E^{(1)}\{dw(t)\} = o, \\ &E^{(1)}\{dw(t)\,dw^T(t)\} = I\,dt \end{aligned} \tag{2}$$

hold for $t \geq 0$, I is the identity matrix. Any realization of the Wiener process $w(t)$ is continuous but not differentiable at each point. Moreover, because $E^{(1)}\{w(t)w^T(t)\} = It$, the process $w(t)$ is a non-stationary stochastic process.

For simplicity, we denote $h(x)\,dw(t) = (h_1\,dw_1, \dots, h_m\,dw_m)^T$ in our consideration. So, the product $h(x)\,dw(t)$ means neither a scalar product nor a vector product, but it is a column vector with components $h_i(x)\,dw_i(t)$, $i = 1, 2, \dots, m$. Using this, system (1) can be rewritten into the vector form

$$dx(t) = f(x)\,dt + h(x)\,dw(t), \quad t \geq 0. \tag{3}$$

We are interested in stability of solutions of the system. There are several various definitions of stability that can be used. Here we recall the mean square stability and the asymptotic mean square stability of the zero solution.

Definition 1 The trivial solution of system (3) is said to be mean square stable on the interval $[0, \infty)$ if, for each $\varepsilon > 0$, there exists $\delta > 0$ such that any solution $x(t)$ corresponding to the initial data $x(0)$ exists for all $t \geq 0$ and the mathematical expectation

$$E^{(1)}\left\{\left\|x(t)\right\|^2\right\} < \varepsilon \quad \text{whenever } t \geq 0 \text{ and } \left\|x(0)\right\| < \delta.$$

The mean stability of the zero solution of system (3) is defined in a very similar way, with only $\|x(t)\|^2$ being replaced by $\|x(t)\|$.

Definition 2 The trivial solution of system (3) is said to be asymptotically mean square stable on the interval $[0, \infty)$ if it is stable and, moreover,

$$\lim_{t \to \infty} E^{(1)}\left\{\left\|x(t)\right\|^2\right\} = 0 \tag{4}$$

is satisfied for each solution of (3).

Remark 1 It is easy to see that (4) is satisfied if and only if the matrix $E^{(2)}\{x(t)\}$ converges to zero matrix Θ,

$$\lim_{t \to \infty} E^{(2)}\left\{x(t)\right\} = \lim_{t \to \infty} E^{(1)}\left\{x(t)x^T(t)\right\} = \Theta.$$

Denote a neighborhood of the point $o \in \mathbb{R}^m$ as $\mathcal{O}(o)$.

Definition 3 The function $g(t, x(t))$ is said to be positive definite on $\mathcal{O}(o)$ if $g(t, x)$ is continuous with respect to t, $t \geq 0$, $\|x(t)\| < \infty$ on $\mathcal{O}(o)$, $g(t, o) = 0$, and there exists a positive definite quadratic form $V(x)$ such that

$$g(t, x) \geq V(x) \quad \text{whenever } t \geq 0, \quad x \in \mathcal{O}(o) \text{ and } \left\|x(t)\right\| < \infty.$$

Recall, if a function $g(t, x(t))$ is positive definite on $\mathcal{O}(o)$, then the function $-g(t, x(t))$ is negative definite on this neighborhood.

Remark 2 The previous definition is equivalent to the following one:

The function $g(t,x(t))$ is said to be positive definite on $\mathcal{O}(o)$ if $g(t,x)$ is continuous with respect to t, $t \geq 0$, $\|x(t)\| < \infty$ on $\mathcal{O}(o)$, $g(t,o) = 0$, and there exists a constant $k > 0$ such that for any positive definite quadratic form $V(x)$,

$$g(t,x) \geq kV(x) \quad \text{whenever } t \geq 0, \quad x \in \mathcal{O}(o) \text{ and } \|x(t)\| < \infty.$$

2 Main results

We define a Lyapunov function v in the form

$$v(x(t)) = x^T(t)Cx(t), \tag{5}$$

where C is an $m \times m$ positive definite symmetric matrix. We also use the diagonal matrix C_d which has the same elements on the diagonal as C.

Theorem 1 *Let there exist a neighborhood $\mathcal{O}(o)$, in which the function*

$$\delta(x) = 2x^T Cf(x) + h^T(x)C_d h(x) \tag{6}$$

is negative definite with respect to system (3).

Then the trivial solution of (3) *is asymptotically mean square stable on the interval* $[0,\infty)$.

Proof First we calculate the differential of Lyapunov function (5) with respect to trajectories of system (3). We have

$$\begin{aligned} dv(x) &= v(x+dx) - v(x) = \left(x^T + dx^T\right)C(x+dx) - x^T Cx \\ &= \left(x^T + f^T(x)\,dt + \left(h(x)\,dw(t)\right)^T\right)C\left(x + f(x)\,dt + \left(h(x)\,dw(t)\right)\right) - x^T Cx \\ &= x^T Cx - x^T Cx + x^T Cf(x)\,dt + x^T C\left(h(x)\,dw(t)\right) + f^T(x)\,dt\,Cx \\ &\quad + f^T(x)\,dt\,Cf(x)\,dt + f^T(x)\,dt\,C\left(h(x)\,dw(t)\right) + \left(h(x)\,dw(t)\right)^T Cx \\ &\quad + \left(h(x)\,dw(t)\right)^T Cf(x)\,dt + \left(h(x)\,dw(t)\right)^T C\left(h(x)\,dw(t)\right). \end{aligned}$$

After modifying the obtained equation, applying operation mathematical expectation $E^{(1)}\{dv(t)\}$, in regard to assumptions (2), and leaving aside the member with dt^2, we get the equation

$$E^{(1)}\left\{dv(x)\right\} = x^T Cf(x)\,dt + f^T(x)Cx\,dt + h^T(x)C_d h(x)\,dt, \tag{7}$$

or, in view of (6), it can be written in the more simple form

$$E^{(1)}\left\{dv(x)\right\} = \delta(x)\,dt. \tag{8}$$

Because $\delta(x)$ is a negative definite deterministic function, the equalities

$$E^{(1)}\left\{\delta(x)\right\} = \delta(x),$$

$$\delta(x) \leq -kv(x), \quad k = \text{const},$$

are true (see Remark 2). By using them and equation (8), we get

$$\frac{d}{dt}E^{(1)}\{v(x)\} \leq -kE^{(1)}\{v(x)\},$$

so

$$E^{(1)}\{v(x)\} \leq e^{-kt}.$$

Therefore,

$$\lim_{t\to\infty} E^{(2)}\{x(t)\} = \lim_{t\to\infty} E^{(1)}\{x(t)x^T(t)\} = \Theta,$$

which implies asymptotic mean square stability of a trivial solution of the considered system. □

Remark 3 Analogous result about instability can be derived in the same way as the result of Theorem 1.

Namely, if there exists a neighborhood $\mathcal{O}(o)$, in which the function

$$\delta(x) = 2x^T Cf(x) + h^T(x)C_d h(x)$$

is positive definite with respect to system (3), then the trivial solution of (3) is unstable on the interval $[0,\infty)$.

Further we will discuss the stability of a system in the more general form

$$dx(t) = f(t,x)\,dt + \sum_{k=1}^{n} h_k(t,x)\,dw_k(t), \quad t \geq 0, \tag{9}$$

where f, h_k, $k = 1,\dots,n$, are m-dimensional column vector-functions continuously differentiable in both variables, such that $f(t,o) = o$, $h_k(t,o) = o$ hold. Expressions $h_k(t,x)\,dw_k(t)$, $k = 1,\dots,n$, are again column vectors with components $h_{ik}(t,x)\,dw_{ik}(t)$, $i = 1,2,\dots,m$, $k = 1,2,\dots,n$. Functions $w_k = (w_{1k},\dots,w_{mk})^T$, $k = 1,2,\dots,n$, are also m-dimensional column vector-functions indicating standard Wiener processes, such that each of them satisfies the following relationships:

$$\begin{aligned} &w_k(0) = o,\\ &E^{(1)}\{dw_k(t)\} = o,\\ &E^{(1)}\{dw_k(t)\,dw_k^T(t)\} = I\,dt, \quad k = 1,2,\dots,n \end{aligned} \tag{10}$$

for $t \geq 0$.

The following theorem can be proved in the same way as Theorem 1.

Theorem 2 *Let there exist a neighborhood $\mathcal{O}(o)$, in which the function*

$$\delta(t,x) = 2x^T Cf(t,x) + \sum_{k=1}^{n} h_k^T(t,x)C_d h_k(t,x) \tag{11}$$

is negative definite with respect to system (9).

Then the trivial solution of (9) *is asymptotically mean square stable on the interval* $[0, \infty)$.

Remark 4 A similar remark as Remark 3 can be formulated in this case. Namely, if there exists a neighborhood $\mathcal{O}(o)$, in which the function

$$\delta(t,x) = 2x^T Cf(t,x) + \sum_{k=1}^{n} h_k^T(t,x) C_d h_k(t,x)$$

is positive definite with respect to system (9), then the trivial solution of (9) is unstable on the interval $[0, \infty)$.

In the last part of this section, let us consider one special case of a stochastic system of differential equations, namely the linear system

$$dx(t) = Ax(t)\,dt + Hx(t)\,dw(t), \quad t \geq 0, \tag{12}$$

where A, H are $m \times m$ constant matrices and w is a standard Wiener process that satisfies (2).

Corollary 1 *Let there exist a neighborhood* $\mathcal{O}(o)$, *in which the function*

$$\delta(x) = x^T A^T Cx + x^T CAx + x^T H^T C_d Hx \tag{13}$$

is negative definite with respect to system (12).

Then the trivial solution of (12) *is asymptotically mean square stable on the interval* $[0, \infty)$.

Proof The proof of this theorem follows immediately from Theorem 1 if $f(x) = Ax$, $h(x) = Hx$. □

Remark 5 If there exists a neighborhood $\mathcal{O}(o)$, in which the function

$$\delta(x) = x^T A^T Cx + x^T CAx + x^T H^T C_d Hx$$

is positive definite with respect to system (12), then the trivial solution of (12) is unstable on the interval $[0, \infty)$.

Remark 6 Stability of the trivial solution of a stochastic system of linear differential equations in the form (12) was studied in the works by Korenevskii in [7]. The following stability criterion was obtained there by different methods: The trivial solution of system (12) is asymptotically mean square stable on the interval $[0, \infty)$ if and only if there exists a positive definite matrix C satisfying the matrix equation

$$A^T C + CA + H^T CH = -G,$$

where matrix G is positive definite.

3 Examples

In the following examples, we illustrate application of Theorem 1 and Remark 3. In these examples, we also construct the corresponding Lyapunov function; however, it is not necessary for verifying the stability or instability condition.

Example 1 We consider stochastic functional differential equation in the form

$$dx(t) = -x(t)\,dt - x^3(t)\,dw(t), \tag{14}$$

where f, h are scalar functions, $f = -x(t)$, $h = -x^3(t)$, x is a scalar state function and w is a standard Wiener process. Stability of the zero solution of this equation can be determined on the basis of Theorem 1.

We define Lyapunov function in the form (5), with $C = 1$,

$$v(x) = x^2. \tag{15}$$

Then equation (7) is in the form

$$E^{(1)}\{dv(t)\} = \left(-x^T x - x^T x + \left(x^3\right)^T x^3\right) dt = \left(-2x^2 + x^6\right) dt.$$

To apply Theorem 1, we have to prove that there exists a neighborhood of the zero point 0, where the function $\delta = -2x^2 + x^6$ is negative definite. This holds if and only if the following inequality is satisfied:

$$x^6 - 2x^2 < 0 \quad \Leftrightarrow \quad 0 < x^4 < 2$$

or

$$-\sqrt[4]{2} < x < \sqrt[4]{2}, \quad x \neq 0.$$

That is, δ is the function of negative values at each point of $(-\sqrt[4]{2}, \sqrt[4]{2})$ except zero, and $\delta(0) = 0$. Therefore, we have the conclusion that there exists a neighborhood $\mathcal{O}(0) = (-\sqrt[4]{2}, \sqrt[4]{2})$ in which the function $\delta = -2x^2 + x^6$ is negative definite. So, the trivial solution $x = 0$ of considered equation is asymptotically mean square stable on the interval $[0, \infty)$.

Example 2 We consider stochastic differential equation in the form

$$dx(t) = ax(t)\,dt + bx(t)\,dw(t),$$

where $a, b \in \mathbb{R}$, $a \neq 0$ and x is a scalar state function.

We again define Lyapunov function in the form (5) with $C = 1$. The function δ, in view of Theorem 1, has the form

$$\delta(x) = \left(2a + b^2\right)x^2.$$

Hence, if $2a + b^2 < 0$, the function δ is negative definite, in the opposite case it is positive definite. Therefore, in accordance with Theorem 1 and Remark 3, if $2a + b^2 < 0$, then the trivial solution of considered equation is asymptotically mean square stable on the interval $[0,\infty)$. In the opposite case, that is, if $2a + b^2 > 0$, then the trivial solution of considered equation is unstable.

Example 3 We consider stochastic functional differential equation in the form

$$dx(t) = -x(t)\,dt + a\sin x(t)\,dw(t), \tag{16}$$

where f, g are scalar functions, $f = -x(t)$, $g = a\sin x(t)$, $a \in \mathbb{R}$, $a \neq 0$ and x is a scalar state function.

We again define Lyapunov function in the form (5) with $C = 1$. Therefore, a function δ from (6) has the form

$$\begin{aligned}\delta(t,x) &= -2x^T x + \left(a\sin x(t)\right)^T a\sin x(t)\\ &= -2x^2 + a^2\sin^2 x.\end{aligned}$$

Because

$$-2x^2 + a^2\sin^2 x > 0 \quad \Leftrightarrow \quad \frac{x^2}{\sin^2 x} < \frac{a^2}{2},$$

in the case $a > \sqrt{2}$, there exists a neighborhood of the zero point 0, where the function δ is positive definite. In the case $a \leq \sqrt{2}$, the function δ is negative definite. So, the zero solution of considered equation is unstable for any $a > \sqrt{2}$ and it is asymptotically mean square stable for any $a \leq \sqrt{2}$.

Example 4 Now we consider two-dimensional stochastic functional differential equation in the form

$$\begin{aligned}dx_1(t) &= x_2(t)\,dt + h_1(x_1,x_2)\,dw(t),\\ dx_2(t) &= -x_1(t)\,dt + h_2(x_1,x_2)\,dw(t),\end{aligned} \tag{17}$$

where $x = (x_1,x_2)^T$ is a two-dimensional vector state function, f, h are two-dimensional vector functions, $f = (x_2(t), -x_1(t))^T$, $h = (h_1(x_1,x_2), h_2(x_1,x_2))^T$, $h \neq (0,0)$ on a neighborhood $\mathcal{O}(0,0)$ with the exception of the zero point $(0,0)$, and w is a standard Wiener process.

We again define Lyapunov function in the form (5) with $C = I$,

$$v(x) = x_1^2 + x_2^2.$$

Then the function $\delta = h^T h$ and it is positive definite on the neighborhood $\mathcal{O}(0,0)$. Thus, by Theorem 1, the trivial zero solution of this equation is unstable on the interval $[0,\infty)$.

Remark 7 If we take $h_1 = 0$ in system (17), then it represents the stochastic second order differential equation

$$d^2x(t) = -x(t)\,dt + h_2\left(x', x\right)dw(t),$$

and the previous example implies that its solution is unstable for any nontrivial h_2. But a function can be added to the right-hand side of this equation such that the trivial solution of the new equation will be asymptotically mean square stable. The following result holds:

If

$$\lim_{(x_1,x_2)\to(0,0)} \frac{g(x_1,x_2)\cdot x_2}{h_2^2(x_1,x_2)} = L > \frac{1}{2} \tag{18}$$

for a continuously differentiable function $g(x_1,x_2)$, then the trivial solution of the system

$$\begin{aligned} dx_1(t) &= x_2(t)\,dt, \\ dx_2(t) &= -x_1(t)\,dt - g(x_1,x_2)\,dt + h_2(x_1,x_2)\,dw(t) \end{aligned} \tag{19}$$

is asymptotically mean square stable.

Condition (18) holds, for example, for $g(x_1,x_2) = h_2(x_1,x_2) = x_2$. Then system (19) is the following:

$$\begin{aligned} dx_1(t) &= x_2(t)\,dt, \\ dx_2(t) &= -x_1(t)\,dt - x_2(t)\,dt + x_2(t)\,dw(t). \end{aligned} \tag{20}$$

Behavior of solutions $x_1(t)$ of this system compared to that of solutions of the unstable system

$$\begin{aligned} dx_1(t) &= x_2(t)\,dt, \\ dx_2(t) &= -x_1(t)\,dt + x_2(t)\,dw(t) \end{aligned} \tag{21}$$

Figure 1 Behavior of solutions of stable system (20).

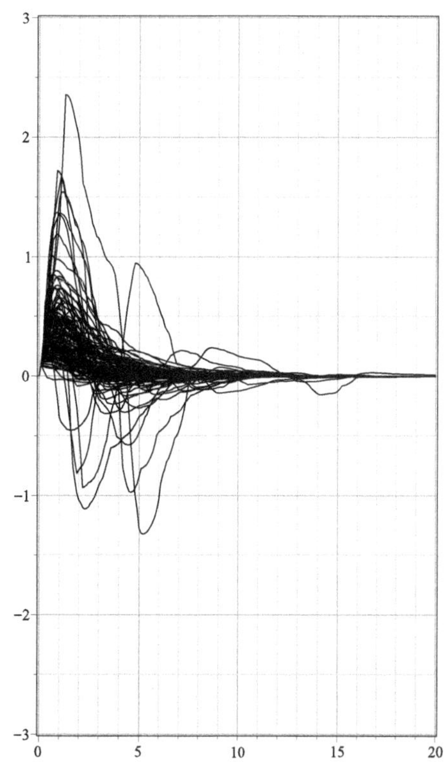

Figure 2 Behavior of solutions of instable system (21).

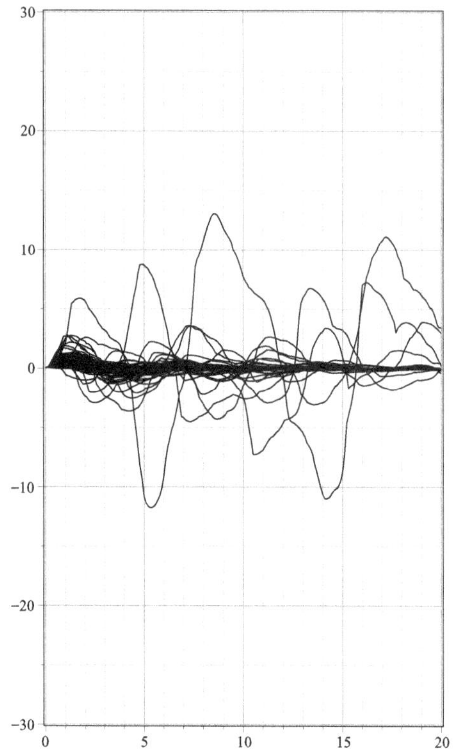

is illustrated in Figures 1 and 2. In both of these figures, there are displayed 120 solutions of the systems (20), (21) with initial conditions $x_1(0) = 0$, $x_2(0) = 1$, respectively. These solutions were computed numerically with the Milstein method on the interval $[0, 20]$.

4 Conclusion

Our concept of studying asymptotic mean square stability of the zero solution of stochastic differential equations is more effective in comparison with previously known methods, and it can also be used in various application problems. For example, by stochastic equation of the type

$$dx(t) = \left[a_1 x(t) - a_2 x(t) \ln x(t)\right] dt + \sigma x(t)\, dw(t)$$

we can describe a tumor growth, where the expected size of the tumor is contaminated with white noise [8]. Such type of equations can be used also in biomedical research [9, 10], epidemic modeling [11], in describing animal motion [12], receiving signals [10, 13] and many others. However, the approach to the study of these models in the cited works is different compared to our method. They use numerical or statistical methods and estimation methods in there.

Competing interests

The authors declare that they have no competing interests.

Authors' contributions

The authors have made the same contribution. All authors read and approved the final manuscript.

Author details
[1]Kiev National Economics Vadym Hetman University, Kyiv, Ukraine. [2]University of Žilina, Žilina, Slovakia. [3]Brno University of Technology, Brno, Czech Republic.

Acknowledgements
The second author was supported by Grant KEGA 004ŽU-4/2014 of Slovak Grant Agency. The third author was supported by the project FSI-S-14-2290 of Brno University of Technology.

References

1. Hasselmann, K: Stochastic climate models. Part I: Theory. Tellus **28**, 473-485 (1976)
2. Arnold, L: Hasselmann's program visited: the analysis of stochasticity in deterministic climate models. In: von Storch, J-S, Imkeller, P (eds.) Stochastic Climate Models, pp. 141-158. Birkhäuser, Boston (2001)
3. Dymnikov, VP, Lykosov, NV, Volodin, EM: Problems of modeling climate and climate change. Izv., Atmos. Ocean. Phys. **42**(5), 568-585 (2006)
4. Benzi, R, Sutera, A, Vulpiani, A: The mechanism of stochastic resonance. J. Phys. A, Math. Gen. **14**(11), L453-L457 (1981)
5. Benzi, R, Parisi, G, Sutera, A, Vulpiani, A: A theory of stochastic resonance in climatic change. SIAM J. Appl. Math. **43**(3), 565-578 (1983)
6. Lipster, R, Shiryaev, A: Statistics of Random Processes I: General Theory, 2nd edn. Springer, Berlin (2001)
7. Korenevskii, DH: The Stability of Solutions to the System of Differential Equations with Coefficients Perturbed by White Noise and Coloured Noises. IM of NASU, Kiev (2012) (in Russian)
8. Lo, CF: Stochastic nonlinear Gompertz model of tumour growth. In: Proceedings of the World Congress on Engineering, London, UK, 1-3 July (2009)
9. Donnet, S, Samson, A: A review on estimation of stochastic differential equations for pharmacokinetic/pharmacodynamic models. Adv. Drug Deliv. Rev. **65**(7), 929-939 (2013)
10. Picchini, U, Ditlevsen, S, De Gaetano, A, Lansky, P: Parameters of the diffusion leaky integrate-and-fire neuronal model for a slowly fluctuating signal. Neural Comput. **20**(11), 2696-2714 (2008)
11. Rao, F: Dynamics analysis of a stochastic SIR epidemic model. Abstr. Appl. Anal. **2014**, Article ID 356013 (2014)
12. Brillinger, DR, Preisler, HK, Ager, AA, Kie, JG, Stewart, BS: Employing stochastic differential equations to model wildlife motion. Bull. Braz. Math. Soc. **33**(3), 385-408 (2002)
13. Gardiner, CW: Handbook of Stochastic Methods for Physics, Chemistry and the Natural Sciences, 2nd edn. Springer, Berlin (1985)

7

The discrete tuberculosis transmission model with treatment of latently infected individuals

Hui Cao[*] and Hongwu Tan

[*]Correspondence:
caohui@sust.edu.cn
Department of Mathematics, Shaanxi University of Science & Technology, Xi'an, 710021, China

Abstract

A discrete tuberculosis model with direct progression and treatment of latently infected individuals is presented. The model does not consider the drug-resistant TB, and it assumes that latently infected individuals develop the active disease only because of being endogenous reactive, and a small fraction of infected individuals is assumed to develop the active disease soon after infection. The global stability of a disease-free equilibrium, the persistence of system, and the local stability of endemic equilibrium are discussed. The basic reproductive numbers with different control measures are determined and analyzed, and we give the critical value of probability of successful detection and treatment of infectious individuals. If a treatment only of infectious individuals cannot control TB transmission, the treatment of latent TB individuals should be carried out, and we give the critical value of the probability of treatment of infectious individuals. Numerical simulations are done to demonstrate the complex dynamics of the model.

Keywords: discrete TB model; direct progression; treatment strategies

1 Introduction

Differential equations and difference equations are widely applied in epidemiological modeling. They are two typical mathematical approaches for modeling infectious diseases. Since the theory and method for dynamical studies of differential equations have developed much more completely than those for difference equations, there are relatively few difference equations in epidemiological modeling compared with differential equations. In recent years, there were increasing interest and research results on discrete epidemic models [1–8]. The fact that the epidemiological data are usually collected in discrete time units, such as days, weeks or months, makes the discrete model a natural choice to describe a disease transmission. The straightforward recurrence relationship of the difference equation models is easier to understand, which is also a prominent advantage over the differential equation models. The direct comparison of the model results with the actual data provides us a fast and simple way to validate the model structure and parameter estimation. The fact that the discrete models exhibit a richer dynamical behavior than the continuous models brings about more challenging problems for researches, and more interesting results can be obtained. For example, the simple logistic model, $x_{n+1} = rx_n(1 - x_n/K)$, the

Ricker model $x_{n+1} = x_n e^{r - x_n/K}$ [9–11], and the Hassell model, $x_{n+1} = \lambda x_n(1 + ax_n)^{-b}$ [12] exhibit a rich dynamical behavior.

There have been increasing interest and more studies on the discrete time epidemic models recently. Various discrete epidemic models have been successfully applied to describe the infectious disease transmission, such as SARS, tuberculosis, HIV/AIDS [13–18]. The theoretical study of discrete epidemic models focuses on the computation of the basic reproduction number [19–21], the existence and the global stability of the disease-free equilibrium [4, 5, 14–16], the existence and the local stability of the endemic equilibrium [22, 23], and the persistence of the disease [14, 15]. Attention has also be paid to various bifurcations of the discrete epidemic models, the equilibrium bifurcation [3, 21, 24–26], the transcritical bifurcation, the flip bifurcation, the saddle-node bifurcation, the Hopf bifurcation, and the bifurcation to chaos [3–5].

There also have been many researches studying mathematical models of the transmission dynamics of TB in human populations. Those researches include slow and fast progression, a variable latent period, MDR TB, multiple strains, exogenous reinfection, generalized households, co-infection with HIV, and the control strategy for TB [27–35]. However, there are a few papers using discrete mathematical models to study the treatment of latent TB as a TB control strategy [3, 36, 37]. The treatment of latent tuberculosis infection is essential to controlling and eliminating TB, because it substantially reduces the risk from latent TB to active TB cases. In this paper, we use the discrete models with direct progression and treatment of latently infected individuals to analyze the impact of the treatment of latently infected individuals on TB transmission.

Our discrete model with treatment of latent TB individuals is presented in the next section. The positivity of solutions is also discussed. The global stability of the disease-free equilibrium and the persistence of the system is discussed in Section 3. The stability of endemic equilibrium is proved in Section 4. The effect of two treatment strategies is studied in Section 5. Numerical simulations are done on the basis of the TB infection data in China to show the effect of treatment strategies. The last section includes concluding remarks and discussions.

2 The discrete TB model

TB is an airborne infectious disease transmitting from person to person via droplets with TB *bacilli*. After being infected, most people become latently infectious, with the bacteria being alive in the body but inactive. Latently infectious individuals do not have symptoms and cannot spread the infection to others. Most people are able to fight the infection with their immune system, but those people with latent infection are at risk of developing the active disease. Based on these characteristics, there have been many continuous models to be used to describe TB transmission. Here, we will build the discrete mathematical model for TB transmission which considers the direct progression, chemoprophylaxis for the latent individuals, and treatment of the infectious individuals.

The total population is epidemiologically divided into susceptible, latent, and infectious classes. Let $S(t)$ be the number of susceptible individuals at time t, $L(t)$ be the number of latent individuals, and $I(t)$ be the number of infectious individuals. $N(t) = S(t) + L(t) + I(t)$ is the total population size. We assume that individuals recovered by successful treatment do not acquire immunity, and they will become member of the susceptible compartment. Hence our model is of SLIS type.

Susceptible individuals may get infected and enter the latent compartment. The surviving latent individuals will experience one of the two mutually exclusive events to leave the latent compartment, naturally progressing into infectious class, or receiving the treatment. These two events happen randomly. The probabilities for receiving the treatment is m, and $1 - m$ is the probabilities for natural progression. Our model framework takes the following form:

$$\begin{aligned} S(t+1) &= \Lambda + p\left(1 - \frac{\beta I(t)}{N(t)}\right)S(t) + pkmL(t) + p\gamma I(t), \\ L(t+1) &= q\frac{p\beta S(t)I(t)}{N(t)} + p(1-\alpha)(1-m)L(t) + p(1-k)mL(t), \\ I(t+1) &= (1-q)\frac{p\beta S(t)I(t)}{N(t)} + p\alpha(1-m)L(t) + p(1-\gamma)I(t), \end{aligned} \tag{1}$$

where Λ is the recruitment rate into the population, p the survival probability; k is the conditional probability that a latent individual is treated successfully given that the individual receives the treatment, and it should rely on the capability of finding latent individuals and budgetary issues. α is the probability that a latent individual becomes an infectious individual that the natural progression happens; γ the probability that an infectious individual recovers successfully; $1 - \frac{\beta I(t)}{N(t)}$ is for the probability of not becoming infected, where β characterizes the disease transmission probability, satisfying $0 < \beta < 1$; q is the probability that the infected ones enter the latent compartment, accordingly, $1-q$ is the probability that the infected ones enter the infectious compartment by the fast development; all parameters are positive and less than one.

Based on biological considerations, system (1) will be studied in the following region:

$$\Omega = \left\{(S,L,I) \in R_+^3 \,\middle|\, S \geq 0, L \geq 0, I \geq 0, S + L + I \leq \frac{\Lambda}{1-p}\right\}.$$

In the following, we first discuss the positivity and boundedness of solutions for system (1). The following theorem holds.

Theorem 2.1 *The solutions $S(t)$, $L(t)$, and $I(t)$ of system* (1) *with initial value $S(0) = S_0 \geq 0$, $L(0) = L_0 \geq 0$, and $I(0) = I_0 \geq 0$, respectively, are non-negative for all $t \geq 0$, $t \in N$. For the system* (1), *the region Ω is positively invariant and all solutions starting in the Ω approach, enter, or stay in Ω.*

Proof Let $S(0) = S_0 \geq 0$, $L(0) = L_0 \geq 0$, and $I(0) = I_0 \geq 0$, by using (1), we have

$$\begin{aligned} S(1) &= \Lambda + p\left(1 - \frac{\beta I_0}{N_0}\right)S_0 + pkmL_0 + p\gamma I_0, \\ L(1) &= q\frac{p\beta S_0 I_0}{N_0} + p(1-\alpha)(1-m)L_0 + p(1-k)mL_0, \\ I(1) &= (1-q)\frac{p\beta S_0 I_0}{N_0} + p\alpha(1-m)L_0 + p(1-\gamma)I_0. \end{aligned}$$

Because of $0 \leq I(t) \leq N(t)$ for any $t \geq 0$, $t \in N$, $0 \leq \beta \leq 1$, we know that $1 - \frac{\beta I_0}{N_0} \geq 0$. It implies that $S(1) > 0$. In addition, $0 < \alpha < 1$, $0 < k < 1$, and $0 < m < 1$ illustrate $L(1) \geq 0$.

Furthermore, $0 < \gamma < 1$ implies that $I(1) \geq 0$ holds. The recurrent procedure implies that $S(t) > 0$, $L(t) \geq 0$, $I(t) \geq 0$ for all $t \geq 0$, $t \in N$. The solutions of (1) are non-negative.

By using $N(t) = S(t) + L(t) + I(t)$, we know $N(t)$ satisfies

$$N(t+1) = \Lambda + pN(t) \quad \text{or} \quad N(t) = \frac{\Lambda(1-p^t)}{1-p} + p^t N(0). \tag{2}$$

The fact $0 < p < 1$ and the last equation in (2) implies that $N^* = \frac{\Lambda}{1-p}$ is the unique equilibrium of (2), and N^* is globally asymptotically stable, *i.e.*, for any solution $N(t)$ of (2) with positive initial value $N(0)$, $\lim_{t\to\infty} N(t) = N^*$ holds. It implies that $N(t)$ is bounded and all solutions starting in the region Ω approach, enter or stay in Ω. □

Using the limiting equations and $S(t) = N^* - L(t) - I(t)$, we reduce the three dimensional system (1) into two dimensional ones:

$$\begin{aligned} L(t+1) &= q\frac{p\beta(N^* - L(t) - I(t))I(t)}{N^*} + p(1-\alpha)(1-m)L(t) + p(1-k)mL(t), \\ I(t+1) &= (1-q)\frac{p\beta(N^* - L(t) - I(t))I(t)}{N^*} + p\alpha(1-m)L(t) + p(1-\gamma)I(t). \end{aligned} \tag{3}$$

In the following, we will study the dynamical behavior of model (3) since model (3) exhibits the same qualitative dynamics as those of system (1) [38]. It is clear that the positively invariant of system (3) is

$$\Omega_1 = \left\{ (L, I) \in R_+^2 \,\middle|\, L \geq 0, I \geq 0, L + I < \frac{\Lambda}{1-p} \right\}.$$

Applying the approach in [39] to our model, we can obtain the basic reproductive number of model (3):

$$R_0 = \frac{(1-m)p\alpha \times qp\beta}{(1-p(1-km) + p\alpha(1-m))(1-p(1-\gamma))} + \frac{(1-q)p\beta}{1-p(1-\gamma)}.$$

The basic reproductive number R_0 is defined mathematically as the spectral radius of the next generation matrix in [39]; in fact, each term in R_0 has a clear epidemiological interpretation. $1/(1-p(1-\gamma))$ is the average infection period. $(1-m)p\alpha/(1-p(1-km)+p\alpha(1-m))$ is the proportion of latent individuals that become infectious by natural progression. $p\beta/(1-p(1-\gamma))$ is the average of new cases generated by a typical infectious member in the entire infection period, where $qp\beta/(1-p(1-\gamma))$ is the average of new cases generated by a typical infectious member who enters the infectious compartment by natural progression in the entire infection period, $(1-q)p\beta/(1-p(1-\gamma))$ is the average of new cases generated by a typical infectious member who enters the infectious compartment by direct progression in the entire infection period.

3 The extinction and persistence for the disease

By directly calculating model (3), we know there exists the disease-free equilibrium $P_0^* = (L_0^*, I_0^*) = (0, 0)$ when $R_0 < 1$, and there is an endemic equilibrium $P_1^* = (L_1^*, I_1^*)$ when $R_0 > 1$, where

$$L_1^* = \frac{qp\beta I_1^*}{R_0(1-p(1-km) + p\alpha(1-m))}, \qquad I_1^* = \frac{N^*(R_0 - 1)}{R_0 + \frac{qp\beta}{1-p(1-km)+p\alpha(1-m)}}.$$

In the following, we will use the linearization matrix to discuss the local stability of equilibrium. The Jacobian matrix of model (3) at equilibrium P^* is

$$J(P^*) = \begin{pmatrix} -\frac{qp\beta}{N^*}I^* + p(1-\alpha)(1-m) + p(1-k)m & \frac{qp\beta}{N^*}(N^* - L^* - 2I^*) \\ -\frac{(1-q)p\beta}{N^*}I^* + p\alpha(1-m) & \frac{(1-q)p\beta}{N^*}(N^* - L^* - 2I^*) + p(1-\gamma) \end{pmatrix}.$$

3.1 The extinction for the disease

Theorem 3.1 *If $R_0 < 1$, then the disease-free equilibrium P_0^* of model (3) is globally asymptotically stable; if $R_0 > 1$, then P_0^* is unstable.*

Proof When $R_0 < 1$, we have $P^* = P_0^*$, and

$$J(P_0^*) = \begin{pmatrix} p(1-\alpha)(1-m) + p(1-k)m & qp\beta \\ p\alpha(1-m) & (1-q)p\beta + p(1-\gamma) \end{pmatrix}.$$

We denote

$$\begin{aligned} f(\lambda) &= \lambda^2 - \big(p(1-\alpha)(1-m) + p(1-k)m + p(1-\gamma) + (1-q)p\beta\big)\lambda \\ &\quad + \big(p(1-\alpha)(1-m) + p(1-k)m\big)\big((1-q)p\beta + p(1-\gamma)\big) - qp\beta p\alpha(1-m). \end{aligned}$$

It is clear that

$$\begin{aligned} f(1) &= \big(1 - p(1-km) + p\alpha(1-m)\big)\big(1 - p(1-\gamma)\big)(1 - R_0) > 0, \\ f(-1) &= 1 + p(1-\alpha)(1-m) + p(1-k)m + p(1-\gamma) + (1-q)p\beta \\ &\quad + \big(p(1-\alpha)(1-m) + p(1-k)m\big)\big((1-q)p\beta + p(1-\gamma)\big) - qp\beta p\alpha(1-m), \\ f(0) &= \big(p(1-\alpha)(1-m) + p(1-k)m\big)\big((1-q)p\beta + p(1-\gamma)\big) - qp\beta p\alpha(1-m). \end{aligned}$$

Since $R_0 < 1$, we have $\frac{(1-m)p\alpha \times qp\beta}{(1-p(1-km)+p\alpha(1-m))(1-p(1-\gamma))} < 1$. Furthermore, we obtain $(1-m)p\alpha \times qp\beta < (1 - p(1-km) + p\alpha(1-m))(1 - p(1-\gamma))$. Therefore,

$$\begin{aligned} f(-1) &> 1 + p(1-\alpha)(1-m) + p(1-k)m + p(1-\gamma) + (1-q)p\beta \\ &\quad + \big(p(1-\alpha)(1-m) + p(1-k)m\big)\big((1-q)p\beta + p(1-\gamma)\big) \\ &\quad - \big(1 - p(1-km) + p\alpha(1-m)\big)\big(1 - p(1-\gamma)\big) \\ &= (1-q)p\beta + \big(p(1-\alpha)(1-m) + p(1-k)m\big)(1-q)p\beta \\ &\quad + 2\big(p(1-\alpha)(1-m) + p(1-k)m\big) + 2p(1-\gamma) > 0. \end{aligned}$$

Similarly, $R_0 < 1$ implies that $\frac{(1-q)p\beta}{1-p(1-\gamma)} < 1$. We obtain $(1-q)p\beta < 1 - p(1-\gamma)$. So, we have

$$\begin{aligned} 1 - f(0) &= 1 - \big(p(1-\alpha)(1-m) + p(1-k)m\big)\big((1-q)p\beta + p(1-\gamma)\big) + qp\beta p\alpha(1-m) \\ &> 1 - p(1-km)\big(1 - p(1-\gamma)\big) + p\alpha(1-m)\big(1 - p(1-\gamma)\big) \\ &\quad + p\alpha(1-m)(1-m)p(1-\gamma) - p(1-km)p(1-\gamma) + qp\beta p\alpha(1-m) \\ &= 1 - p(1-km) + p\alpha(1-m) + qp\beta p\alpha(1-m) > 0, \end{aligned}$$

which illustrates that $f(0) < 1$ as $R_0 < 1$. By using of the Jury criterion, we know the disease-free equilibrium P_0^* is of local stability. If $R_0 > 1$, we have $f(1) < 0$. The Jury criterion implies that the disease-free equilibrium P_0^* is unstable. □

In the following, we use the Lyapunov function to prove the globally asymptotically stability of P_0^*. We define $F : [0, N^*] \times [0, N^*] \to [0, N^*] \times [0, N^*]$ by

$$F(L, I) = p\Bigg(q\frac{\beta(N^* - L - I)I}{N^*} + (1-\alpha)(1-m)L + (1-k)mL,$$
$$(1-q)\frac{\beta(N^* - L - I)I}{N^*} + \alpha(1-m)L(t) + (1-\gamma)I \Bigg).$$

Obviously, F is the mapping derived by system (3), and $(0, 0)$ is a fixed point of F. The linear function

$$V(L, I) = \frac{(1-m)p\alpha}{(1-p(1-km) + p\alpha(1-m))(1-p(1-\gamma))}L + \frac{1}{1-p(1-\gamma)}I$$

on $[0, N^*] \times [0, N^*]$ is continuous and positive definite with respect to $(0, 0)$. Therefore, V is a Lyapunov function on the domain of F. For any $(L, I) \in [0, N^*] \times [0, N^*]$,

$$\begin{aligned}
\Delta V(L, I) &= \frac{(1-m)p\alpha q p\beta(N^* - L - I)I}{(1-p(1-km) + p\alpha(1-m))(1-p(1-\gamma))N^*} \\
&\quad + \frac{(1-m)p\alpha p(1-\alpha)(1-m)L}{(1-p(1-km) + p\alpha(1-m))(1-p(1-\gamma))} \\
&\quad + \frac{(1-m)p\alpha p(1-k)mL}{(1-p(1-km) + p\alpha(1-m))(1-p(1-\gamma))} \\
&\quad + \frac{(1-q)p\beta(N^* - L - I)I}{(1-p(1-\gamma))N^*} + \frac{p\alpha(1-m)L}{1-p(1-\gamma)} + \frac{p(1-\gamma)I}{1-p(1-\gamma)} \\
&\quad - \frac{p\alpha(1-m)L}{(1-p(1-km) + p\alpha(1-m))(1-p(1-\gamma))} - \frac{I}{1-p(1-\gamma)} \\
&\le \frac{(1-m)p\alpha q p\beta I}{(1-p(1-km) + p\alpha(1-m))(1-p(1-\gamma))} \\
&\quad + \frac{(1-m)p\alpha(p(1-km) - p\alpha(1-m))L}{(1-p(1-km) + p\alpha(1-m))(1-p(1-\gamma))} - I + \frac{(1-q)p\beta I}{(1-p(1-\gamma))} \\
&\quad + \frac{p\alpha(1-m)L}{1-p(1-\gamma)} - \frac{p\alpha(1-m)L}{(1-p(1-km) + p\alpha(1-m))(1-p(1-\gamma))} \\
&= \frac{(1-m)p\alpha q p\beta I}{(1-p(1-km) + p\alpha(1-m))(1-p(1-\gamma))} - \frac{p\alpha(1-m)L}{1-p(1-\gamma)} \\
&\quad - I + \frac{(1-q)p\beta I}{(1-p(1-\gamma))} + \frac{p\alpha(1-m)L}{1-p(1-\gamma)} \\
&= \frac{(1-m)p\alpha q p\beta I}{(1-p(1-km) + p\alpha(1-m))(1-p(1-\gamma))} + \frac{(1-q)p\beta I}{(1-p(1-\gamma))} - I \\
&= (R_0 - 1)I.
\end{aligned}$$

Hence, if $R_0 < 1$, then $\Delta V(L, I) < 0$ holds for $(L, I) \in [0, N^*] \times [0, N^*]$. It follows from Theorem 4.22 in [40] that P_0^* is globally asymptotically stable.

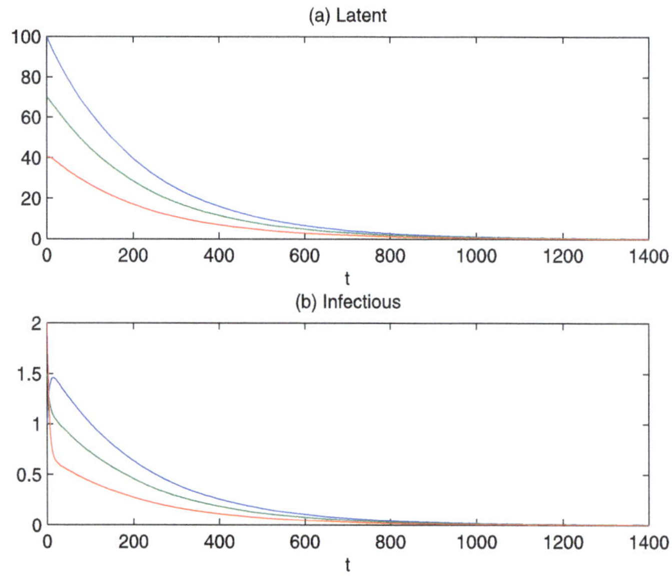

Figure 1 The global stability of disease-free equilibrium P_0^* of model (3) as $R_0 < 1$.

We illustrate our theoretical results on disease extinction by numerical simulation. Taking $\Lambda = 10$, $p = 0.994$, $q = 0.95$, $\beta = 0.3$, $\alpha = 0.003$, $m = 0.001$, $k = 0.001$, and $\gamma = 0.2$, we obtain $R_0 = 0.5298 < 1$, and the disease-free equilibrium $P_0^* = (0,0)$ is globally asymptotically stable. The solution curves of the latently infected individuals $L(t)$ and the infectious individuals $I(t)$ are given in Figure 1(a) and (b), respectively. The initial values at $t = 0$ are $(100,1)$, $(70,1.5)$, and $(40,2)$, respectively. We observe that solutions with positive initial values will converge to the disease-free equilibrium P_0^*.

3.2 The persistence for the disease

Theorem 3.2 *If* $R_0 > 1$, *then the disease will have persistence in the population, i.e., there exists a positive* ϵ, *such that the solution of model* (3) *with the initial value* $L(0) > 0$ *and* $I(0) > 0$ *satisfies*

$$\liminf_{t\to\infty}\{L(t), I(t)\} > \epsilon.$$

Proof We denote $\mathcal{X} = \Omega$, $\mathcal{X}_0 = \{(L,I) \in \mathcal{X} \mid L > 0, I > 0\}$, and $\partial\mathcal{X}_0 = \mathcal{X}\backslash\mathcal{X}_0$. Let $\Phi : \mathcal{X} \to \mathcal{X}$, $\Phi_t(x_0) = \phi(t,x_0)$ be the solution map of model (3) with $\phi(0,x_0) = x_0$, and $x_0 = (L(0), I(0))$.

Define $\mathcal{M} = \{P_0^*\}$, and

$$M_\partial = \{(L,I) \in \partial\mathcal{X}_0 : \phi(t,(L,I)) \in \partial\mathcal{X}_0, \forall t \geq 0\}.$$

It is clear that $M_\partial = \{(0,0)\}$. Furthermore, there is exactly one fixed point $P_0^* = (0,0)$ in M_∂. Because N^* is globally attractive in ∂X_0 and due to Lemma 5.9 in [41], we know that no subset of $\mathcal{M}$ forms a cycle in $\partial\mathcal{X}_0$. The definition of M_∂ implies that M_∂ is the maximum positive invariant set in $\partial\mathcal{X}_0$, that is, $\Phi_t(M_\partial) \subset M_\partial$. Therefore, $\bigcup_{(L(0),I(0))\in M_\partial} = P_0^*$.

For any solutions $(L(t), I(t))$ of model (3) with initial value $(L(0), I(0)) \in \mathcal{X}_0$, we further claim that

$$\limsup_{t\to\infty} d(\Phi_t(L(0), I(0)), P_0^*) > \epsilon. \tag{4}$$

If it is not true, there exists $t_1 > 0$, $t_1 \in N$, such that $L(t) \leq \epsilon$, $I(t) \leq \epsilon$ for all $t \geq t_1$, $t \in N$. Let $H(t) = L(t) + I(t)$, then $H(t) \leq 2\epsilon$, $t \geq t_1$, $t \in N$, and we have

$$H(t+1) = \frac{p\beta(N^* - H(t))}{N^*} I(t) + pH(t) - pkmL(t) - p\gamma I(t).$$

Furthermore, we have

$$\begin{aligned}
H(t+1) &= p\beta I(t) - \frac{p\beta H(t)I(t)}{N^*} + pH(t) - pkmL(t) - p\gamma I(t) \\
&= p\beta\big(H(t) - L(t)\big) - \frac{p\beta H(t)I(t)}{N^*} + pH(t) - pkmL(t) - p\gamma I(t) \\
&\geq p\beta H(t) - p\beta L(t) - \frac{p\beta H^2(t)}{N^*} + p(1-\gamma)H(t) - pkmL(t) \\
&= \left(p\beta + p(1-\gamma) - \frac{p\beta H(t)}{N^*}\right) H(t) - p(\beta + km)\epsilon \\
&\geq \left(p\beta + p(1-\gamma) - \frac{p\beta\epsilon}{N^*}\right) H(t) - p(\beta + km)\epsilon, \quad t \geq t_1.
\end{aligned}$$

It is clear that the perturbed equation

$$H_1(t+1) = \left(p\beta + p(1-\gamma) - \frac{p\beta\epsilon}{N^*}\right) H_1(t) - p(\beta + km)\epsilon, \quad t \geq t_1, t \in N,$$

or

$$\begin{aligned}
H_1(t) &= \big(-p(\beta + km)\epsilon\big) \frac{1 - (p\beta + p(1-\gamma) - \frac{p\beta\epsilon}{N^*})^{t-t_1}}{1 - (p\beta + p(1-\gamma) - \frac{p\beta\epsilon}{N^*})} \\
&\quad + \left(p\beta + p(1-\gamma) - \frac{p\beta\epsilon}{N^*}\right)^{t-t_1} H_1(t_1), \quad t \geq t_1, t \in N.
\end{aligned}$$

$R_0 > 1$ implies that $p\beta + p(1-\gamma) > 1$, namely, $p\beta + p(1-\gamma) - \frac{p\beta\epsilon}{N^*} > 1$ holds for a small $\epsilon > 0$. Therefore, $H_1(t) \to \infty$ as $t \to \infty$ and $\epsilon \to 0$. The comparison principle implies that $H(t) \to \infty$ as $t \to \infty$ and $\epsilon \to 0$. In fact, $H(t) \leq 2\epsilon$ for all $t \geq t_1$, $t \in N$, a contradiction. It implies that $L(t) + I(t) \to \infty$, for all $t \geq t_1$, $t \in N$. On the other hand, $0 \leq L(t) \leq \epsilon$, $0 \leq I(t) \leq \epsilon$ for all $t \geq t_1$, $t \in N$. It implies that $L(t) \to \infty$ and $I(t) \to \infty$ at least established for all $t \geq t_1$, $t \in N$. There is a contradiction, that is, the conclusion in (4) holds.

Equation (4) implies that P_0^* is isolated in $\mathcal{X}_0$, and $W^s(P_0^*) \cap \mathcal{X}_0 = \emptyset$. In fact, P_0^* is also isolated in $\partial\mathcal{X}_0$ because N^* is globally attractive in $\partial\mathcal{X}_0$. Thus, P_0^* is isolated in $\mathcal{X}$. From Theorem 1.3.1 and Remark 1.3.1 in [42], it follows that Φ is uniformly persistent with respect to $(\mathcal{X}_0, \partial\mathcal{X}_0)$. Furthermore, Theorem 1.3.4 in [42] implies that the solutions of model (3) are uniformly persistent with respect to $(\mathcal{X}_0, \partial\mathcal{X}_0)$ when $R_0 > 1$. That is, there exists an $\epsilon > 0$, such that

$$\liminf_{t\to\infty} L(t) > \epsilon > 0, \qquad \liminf_{t\to\infty} I(t) > \epsilon > 0. \qquad \square$$

4 The stability of the endemic equilibrium

Theorem 4.1 *If $R_0 > 1$, then the endemic equilibrium P_1^* of model* (3) *is locally asymptotically stable.*

Proof In the case where $P^* = P_1^*$, we have

$$J(P_1^*) = \begin{pmatrix} -\frac{qp\beta}{N^*}I_1^* + p(1-\alpha)(1-m) + p(1-k)m & \frac{qp\beta}{N^*}(N^* - L_1^* - 2I_1^*) \\ -\frac{(1-q)p\beta}{N^*}I_1^* + p\alpha(1-m) & \frac{(1-q)p\beta}{N^*}(N^* - L_1^* - 2I_1^*) + p(1-\gamma) \end{pmatrix}.$$

We denote

$$\begin{aligned} g(\lambda) = \lambda^2 &- \Big(p(1-\alpha)(1-m) + p(1-k)m + p(1-\gamma) - \frac{qp\beta}{N^*}I_1^* \\ &+ \frac{(1-q)p\beta}{N^*}(N^* - L_1^* - 2I_1^*)\Big)\lambda + \big(p(1-\alpha)(1-m) + p(1-k)m\big)p(1-\gamma) \\ &+ \big(p(1-\alpha)(1-m) + p(1-k)m\big)\frac{(1-q)p\beta}{N^*}(N^* - L_1^* - 2I_1^*) \\ &- p(1-\gamma)\frac{qp\beta}{N^*}I_1^* - p\alpha\frac{qp\beta}{N^*}(N^* - L_1^* - 2I_1^*). \end{aligned}$$

By directly calculating, we obtain

$$\begin{aligned} g(1) &= R_0\frac{L_1^* + I_1^*}{N^*}\big(1 - p(1-\alpha-k)\big)\big(1 - p(1-\gamma)\big) > 0, \\ g(-1) &= 2 + 2\big(p(1-\alpha)(1-m) + p(1-k)m\big) - 2(1-m)p\alpha\frac{L_1^*}{I_1^*} \\ &\quad - \big(1 + \big(p(1-\alpha)(1-m) + p(1-k)m\big)(1-q) + p(1-\gamma)q \\ &\quad - (1-m)p\alpha q\big)\frac{p\beta}{N^*}I_1^*, \\ g(0) &= p(1-km) - p\alpha(1-m) - \big(p(1-km) - p\alpha(1-m)\big)\frac{(1-q)p\beta I_1^*}{N^*} \\ &\quad - p(1-\gamma)\frac{qp\beta I_1^*}{N^*} - (1-m)p\alpha\frac{L_1^*}{I_1^*} + (1-m)p\alpha\frac{qp\beta I_1^*}{N^*}. \end{aligned}$$

Rearranging the expression of $g(-1)$, we have

$$g(-1) = \frac{a_2R_0^2 + a_1R_0 + a_0}{R_0(1 - p(1-km) + p\alpha(1-m))(qp\beta + R_0(1 - p(1-km) + p\alpha(1-m)))},$$

where

$$\begin{aligned} a_2 &= 2\big(1 + p(1-km) - p\alpha(1-m)\big)\big(1 - p(1-km) + p\alpha(1-m)\big)^2 \\ &\quad - \big(1 - p(1-km) + p\alpha(1-m)\big)^2 p\beta\big(1 + p\big(1 - km - \alpha(1-m)\big)(1-q) \\ &\quad + p(1-\gamma)q - (1-m)p\alpha q\big) \\ &= \big(1 - p(1-km) + p\alpha(1-m)\big)^2\big\{1 - p\beta + 1 - p(1-\gamma)qp\beta \\ &\quad + p\big((1-\alpha)(1-m) + (1-k)m\big)\big(2 - (1-q)p\beta\big) + (1-m)p\alpha qp\beta\big\} > 0, \\ a_1 &= 2\big(1 + p(1-km) - p\alpha(1-m)\big)\big(1 - p(1-km) + p\alpha(1-m)\big)qp\beta \\ &\quad - 2(1-m)p\alpha qp\beta\big(1 - p(1-km) + p\alpha(1-m)\big) \\ &\quad + \big(1 - p(1-km) + p\alpha(1-m)\big)^2 p\beta\big(1 + p\big(1 - km - \alpha(1-m)\big)(1-q) \end{aligned}$$

$$+p(1-\gamma)q-(1-m)p\alpha q),$$

$$a_0 = -2(1-m)p\alpha qp\beta \times qp\beta < 0.$$

$a_2 > 0$ implies that the quadratic function curve opens upward. $a_0 < 0$ implies that the quadratic function equation has two real roots, which are positive and negative, respectively. Because of $a_2 + a_1 > 0$, we obtain $-\frac{a_1}{2a_2} < \frac{1}{2}$, which implies that the axis of symmetry of the quadratic function curve is on the left of $R_0 = 1$. In addition,

$$a_2R_0^2 + a_1R_0 + a_0 = 2\big(1-p(1-km)+p\alpha(1-m)+qp\beta\big)\big\{\big(1+p(1-km) - p\alpha(1-m)\big)\big(1-p(1-km)+p\alpha(1-m)\big)-(1-m)p\alpha qp\beta\big\}$$

with $R_0 = 1$. Because of $1-p(1-km)+p\alpha(1-m) > (1-m)p\alpha qp\beta$, and $1+p(1-km)-p\alpha(1-m) = 1+p(1-\alpha)(1-m)+p(1-k)m > 1$, we obtain $a_2R_0^2 + a_1R_0 + a_0 > 0$ as $R_0 = 1$. Therefore, $a_2R_0^2 + a_1R_0 + a_0 > 0$ as $R_0 > 1$, that is, $g(-1) > 0$.

Similarly, we have

$$1-g(0) = \frac{b_2R_0^2 + b_1R_0 + b_0}{R_0(1-p(1-km)+p\alpha(1-m))(qp\beta + R_0(1-p(1-km)+p\alpha(1-m)))},$$

where

$$\begin{aligned} b_2 &= \big(1-p(1-km)+p\alpha(1-m)\big)^3 + \big(1-p(1-km)+p\alpha(1-m)\big)^2 \\ &\quad \times \big(p(1-km-\alpha(1-m))(1-q)+p(1-\gamma)q-(1-m)p\alpha q\big)p\beta \\ &= \big(1-p(1-km)+p\alpha(1-m)\big)^2\big\{1-p(1-km)\big(1-p\beta(1-q)\big) \\ &\quad + p\alpha(1-m)(1-p\beta)+p(1-\gamma)qp\beta\big\} > 0, \\ b_1 &= \big(1-p(1-km)+p\alpha(1-m)\big)^2 qp\beta + \big(1-p(1-km)+p\alpha(1-m)\big)(1-m)p\alpha qp\beta \\ &\quad - \big(1-p(1-km)+p\alpha(1-m)\big)^2\big(p\big(1-km-\alpha(1-m)\big)(1-q)+p(1-\gamma)q \\ &\quad - (1-m)p\alpha q\big)p\beta, \\ b_0 &= (1-m)p\alpha qp\beta \times qp\beta > 0. \end{aligned}$$

$b_2 > 0$ implies that the quadratic function curve opens upward. $b_0 > 0$ implies that the quadratic function equation has roots which are of the same sign. Because of $b_2 + b_1 > 0$, we obtain $-\frac{b_1}{2b_2} < \frac{1}{2}$, which implies that the axis of symmetry of the quadratic function curve is on the left of $R_0 = 1$. In addition, $b_2R_0^2 + b_1R_0 + b_0 > 0$ as $R_0 = 1$. Therefore, we obtain $b_2R_0^2 + b_1R_0 + b_0 > 0$ as $R_0 > 1$, that is, $g(0) < 1$. The Jury criterion implies that the endemic equilibrium P_1^* is locally stable as $R_0 > 1$. □

The global stability of the endemic equilibrium of (3) is very difficult, though Theorems 3.2 and 4.1 established the persistence and local stability as $R_0 > 1$. We investigate the global stability of the endemic equilibrium by numerical simulations. Taking $\Lambda = 10$, $p = 0.994$, $q = 0.95$, $\beta = 0.7$, $\alpha = 0.003$, $m = 0.001$, $k = 0.001$, and $\gamma = 0.2$, we obtain $R_0 = 1.2363 > 1$. In this case, the numerical simulation shows that the endemic equilibrium $P_1^* = (313.3017, 5.2621)$ may be globally asymptotically stable. The solution curves of the

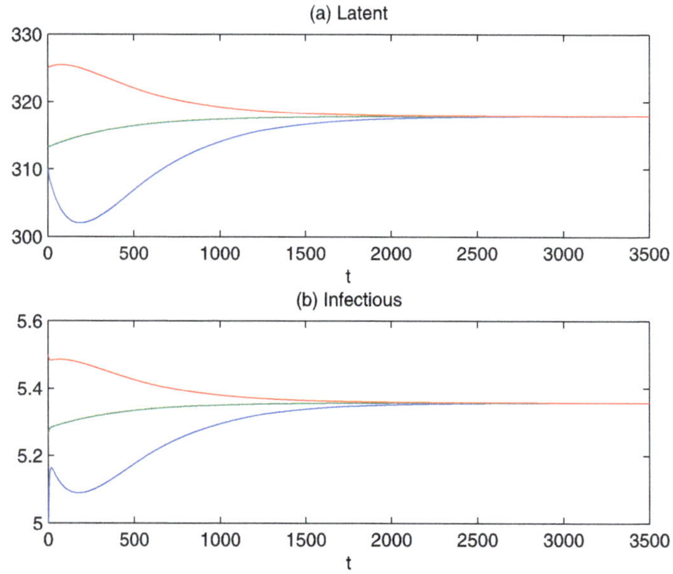

Figure 2 **The global stability of the endemic equilibrium P_1^* of (3) as $R_0 > 1$.**

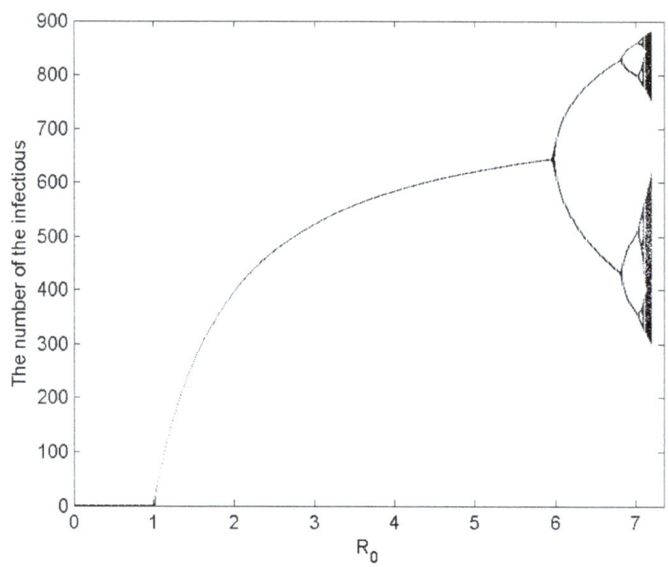

Figure 3 **The chaos when endemic equilibrium P_1^* loses the global stability.**

latently infected individuals $L(t)$ and the infectious individuals $I(t)$ are given in Figure 2(a) and (b), respectively. The initial values at $t = 0$ are $(310, 5)$, $(313, 5.3)$, and $(325, 5.5)$, respectively. We observe that solutions with positive initial values will converge to the endemic equilibrium P_1^*.

We always assume that the parameter obeys $0 < \beta < 1$ in theoretical discussions so as to guarantee the positivity of the solution of (3). In fact, if $\beta > 1$, and β is not large enough, the numerical simulation shows that solutions of (3) are still positive, and the endemic equilibrium P_1^* may lose the global stability and there exists chaos as β increases, which can lead to increasing R_0 (see Figure 3). In Figure 3, we fix $\Lambda = 10$, $p = 0.994$, $q = 0.01$,

$\alpha = 0.001$, $m = 0.01$, $k = 0.1$, $\gamma = 0.8$, and $\beta \in [0.01, 6.00)$, furthermore, $0 < R_0 < 7.3701$. Figure 3 illustrates that one only has a disease-free equilibrium as $R_0 < 1$, and an endemic equilibrium exists as $1 < R_0 < 6$. In addition, the endemic equilibrium loses the global stability as $R_0 > 6$, and exhibits chaotic behavior.

5 Effect of control strategies

In our model, we consider the direct progression to active TB once one is infected, and both treatment of latently infected individuals and infectious individuals. By directly calculating, we obtain

$$\begin{aligned} \frac{\partial R_0}{\partial k} &= -\frac{pm \times (1-m)p\alpha \times qp\beta}{(1-p(1-km)+p\alpha(1-m))^2(1-p(1-\gamma))} < 0, \\ \frac{\partial R_0}{\partial m} &= -\frac{p\alpha \times qp\beta(1-p(1-k))}{(1-p(1-km)+p\alpha(1-m))^2(1-p(1-\gamma))} < 0. \end{aligned} \tag{5}$$

The two equations of (5) illustrate that the basic reproductive number R_0 decreases as the proportion of m and the probability k of successful detection and effective treatment of latently infected individuals. The treatment of latent tuberculosis infection contributes to slowing down TB transmission because it substantially reduces the risk that TB infection will progress to TB disease. Therefore, we consider both treatment of the latently infected individuals and for the infectious individuals. We assume that the word 'control' would mean bringing down the number of infectious TB cases.

To see the effect of these interventions, we need to consider the basic reproduction number of the model without treatment of the latent and infectious individuals. This is given by

$$R_1 = \frac{p\alpha \times qp\beta}{(1-p(1-\alpha))(1-p)} + \frac{(1-q)p\beta}{1-p}.$$

$R_1 < 1$ implies that TB will die out without any treatment. In fact, $R_1 > 1$ is the usual case for the current TB epidemic in many countries. Taking China as an example, large scale national sampling surveys of TB epidemiology were carried out in 1979, 1984/85, 2000, and 2010, respectively. The 1979 and 2000 national surveys were more comprehensive and more information was obtained. The tuberculin skin test was done in those two sampling surveys. The result of those two surveys show that PPD (purified protein derivative) positive rates in 1979 and 2000 are 27.9% and 44.5%, respectively [43]. The 2010 national TB epidemiological survey shows that the prevalence of active TB cases in the population 15 years of age and older is 459 cases per 100,000, a little lower than the 466 cases per 100,000 in 2000. The estimated active TB prevalence in the Chinese population in 2010 is 392 cases per 100,000, a little higher than the 367 cases per 100,000 found in 2000 [44]. This information indicates that TB infection remains a serious public health challenge in China.

In China, the population recruitment mainly depends on births. Therefore, we use the average natural death from 2000-2009 as the natural death rate, namely, $p = 0.994$. It is estimated that about 5%-10% of latent TB infection will progress to TB disease [45]. We assume that the lifespan of the population is 75, then the age of the infected individual may be from 1 year old to 75 years old. The individuals may survive 1 year, 2 years, ...,

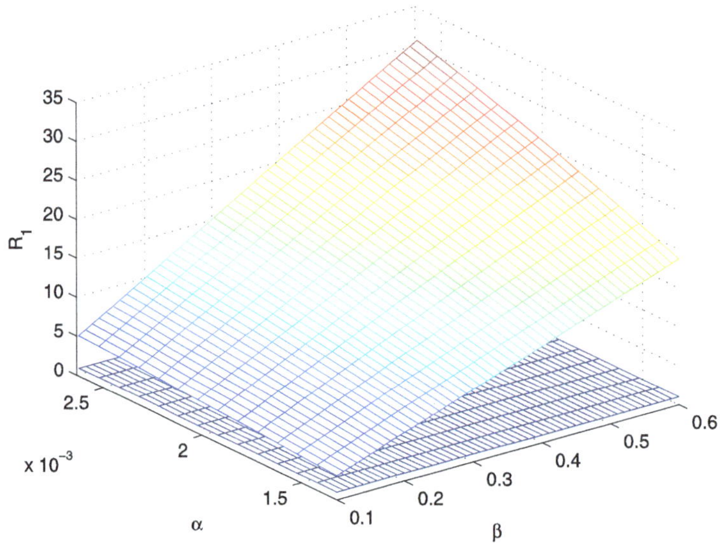

Figure 4 The basic reproductive number R_1.

or 75 years after becoming TB latent. We choose the average survival time to be $\frac{75}{2} \approx 38$. The estimate of the annual progression rate from latent to infectious is $\alpha \in [\frac{0.05}{38}, \frac{0.1}{38}] = [0.00132, 0.00263]$. We take $q = 0.99$, and $\beta \in [0.1, 0.6]$ [46]. The numerical calculation shows that $R_1 > 1$ (see Figure 4). In short, we always assume that $R_1 > 1$ in this section.

In addition, as $m \to 0$, the system (3) reduces to a model in which there is no treatment of latent TB individuals. In this case, every TB infected individual leaves the latent compartment only by the natural progression and the reproductive number becomes

$$\lim_{m \to 0} R_0 = \frac{p\alpha \times qp\beta}{(1-p(1-\alpha))(1-p(1-\gamma))} + \frac{(1-q)p\beta}{1-p(1-\gamma)} := R_2.$$

We first consider the effect of both treatment of latently infected individuals and for infectious individuals on control TB transmission. The difference between R_1 and R_0 is

$$\begin{aligned}\Delta_1 &= R_1 - R_0\\ &= \frac{m(1-p(1-k))p\alpha qp\beta + p\gamma(1-p(1-km)+p\alpha(1-m))p\alpha qp\beta}{(1-p(1-\alpha))(1-p)(1-p(1-km)+p\alpha(1-m))(1-p(1-\gamma))}\\ &\quad + \frac{p\gamma(1-q)p\beta}{(1-p)(1-p(1-\gamma))}.\end{aligned}$$

It easy to see that $\Delta_1 > 0$. It implies that both the treatment of latently infected individuals and for infectious individuals can lead to a decrease of the basic reproductive number. That is, treatment is effective at the population level so as to slow down the spread of TB, while having treatment of the individuals in the latent and the infectious compartment.

5.1 Effect of treatment of the infectious individuals

Because the latently TB infected individuals cannot infect others, one does not take to treatment of latently infected individuals in developing and undeveloped countries so as to save expense. Therefore, we only consider treatment of the TB infectious individuals. We will give the critical value γ^* of the probability γ of successful detection and treatment of infectious individuals so that the basic reproductive number of the model is less than one.

When $m = 0$ and $\gamma > 0$ (with only treatment for the infectious individuals), the basic reproductive number is R_2. With the application of only treatment of the population in the infectious compartment, we can obtain the difference between R_1 and R_2:

$$\Delta_2 = R_1 - R_2 = \frac{p\gamma}{(1-p)(1-p(1-\gamma))}\left(\frac{p\alpha \times qp\beta}{1-p(1-\alpha)} + (1-q)p\beta\right) > 0. \tag{6}$$

$\Delta_2 > 0$ implies that the treatment is effective at the population level so as to slow down the spread of TB, while having treatment of the individuals in the infectious compartment.

Using (6), we can calculate the critical values γ^* of the disease control parameters γ to have $R_2 < 1$. Namely,

$$R_1 - \frac{p\gamma}{(1-p)(1-p(1-\gamma))}\left(\frac{p\alpha \times qp\beta}{1-p(1-\alpha)} + (1-q)p\beta\right) < 1. \tag{7}$$

Rearranging (7), we can obtain

$$\gamma > \gamma^* := \frac{(1-p(1-\alpha))(1-p)^2(R_1-1)}{p(1-p)(1-p(1-\alpha))} = \left(\frac{1}{p}-1\right)(R_1-1). \tag{8}$$

Equation (8) shows that the basic reproductive number $R_2 < 1$, while the successful detection and effective treatment probability γ of infectious individuals is more than γ^*. However, $\gamma < \gamma^*$ will lead to $R_2 > 1$, which implies that successful detection and effective treatment of infectious individuals is not enough, TB transmission will be in persistence. Therefore, we should strengthen the successful detection and effective treatment of infectious individuals so as to eradicate TB transmission in the population. Taking the same parameter values in Figure 4, the function of the basic reproductive number, R_2, is shown in Figure 5 with $m = 0$ and for $\gamma = 0.05$, $\gamma = 0.1$, and $\gamma = 0.2$, respectively.

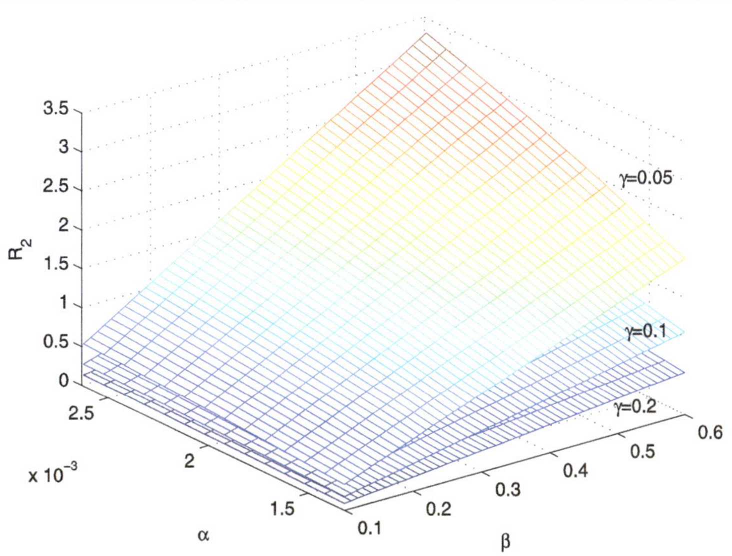

Figure 5 The basic reproductive number R_2.

5.2 Effect of treatment of the latent individuals

When the probability γ of successful detection and treatment of infectious individuals does not reach the critical value γ^*, we also take the treatment of latently TB individuals, and give the critical value m^* of the probability m of the treatment of latently individuals, so that $R_0 < 1$.

In this case, we assume that $\gamma < \gamma^*$, or $R_2 > 1$. Similarly, we can obtain the difference between R_2 and R_0:

$$\Delta_3 = R_2 - R_0 = \frac{m(1-p(1-k))p\alpha qp\beta}{(1-p(1-\alpha))(1-p(1-km)+p\alpha(1-m))(1-p(1-\gamma))} > 0. \tag{9}$$

$\Delta_3 > 0$ implies that the treatment is effective at the population level so as to slow down the spread of the disease, while having treatment of the individuals in the latent and the infectious compartment.

Using (9), we can calculate the critical values m^* of the disease control parameters m to have $R_0 < 1$. Namely,

$$R_2 - \frac{m(1-p(1-k))p\alpha qp\beta}{(1-p(1-\alpha))(1-p(1-km)+p\alpha(1-m))(1-p(1-\gamma))} < 1. \tag{10}$$

Rearranging (10), we can obtain

$$m > m^* := \frac{p\alpha qp\beta + (1-p(1-\alpha))((1-q)p\beta - (1-p(1-\gamma)))}{p\alpha qp\beta + (p(1-k) - p(1-\alpha))((1-q)p\beta - (1-p(1-\gamma)))}. \tag{11}$$

Because of $R_2 > 1$, we obtain $p\alpha qp\beta > (1-p(1-\alpha))(1-p(1-\gamma)-(1-q)p\beta)$. It is clear that $(1-p(1-\alpha))(1-p(1-\gamma)-(1-q)p\beta) > (p(1-k)-p(1-\alpha))((1-q)p\beta-(1-p(1-\gamma)))$, which implies $m^* > 0$. By (11), we know that if the effective treatment of the individuals in the infectious compartment satisfies $0 < \gamma < \gamma^*$, we should strengthen control measures so as to have $R_0 < 1$. Therefore, we also take the treatment of latent TB individuals, and the probability for receiving the treatment m satisfies $m > m^*$.

Taking the same parameter values in Figure 4, by calculating, we can obtain $\gamma^* = (\frac{1}{p} - 1)(R_1 - 1) \in [0.0127, 0.1798]$, furthermore, we fix $\gamma = 0.01$, satisfying $0 < \gamma < \gamma^*$. In addition, taking $k = 0.9$, we use the same parameter values to obtain $m^* \in [0.0015, 0.1391]$. The numerical method shows that R_0 with $\gamma = 0.01$ and $k = 0.9$ for $m = 0.001$, $m = 0.01$, $m = 0.1$, and $m = 0.2$, respectively (see Figure 6). Figure 6 shows that, in the case where $\gamma < \gamma^*$, we can control the probability $m > m^*$ of the treatment of latent individuals, so that $R_0 < 1$.

6 Conclusion and discussion

In this paper, we analyze a class of discrete SLIS models with direct progression and chemoprophylaxis for latent TB individuals. The basic reproductive numbers R_0 with both treatment of latently infected individuals and infectious individuals are determined. Furthermore, we study the global stability of the disease-free equilibrium as $R_0 < 1$, the persistence of the system, and the local stability of the endemic equilibrium as $R_0 > 1$. The numerical simulation shows that when $\beta > 1$, and β is not large enough, the solutions of the model are still positive, and the endemic equilibrium may lose the stability and there exists chaos as β increases, which can lead to R_0 increasing. By analyzing R_0, we learn that

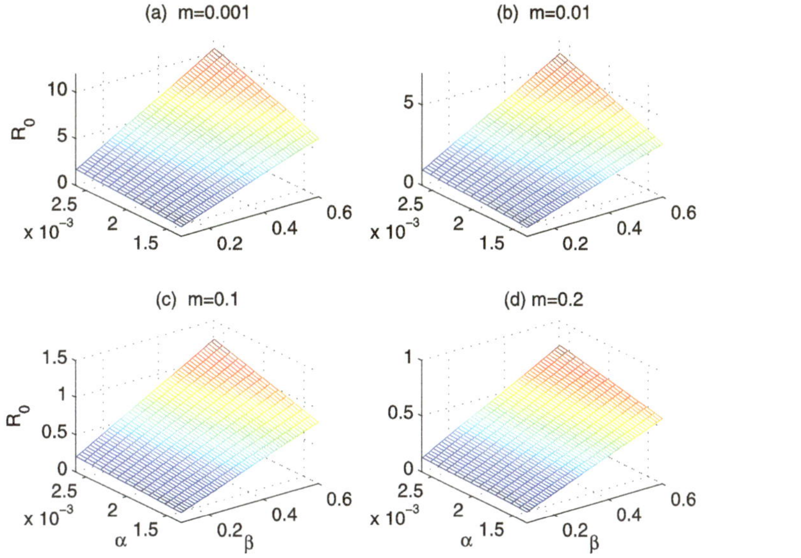

Figure 6 The basic reproductive number R_0 with different *m*.

the treatment of latent TB individuals contributes to a decrease of the basic reproductive number, namely, a reduction of the speed of TB transmission.

In addition, the basic reproductive number R_1 without treatment, and the basic reproductive number R_2 with treatment of infectious individuals are determined, respectively. We also discuss the effect of the treatment of the infectious individuals on TB transmission, and give the critical value γ^* of the probability γ of successful detection and treatment of infectious individuals so that the basic reproductive number of the model is $R_2 < 1$. Once $R_2 > 1$, that is, $\gamma < \gamma^*$, we need to strengthen control measures so as to slow down the TB transmission. In this situation, we take both the treatment of latently infected individuals and infectious individuals, and give the critical value m^* of probability m of the treatment of latently infected individuals so that the basic reproductive number of the model obeys $R_0 < 1$.

By using the TB data in China, we compare the effect of these different treatment strategies for the control of TB. The numerical simulations totally support our theoretical results. Moreover, our conclusions show that, if the probability γ of successful detection and treatment of infectious individuals cannot reach the critical value γ^*, only treatment of infectious individuals cannot effectively control the TB transmission, which implies that we should strengthen control measures so as to slow down the TB transmission. That is, we should take both the treatment of latently infected individuals and of infectious individuals.

Competing interests

The authors declare that no competing interests exist.

Authors' contributions

HC is responsible for the model formulation, the calculation, and the proof. HT has done the simulation and the application. All authors have read and approved the final manuscript.

Acknowledgements
This research was supported by National Nature Science Foundation of China Grant 11301314; by Natural Science Basic Research Plan in Shaanxi Province of China Grant 2014JQ1025; by Backbone Youth Training Scheme of Shaanxi University of Science & Technology Grant XSG(4)014; and by National Nature Science Foundation of China Grant 11400856.

References

1. Allen, L: Some discrete-time SI, SIR, and SIS epidemic models. Math. Biosci. **124**, 83-105 (1994)
2. Allen, L, Burgin, A: Comparison of deterministic and stochastic SIS and SIR models in discrete time. Math. Biosci. **163**, 1-33 (2000)
3. Cao, H, Zhou, YC, Song, BJ: Complex dynamics of discrete SEIS models with simple demography. Discrete Dyn. Nat. Soc. **2011**, Article ID 653937 (2011). doi:10.1155/2011/653937
4. Castillo-Chavez, C, Yakubu, AA: Discrete-time SIS models with complex dynamics. Nonlinear Anal. TMA **47**, 4753-4762 (2001)
5. Castillo-Chavez, C, Yakubu, AA: Discrete-time SIS models with simple and complex population dynamics. In: Castillo-Chavez, C, Blower, S, van den Driessche, P, Kirschner, D, Yakubu, AA (eds.) Mathematical Approaches for Emerging and Reemerging Infectious Diseases: An Introduction, pp. 153-163. Springer, New York (2002)
6. Franke, JE, Yakubu, AA: Discrete-time SIS epidemic model in a seasonal environment. SIAM J. Appl. Math. **66**, 1563-1587 (2006)
7. Zhou, YC, Ma, ZE: Global stability of a class of discrete age-structured SIS models with immigration. Math. Biosci. Eng. **6**, 409-425 (2009)
8. Zhou, YC, Paolo, F: Dynamics of a discrete age-structured SIS models. Discrete Contin. Dyn. Syst., Ser. B **4**, 843-852 (2004)
9. May, RM: Biological population obeying difference equations: stable points, stable cycles, and chaos. J. Theor. Biol. **51**, 511-524 (1975)
10. May, RM: Deterministic models with chaotic dynamics. Nature **256**, 165-166 (1975)
11. May, RM: Simple mathematical models with very complicated dynamics. Nature **261**, 459-467 (1976)
12. Hassell, MP: Density dependence in single-species populations. J. Anim. Ecol. **44**, 283-289 (1975)
13. Cao, H, Dou, ZH, Liu, X, Zhang, FJ, Zhou, YC, Ma, ZE: The impact of antiretroviral therapy on the basic reproductive number of HIV transmission. Math. Model. Appl. **1**, 33-37 (2012)
14. Cao, H, Xiao, YN, Zhou, YC: The dynamics of a discrete SEIT model with age and infection-age structures. Int. J. Biomath. **5**, 61-76 (2012)
15. Cao, H, Zhou, YC: The discrete age-structured SEIT model with application to tuberculosis transmission in China. Math. Comput. Model. **55**, 385-395 (2012)
16. Zhou, YC, Cao, H: Discrete tuberculosis transmission models and their application. In: Sivaloganathan, SA (ed.) Survey of Mathematical Biology. Fields Communications Series, vol. 57, pp. 83-112. A co-publication of the AMS and Fields Institute, Canada (2010)
17. Zhou, YC, Khan, K, Feng, ZL, Wu, JH: Projection of tuberculosis incidence with increasing immigration trends. J. Theor. Biol. **254**, 215-228 (2008)
18. Zhou, YC, Ma, ZE, Brauer, F: A discrete epidemic model for SARS transmission and control in China. Math. Comput. Model. **40**, 1491-1506 (2004)
19. Allen, L, van den Driessche, P: The basic reproduction number in some discrete-time epidemic models. J. Differ. Equ. Appl. **14**, 1127-1147 (2008)
20. Cao, H, Zhou, YC: The basic reproduction number of discrete SIR and SEIS models with periodic parameters. Discrete Contin. Dyn. Syst., Ser. B **18**, 37-56 (2013)
21. Li, X, Wang, W: A discrete epidemic model with stage structure. Chaos Solitons Fractals **26**, 947-958 (2005)
22. Arreola, R, Crossa, A, Velasco, MC, Yakubu, AA: Discrete-time SEIS models with exogenous re-infection and dispersal between two patches. http://mtbi.asu.edu/files/0835_001.pdf (2000). Accessed 15 Apr 2015
23. Gonzalez, PA, Saenz, RA, Sanchez, BN, Castillo-Chavez, C, Yakubu, AA: Dispersal between two patches in a discrete time SEIS model. http://www.researchgate.net/publication/221711667_Dispersal_between_two_patches_in_a_discrete_time_SEIS_model (2000). Accessed 15 Apr 2015
24. Hu, Z, Teng, ZD, Jiang, H: Stability analysis in a class of discrete SIRS epidemic models. Nonlinear Anal., Real World Appl. **13**, 2017-2033 (2012)
25. Hu, Z, Teng, ZD, Zhang, L: Stability and bifurcation analysis of a discrete predator-prey model with nonmonotonic functional response. Nonlinear Anal., Real World Appl. **12**, 2356-2377 (2011)
26. Li, L, Sun, GQ, Jin, Z: Bifurcation and chaos in an epidemic model with nonlinear incidence rates. Appl. Math. Comput. **216**, 1226-1234 (2010)
27. Aparicio, JP, Capurro, AF, Castillo-Chavez, C: Transmission and dynamics of tuberculosis on generalized households. J. Theor. Biol. **206**, 327-341 (2000)
28. Blower, SM, McLean, AR, Porco, TC, Small, PM, Hopwell, PC: The intrinsic transmission dynamics of tuberculosis epidemics. Nat. Med. **1**, 815-821 (1995)
29. Blower, SM, Small, PM, Hopwell, PC: Control strategies for tuberculosis epidemics: new models for old problems. Science **273**, 497-500 (1996)
30. Castillo-Chavez, C, Feng, Z: To treat or not to treat: the case of tuberculosis. J. Math. Biol. **35**, 629-656 (1997)
31. Castillo-Chavez, C, Feng, Z: Global stability of an age-structure model for TB and its application to optimal vaccination strategies. Math. Biosci. **151**, 135-154 (1998)
32. Castillo-Chavez, C, Feng, Z: Mathematical models for the disease dynamics of tuberculosis. In: Arino, O, Axelrod, D, Kimmel, M (eds.) Advances in Mathematical Population Dynamics: Molecules, Cells and Man, pp. 629-656. World Scientific, Singapore (1998)
33. Feng, Z, Huang, W, Castillo-Chavez, C: On the role of variable latent periods in mathematical models for tuberculosis. J. Dyn. Differ. Equ. **2**, 425-452 (2001)

34. Feng, Z, Castillo-Chavez, C, Capurro, AF: A model for tuberculosis with exogenous reinfection. Theor. Popul. Biol. **57**, 235-247 (2000)
35. Kirschner, D: Dynamics of co-infection with *M. tuberculosis* and HIV-1. Theor. Popul. Biol. **55**, 94-109 (1999)
36. Song, BJ, Castillo-Chavez, C: Tuberculosis control in the U.S.A.: A strategy to meet CDC's goal. Department of Biometrics Technical Report Series BU-1562-M, Biometrics Department, Cornell University
37. Ziv, E, Daley, CL, Blower, SM: Early therapy for latent tuberculosis infection. Am. J. Epidemiol. **153**, 381-385 (2001)
38. Zhao, XQ: Asymptotic behavior for asymptotically periodic semiflows with applications. Commun. Appl. Nonlinear Anal. **3**, 43-66 (1996)
39. Allen, L, van den Driessche, P: The basic reproduction number in some discrete-time epidemic models. J. Differ. Equ. Appl. **14**, 1127-1147 (2008)
40. Elaydi, SN: An Introduction to Difference Equations. Undergraduate Texts in Mathematics. Springer, New York (1999)
41. Elaydi, SN: Persistence in a discrete-time stage-structured epidemic model. J. Differ. Equ. Appl. **16**, 73-103 (2010)
42. Zhao, XQ: Dynamical Systems in Population Biology. Springer, New York (2003)
43. National Technical Steering Group of the Epidemiological Sampling Survey for Tuberculosis, Office of the Nationwide Epidemiological Sampling Survey for Tuberculosis: On nationwide random survey for the epidemiology of tuberculosis in 2000. J. Chin. Antituberc. Assoc. **24**, 65-108 (2002)
44. National Technical Steering Group of the Epidemiological Sampling Survey for Tuberculosis, Office of the Nationwide Epidemiological Sampling Survey for Tuberculosis: On nationwide random survey for the epidemiology of tuberculosis in 2010. J. Chin. Antituberc. Assoc. **34**, 485-508 (2012)
45. Kumar, V, Abbas, AK, Fausto, N, Robbins, RN: Basic Pathology, 8th edn. Elsevier, Amsterdam (2007)
46. Bhunu, CP, Garira, W, Mukandavire, Z, Zimba, M: Tuberculosis transmission model with chemoprophylaxis and treatment. Bull. Math. Biol. **70**, 1161-1191 (2008)

8

Global behavior of a plant-herbivore model

Qamar Din*

*Correspondence:
qamar.sms@gmail.com
Department of Mathematics,
Faculty of Basic and Applied
Sciences, University of Poonch
Rawalakot, Rawalakot, Pakistan

Abstract

The present work deals with an analysis of the local asymptotic stability and global behavior of the unique positive equilibrium point of the following discrete-time plant-herbivore model: $x_{n+1} = \frac{\alpha x_n}{\beta x_n + e^{y_n}}$, $y_{n+1} = \gamma(x_n + 1)y_n$, where $\alpha \in (1, \infty)$, $\beta \in (0, \infty)$, and $\gamma \in (0, 1)$ with $\alpha + \beta > 1 + \frac{\beta}{\gamma}$ and initial conditions x_0, y_0 are positive real numbers. Moreover, the rate of convergence of positive solutions that converge to the unique positive equilibrium point of this model is also discussed. In particular, our results solve an open problem and a conjecture proposed by Kulenović and Ladas in their monograph (Dynamics of Second Order Rational Difference Equations: With Open Problems and Conjectures, 2002). Some numerical examples are given to verify our theoretical results.

MSC: 39A10; 40A05

Keywords: plant-herbivore system; steady-states; local stability; global behavior; rate of convergence

1 Introduction and preliminaries

In mathematical biology, the model of the plant-herbivore has attracted many researchers during the last few decades. There are several theoretical results for the mutual dependence between plant and herbivore populations. We consider here an open problem proposed in [1] related to the global character of solutions of the plant-herbivore model

$$x_{n+1} = \frac{\alpha x_n}{\beta x_n + e^{y_n}}, \qquad y_{n+1} = \gamma(x_n + 1)y_n, \tag{1}$$

where x_n and y_n are the population biomasses of the plant and the herbivore in successive generations n and $n + 1$, respectively. Moreover, $\alpha \in (1, \infty)$, $\beta \in (0, \infty)$, and $\gamma \in (0, 1)$ with $\alpha + \beta > 1 + \frac{\beta}{\gamma}$ and initial conditions x_0, y_0 are positive real numbers. Furthermore, (1) has a unique positive equilibrium point under the following inequalities:

$$0 < \gamma < 1, \quad \alpha + \beta > 1 + \frac{\beta}{\gamma}.$$

It is pointed out in [2] that the system (1) has a very complex behavior, which has been investigated mainly numerically and visually. Moreover, there is a range of the values of parameters for which the equilibrium points on the x-axis are globally asymptotically stable. The computer simulations indicate that for some parameter values there is chaotic behavior, but proofs are yet to be completed.

In 2002, Kulenović and Ladas [1] proposed the following open problem:

Open Problem 6.10.13 ([1], p.128) (a plant-herbivore system) Assume that $\alpha \in (1,\infty)$, $\beta \in (0,\infty)$, and $\gamma \in (0,1)$ with $\alpha + \beta > 1 + \frac{\beta}{\gamma}$. Obtain conditions for the global asymptotic stability of the positive equilibrium of the system

$$x_{n+1} = \frac{\alpha x_n}{\beta x_n + e^{y_n}}, \qquad y_{n+1} = \gamma(x_n + 1)y_n.$$

The above model is for the interaction of the apple twig borer (an insect pest of the grape vine) and grapes in the Texas High Plains was developed and studied by Allen *et al.* [3]. See also Allen *et al.* [4] and Allen *et al.* [5].

In this paper, our aim is to investigate the local asymptotic stability, the global asymptotic character of the unique positive equilibrium point, and the rate of convergence of positive solutions of the system (1). For some interesting results related to the qualitative behavior of discrete dynamical systems, we refer to [6–14].

Let $(\bar{x},\bar{y})$ be an equilibrium point of the system (1), then

$$\bar{x} = \frac{\alpha\bar{x}}{\beta\bar{x} + e^{\bar{y}}}, \qquad \bar{y} = \gamma(\bar{x} + 1)\bar{y}.$$

Hence, $P_0 = (0,0)$, $P_1 = (\frac{\alpha-1}{\beta}, 0)$ and $P_2 = (\frac{1-\gamma}{\gamma}, \ln(\frac{\alpha\gamma-\beta+\beta\gamma}{\gamma}))$ are equilibrium points of the system (1). Furthermore, $P_2 = (\frac{1-\gamma}{\gamma}, \ln(\frac{\alpha\gamma-\beta+\beta\gamma}{\gamma}))$ is a unique positive equilibrium point of the system (1) if and only if

$$0 < \gamma < 1, \quad \alpha + \beta > 1 + \frac{\beta}{\gamma}.$$

The Jacobian matrix of linearized system of (1) about the fixed point $(\bar{x},\bar{y})$ is given by

$$F_J(\bar{x},\bar{y}) = \begin{bmatrix} \frac{\alpha e^{\bar{y}}}{(\beta\bar{x}+e^{\bar{y}})^2} & -\frac{\alpha\bar{x}e^{\bar{y}}}{(\beta\bar{x}+e^{\bar{y}})^2} \\ \gamma\bar{y} & \gamma\bar{x} + \gamma \end{bmatrix}.$$

Lemma 1.1 (Jury condition) *Consider the second-degree polynomial equation*

$$\lambda^2 + p\lambda + q = 0, \tag{2}$$

where p and q are real numbers. Then the necessary and sufficient condition for both roots of (2) *to lie inside the open disk* $|\lambda| < 1$ *is*

$$|p| < 1 + q < 2.$$

2 Boundedness

The following theorem shows that every positive solution of the system (1) is bounded.

Theorem 2.1 *Assume that* $x_n \leq x_{n+1}$ *for all* $n = 0,1,\ldots,$ *then every positive solution* $\{(x_n,y_n)\}$ *of the system* (1) *is bounded.*

Proof Assume that the initial conditions x_0, y_0 of (1) are positive, then every solution of (1) is positive. Assume that $\{(x_n,y_n)\}$ is an arbitrary positive solution of the system (1). Then, from the first part of the system (1), one has

$$x_{n+1} = \frac{\alpha x_n}{\beta x_n + e^{y_n}} \leq \frac{\alpha}{\beta},$$

for all $n = 0, 1, 2, \ldots$. Assume that $x_n \leq x_{n+1}$ for all $n = 0, 1, \ldots$. Moreover, from the first equation of the system (1), we obtain

$$\begin{aligned} y_n &= \ln\left(\frac{x_n(\alpha - \beta x_{n+1})}{x_{n+1}}\right) \\ &\leq \ln(\alpha - \beta x_{n+1}). \end{aligned} \tag{3}$$

From (3) and the second part of the system (1), we have

$$\begin{aligned} y_{n+1} &= \gamma(x_n + 1)y_n \\ &= \gamma(x_n + 1)\ln\left(\frac{\alpha x_n - \beta x_n x_{n+1}}{x_{n+1}}\right) \\ &\leq \gamma\left(\frac{\alpha}{\beta} + 1\right)\ln\left(\frac{\alpha x_n - \beta x_n x_{n+1}}{x_{n+1}}\right) \\ &\leq \gamma\left(\frac{\alpha}{\beta} + 1\right)\ln(\alpha - \beta x_{n+1}), \end{aligned}$$

for all $n = 0, 1, \ldots$. Finally, we let $F(x) = \ln(\alpha - \beta x)$ such that $0 \leq x < \frac{\alpha}{\beta}$. Then it follows that $F'(x) = -\frac{\beta}{\alpha - \beta x} < 0$ for all $0 \leq x < \frac{\alpha}{\beta}$. Hence, $F(x) \leq F(0)$ for all $0 \leq x < \frac{\alpha}{\beta}$, *i.e.*, $\ln(\alpha - \beta x) \leq \ln(\alpha)$ for all $0 \leq x < \frac{\alpha}{\beta}$. Thus for any positive solution $\{(x_n, y_n)\}$ of the system (1), we have

$$0 \leq x_n \leq \frac{\alpha}{\beta}, \qquad 0 \leq y_n \leq \gamma\left(\frac{\alpha}{\beta} + 1\right)\ln(\alpha),$$

for all $n = 1, 2, \ldots$. The proof is therefore completed. □

3 Linearized stability

Theorem 3.1 *Assume that* $\alpha \in (1, \infty)$ *and* $\gamma \in (0, 1)$, *then the following statements are true*:

(i) *The equilibrium point* $P_0 = (0, 0)$ *of the system* (1) *is a saddle point.*

(ii) *The equilibrium point* $P_1 = (\frac{\alpha - 1}{\beta}, 0)$ *of the system* (1) *is locally asymptotically stable if and only if* $\beta > \frac{\gamma - \alpha\gamma}{\gamma - 1}$.

Proof (i) The proof follows from the fact that the Jacobian matrix of linearized system of (1) about the equilibrium point $(0, 0)$ is given by

$$F_J(0, 0) = \begin{bmatrix} \alpha & 0 \\ 0 & \gamma \end{bmatrix}.$$

(ii) The Jacobian matrix of linearized system of (1) about the fixed point $(\frac{\alpha - 1}{\beta}, 0)$ is given by

$$F_J\left(\frac{\alpha - 1}{\beta}, 0\right) = \begin{bmatrix} \frac{1}{\alpha} & \frac{1 - \alpha}{\alpha\beta} \\ 0 & \frac{\gamma(\alpha + \beta - 1)}{\beta} \end{bmatrix}.$$

It is obvious that the roots of the characteristic polynomial of $F_J(\frac{\alpha-1}{\beta},0)$ about $(\frac{\alpha-1}{\beta},0)$ are $\lambda_1 = \frac{1}{\alpha} < 1$ for $\alpha \in (1,\infty)$ and $\lambda_2 = \frac{\alpha\gamma+\beta\gamma-\gamma}{\beta} < 1$ if and only if $\beta > \frac{\gamma-\alpha\gamma}{\gamma-1}$. □

The following theorem shows a necessary and sufficient condition for local asymptotic stability of unique positive equilibrium point of the system (1).

Theorem 3.2 *The unique positive equilibrium point* $P_2 = (\frac{1-\gamma}{\gamma}, \ln(\frac{\alpha\gamma-\beta+\beta\gamma}{\gamma}))$ *of the system* (1) *is locally asymptotically stable if and only if the following holds*:

$$\ln\left(\alpha+\beta-\frac{\beta}{\gamma}\right) < \frac{\beta}{\gamma(\alpha+\beta)-\beta}. \tag{4}$$

Proof The Jacobian matrix of linearized system of (1) about the positive equilibrium point $P_2 = (\frac{1-\gamma}{\gamma}, \ln(\frac{\alpha\gamma-\beta+\beta\gamma}{\gamma}))$ is given by

$$F_J(P_2) = \begin{bmatrix} \frac{\alpha\gamma+\beta(\gamma-1)}{\alpha\gamma} & \frac{(\gamma-1)(\alpha\gamma+\beta(\gamma-1))}{\alpha\gamma^2} \\ \gamma\ln(\alpha-\frac{\beta}{\gamma}+\beta) & 1 \end{bmatrix}.$$

The characteristic polynomial of $F_J(P_2)$ about positive equilibrium point $P_2 = (\frac{1-\gamma}{\gamma}, \ln(\frac{\alpha\gamma-\beta+\beta\gamma}{\gamma}))$ is given by

$$P(\lambda) = \lambda^2 + \left(\frac{\beta}{\alpha\gamma} - \frac{\beta}{\alpha} - 2\right)\lambda + 1 + \frac{\beta}{\alpha} - \frac{\beta}{\alpha\gamma} + \frac{\delta}{\alpha\gamma}, \tag{5}$$

where $\delta = (1-\gamma)(\beta(\gamma-1)+\alpha\gamma)\ln(\alpha+\beta-\frac{\beta}{\gamma})$. Set

$$p = \frac{\beta}{\alpha\gamma} - \frac{\beta}{\alpha} - 2, \qquad q = 1 + \frac{\beta}{\alpha} - \frac{\beta}{\alpha\gamma} + \frac{\delta}{\alpha\gamma}.$$

Then (5) can be written as

$$P(\lambda) = \lambda^2 + p\lambda + q. \tag{6}$$

Then it follows from Lemma 1.1 that both eigenvalues of $F_J(P_2)$ lie inside the unit disk $|\lambda| < 1$, if and only if

$$|p| < 1 + q < 2. \tag{7}$$

Inequality (7) is equivalent to the following three inequalities:

$$\frac{\beta}{\alpha} - \frac{\beta}{\alpha\gamma} + \frac{\delta}{\alpha\gamma} < 0, \tag{8}$$

$$p < 1 + q, \tag{9}$$

$$-1 - q < p. \tag{10}$$

First, we will prove (8). This inequality is equivalent to

$$\beta(\gamma-1) + \delta = (1-\gamma)\left[-\beta + ((\alpha+\beta)\gamma - \beta)\ln\left(\alpha+\beta-\frac{\beta}{\gamma}\right)\right] < 0.$$

Assume that $\ln(\alpha+\beta-\frac{\beta}{\gamma})<\frac{\beta}{\gamma(\alpha+\beta)-\beta}$, then it follows that $\beta(\gamma-1)+\delta<0$. Next, we consider (9). This inequality is equivalent to

$$2\beta(1-\gamma)-4\alpha\gamma-\delta<0.$$

It is easy to see that under condition (4), one has $2\beta(1-\gamma)-4\alpha\gamma-\delta<0$. Finally, (10) is equivalent to the following inequality:

$$(1-\gamma)\big(\beta(\gamma-1)+\alpha\gamma\big)\ln\left(\alpha+\beta-\frac{\beta}{\gamma}\right)>0,$$

and this obviously holds due to the assumption that $\alpha+\beta>1+\frac{\beta}{\gamma}$. Then it follows from Lemma 1.1 that the unique positive equilibrium point $P_2=(\frac{1-\gamma}{\gamma},\ln(\frac{\alpha\gamma-\beta+\beta\gamma}{\gamma}))$ of the system (1) is locally asymptotically stable if and only if $\ln(\alpha+\beta-\frac{\beta}{\gamma})<\frac{\beta}{\gamma(\alpha+\beta)-\beta}$. □

4 Global behavior

Lemma 4.1 [7] *Let $I=[a,b]$ and $J=[c,d]$ be real intervals, and let $f:I\times J\to I$ and $g:I\times J\to J$ be continuous functions. Consider the following system:*

$$\begin{aligned} x_{n+1}&=f(x_n,y_n),\\ y_{n+1}&=g(x_n,y_n),\quad n=0,1,\ldots, \end{aligned}\tag{11}$$

with initial conditions $(x_0,y_0)\in I\times J$. Suppose that the following statements are true:

(i) *$f(x,y)$ is non-decreasing in x and non-increasing in y.*
(ii) *$g(x,y)$ is non-decreasing in both arguments.*
(iii) *If $(m_1,M_1,m_2,M_2)\in I^2\times J^2$ is a solution of the system*

$$\begin{aligned} m_1&=f(m_1,M_2), & M_1&=f(M_1,m_2),\\ m_2&=g(m_1,m_2), & M_2&=g(M_1,M_2), \end{aligned}$$

such that $m_1=M_1$ and $m_2=M_2$. Then there exists exactly one positive equilibrium point $(\bar{x},\bar{y})$ of the system (11) such that $\lim_{n\to\infty}(x_n,y_n)=(\bar{x},\bar{y})$.

Theorem 4.1 *The positive equilibrium point P_2 of the system (1) is a global attractor.*

Proof Let $f(x,y)=\frac{\alpha x}{\beta x+e^y}$ and $g(x,y)=\gamma(x+1)y$. Then it is easy to see that $f(x,y)$ is non-decreasing in x and non-increasing in y. Moreover, $g(x,y)$ is non-decreasing in both x and y. Let (m_1,M_1,m_2,M_2) be a positive solution of the system

$$\begin{aligned} m_1&=f(m_1,M_2), & M_1&=f(M_1,m_2),\\ m_2&=g(m_1,m_2), & M_2&=g(M_1,M_2). \end{aligned}$$

Then one has

$$m_1=\frac{\alpha m_1}{\beta m_1+e^{M_2}},\qquad M_1=\frac{\alpha M_1}{\beta M_1+e^{m_2}},\tag{12}$$

and

$$m_2 = \gamma(m_1 + 1)m_2, \qquad M_2 = \gamma(M_1 + 1)M_2. \tag{13}$$

Furthermore, we make the assumption as in the proof of Theorem 1.16 of [15]; it suffices to suppose that

$$0 < m_1 \leq M_1, \qquad 0 < m_2 \leq M_2.$$

From (12), one has

$$\beta m_1 + e^{M_2} = \alpha, \qquad \beta M_1 + e^{m_2} = \alpha. \tag{14}$$

From (13), one has

$$\gamma(m_1 + 1) = 1, \qquad \gamma(M_1 + 1) = 1. \tag{15}$$

It follows from (15) that

$$m_1 = \frac{1-\gamma}{\gamma} = M_1. \tag{16}$$

Moreover, from (14) and (16), we obtain

$$m_2 = \ln\left(\frac{\alpha\gamma + \beta\gamma - \beta}{\gamma}\right) = M_2.$$

Hence, from Lemma 4.1 the equilibrium point P_2 is a global attractor. □

Lemma 4.2 *Under the condition* (4) *the unique positive equilibrium point* P_2 *of the system* (1) *is globally asymptotically stable.*

Proof The proof follows from Theorem 3.2 and Theorem 4.1. □

5 Conjecture

Now we consider the following conjecture related to special form of the system (1).

Conjecture 6.10.6 ([1], p.128) Assume that $x_0, y_0 \in (0, \infty)$ and that

$$\alpha \in (2, 3), \qquad \beta = 1, \qquad \gamma = 0.5.$$

Show that the positive equilibrium of the system (1) is globally asymptotically stable.

In this conjecture, we modify the interval of α for which the unique positive equilibrium point of the system (1) is globally asymptotically stable. In this case the system (1) can be written as

$$x_{n+1} = \frac{\alpha x_n}{x_n + e^{y_n}}, \qquad y_{n+1} = 0.5(x_n + 1)y_n, \quad n = 0, 1, 2, \ldots, \tag{17}$$

where $\alpha \in (2, R_0)$ such that R_0 is a root of the function defined by

$$F(\alpha) := \frac{(\frac{\alpha}{2} - \frac{1}{2})\ln(\alpha - 1)}{\alpha} - \frac{1}{\alpha}. \tag{18}$$

Theorem 5.1 *The unique positive equilibrium point* $(1, \ln(\alpha - 1))$ *of the system* (17) *is globally asymptotically stable if and only if* $2 < \alpha < R_0$, *where* R_0 *is a root of the function defined in* (18) *and* $R_0 \approx 3.3457507549227654$.

Proof The characteristic polynomial of $F_J(\bar{x}, \bar{y})$ about positive equilibrium point $(\bar{x}, \bar{y}) = (1, \ln(\alpha - 1))$ for the system (17) is given by

$$P(\lambda) = \lambda^2 + \left(\frac{1}{\alpha} - 2\right)\lambda + 1 - \frac{1}{\alpha} + \frac{(\frac{\alpha}{2} - \frac{1}{2})\ln(\alpha - 1)}{\alpha}. \tag{19}$$

Then it follows from Lemma 1.1 that the unique positive equilibrium point $(1, \ln(\alpha - 1))$ of the system (17) is locally asymptotically stable if and only if

$$\left|\frac{1}{\alpha} - 2\right| < 2 - \frac{1}{\alpha} + \frac{(\frac{\alpha}{2} - \frac{1}{2})\ln(\alpha - 1)}{\alpha} < 2.$$

It is easy to see that $|\frac{1}{\alpha} - 2| < 2 - \frac{1}{\alpha} + \frac{(\frac{\alpha}{2} - \frac{1}{2})\ln(\alpha-1)}{\alpha}$ for all $\alpha > 2$. Hence, it is enough to show that $2 - \frac{1}{\alpha} + \frac{(\frac{\alpha}{2} - \frac{1}{2})\ln(\alpha-1)}{\alpha} < 2$, or equivalently

$$F(\alpha) = \frac{(\frac{\alpha}{2} - \frac{1}{2})\ln(\alpha - 1)}{\alpha} - \frac{1}{\alpha} < 0.$$

Moreover, $F(2) = -\frac{1}{2}$, $F(3.4) = 0.0148713$, and $F'(\alpha) = \frac{\alpha + \ln(\alpha-1) + 2}{2\alpha^2} > 0$ for all $\alpha > 2$. From this it follows that $F(\alpha)$ has the unique root R_0 in $(2, 3.4)$. With the help of Mathematica this unique root of $F(\alpha)$ is approximated by $R_0 \approx 3.3457507549227654$. It is easy to see that $F(\alpha) < 0$ if and only if $2 < \alpha < R_0$. Hence, the unique positive equilibrium point $(1, \ln(\alpha - 1))$ of the system (17) is locally asymptotically stable if and only if $\alpha \in (2, R_0)$. Moreover, Theorem 4.1 shows that the unique positive equilibrium point $(1, \ln(\alpha - 1))$ of the system (17) is a global attractor. Hence, the proof is completed. □

6 Rate of convergence

In this section we will determine the rate of convergence of a solution that converges to the unique positive equilibrium point of the system (1).

The following result gives the rate of convergence of solutions of a system of difference equations:

$$X_{n+1} = \big(A + B(n)\big)X_n, \tag{20}$$

where X_n is an m-dimensional vector, $A \in C^{m\times m}$ is a constant matrix, and $B : \mathbb{Z}^+ \to C^{m\times m}$ is a matrix function satisfying

$$\|B(n)\| \to 0 \tag{21}$$

as $n \to \infty$, where $\| \cdot \|$ denotes any matrix norm which is associated with the vector norm

$$\|(x, y)\| = \sqrt{x^2 + y^2}.$$

Proposition 6.1 (Perron's theorem) [16] *Suppose that condition* (21) *holds. If X_n is a solution of* (20), *then either $X_n = 0$ for all large n or*

$$\rho = \lim_{n\to\infty} \left(\|X_n\|\right)^{1/n} \tag{22}$$

exists and is equal to the modulus of one of the eigenvalues of matrix A.

Proposition 6.2 [16] *Suppose that condition* (21) *holds. If X_n is a solution of* (20), *then either $X_n = 0$ for all large n or*

$$\rho = \lim_{n\to\infty} \frac{\|X_{n+1}\|}{\|X_n\|} \tag{23}$$

exists and is equal to the modulus of one of the eigenvalues of matrix A.

Let $\{(x_n, y_n)\}$ be any solution of the system (1) such that $\lim_{n\to\infty} x_n = \bar{x}$ and $\lim_{n\to\infty} y_n = \bar{y}$. To find the error terms, one has the following system:

$$\begin{aligned} x_{n+1} - \bar{x} &= \frac{\alpha x_n}{\beta x_n + e^{y_n}} - \frac{\alpha \bar{x}}{\beta \bar{x} + e^{\bar{y}}} \\ &= \frac{\alpha e^{\bar{y}}}{(\beta x_n + e^{y_n})(\beta \bar{x} + e^{\bar{y}})}(x_n - \bar{x}) - \frac{\alpha \bar{x}(e^{y_n} - e^{\bar{y}})}{(\beta x_n + e^{y_n})(\beta \bar{x} + e^{\bar{y}})(y_n - \bar{y})}(y_n - \bar{y}) \end{aligned}$$

and

$$\begin{aligned} y_{n+1} - \bar{y} &= \gamma(x_n + 1)y_n - \gamma(\bar{x} + 1)\bar{y} \\ &= \gamma y_n(x_n - \bar{x}) + (\gamma \bar{x} + \gamma)(y_n - \bar{y}). \end{aligned}$$

Let $e_n^1 = x_n - \bar{x}$ and $e_n^2 = y_n - \bar{y}$, then one has

$$e_{n+1}^1 = a_n e_n^1 + b_n e_n^2,$$

and

$$e_{n+1}^2 = c_n e_n^1 + d_n e_n^2,$$

where

$$a_n = \frac{\alpha e^{\bar{y}}}{(\beta x_n + e^{y_n})(\beta \bar{x} + e^{\bar{y}})}, \qquad b_n = -\frac{\alpha \bar{x}(e^{y_n} - e^{\bar{y}})}{(\beta x_n + e^{y_n})(\beta \bar{x} + e^{\bar{y}})(y_n - \bar{y})},$$

$$c_n = \gamma y_n, \qquad d_n = \gamma \bar{x} + \gamma.$$

Moreover,

$$\lim_{n\to\infty} a_n = \frac{\alpha e^{\bar{y}}}{(\beta\bar{x} + e^{\bar{y}})^2}, \qquad \lim_{n\to\infty} b_n = -\frac{\alpha e^{\bar{y}}\bar{x}}{(\beta\bar{x} + e^{\bar{y}})^2},$$

$$\lim_{n\to\infty} c_n = \gamma\bar{y}, \qquad \lim_{n\to\infty} d_n = \gamma\bar{x} + \gamma.$$

Now the limiting system of error terms can be written as

$$\begin{bmatrix} e^1_{n+1} \\ e^2_{n+1} \end{bmatrix} = \begin{bmatrix} \frac{\alpha e^{\bar{y}}}{(\beta\bar{x}+e^{\bar{y}})^2} & -\frac{\alpha\bar{x}e^{\bar{y}}}{(\beta\bar{x}+e^{\bar{y}})^2} \\ \gamma\bar{y} & \gamma\bar{x}+\gamma \end{bmatrix} \begin{bmatrix} e^1_n \\ e^2_n \end{bmatrix},$$

which is similar to the linearized system of (1) about the equilibrium point P_2.

Using Proposition 6.1, one has the following result.

Theorem 6.1 *Assume that* $\{(x_n, y_n)\}$ *is a positive solution of the system* (1) *such that* $\lim_{n\to\infty} x_n = \bar{x}$ *and* $\lim_{n\to\infty} y_n = \bar{y}$, *where*

$$(\bar{x}, \bar{y}) = \left(\frac{1-\gamma}{\gamma}, \ln\left(\frac{\alpha\gamma - \beta + \beta\gamma}{\gamma}\right)\right).$$

Then the error vector $e_n = \binom{e^1_n}{e^2_n}$ *of every solution of* (1) *satisfies both of the following asymptotic relations*:

$$\lim_{n\to\infty} \left(\|e_n\|\right)^{\frac{1}{n}} = \left|\lambda_{1,2}F_J(\bar{x},\bar{y})\right|,$$

$$\lim_{n\to\infty} \frac{\|e_{n+1}\|}{\|e_n\|} = \left|\lambda_{1,2}F_J(\bar{x},\bar{y})\right|,$$

where $\lambda_{1,2}F_J(\bar{x},\bar{y})$ *are the characteristic roots of Jacobian matrix* $F_J(\bar{x},\bar{y})$.

7 Examples

In order to verify our theoretical results we consider some interesting numerical examples in this section. These examples represent different types of qualitative behavior of the system (1). First two examples show that positive equilibrium point of the system (1) is locally asymptotically stable, *i.e.*, condition (4) of Theorem 3.2 is satisfied. Meanwhile, the third example shows that the positive equilibrium point of the system (1) is unstable, *i.e.*, condition (4) of Theorem 3.2 does not hold. All plots in this section are drawn with Mathematica.

Example 7.1 Let $\alpha = 24$, $\beta = 39$, $\gamma = 0.81$. Then the system (1) can be written as

$$x_{n+1} = \frac{24x_n}{39x_n + e^{y_n}}, \qquad y_{n+1} = 0.81(x_n+1)y_n, \quad n = 0, 1, \ldots, \tag{24}$$

with initial conditions $x_0 = 0.23$, $y_0 = 2.7$. In Figure 1, a plot of x_n is shown in Figure 1(a), a plot of y_n is shown in Figure 1(b), and an attractor of the system (24) is shown in Figure 1(c). In this case, $\ln(\alpha + \beta - \frac{\beta}{\gamma}) = 2.69812$ and $\frac{\beta}{\gamma(\alpha+\beta)-\beta} = 3.2419$. Hence, condition (4) of Theorem 3.2 is satisfied, *i.e.*, $\ln(\alpha + \beta - \frac{\beta}{\gamma}) < \frac{\beta}{\gamma(\alpha+\beta)-\beta}$. The unique positive equilibrium point of the system (24) is given by $(\frac{1-\gamma}{\gamma}, \ln(\frac{\alpha\gamma-\beta+\beta\gamma}{\gamma})) = (0.234568, 2.69812)$.

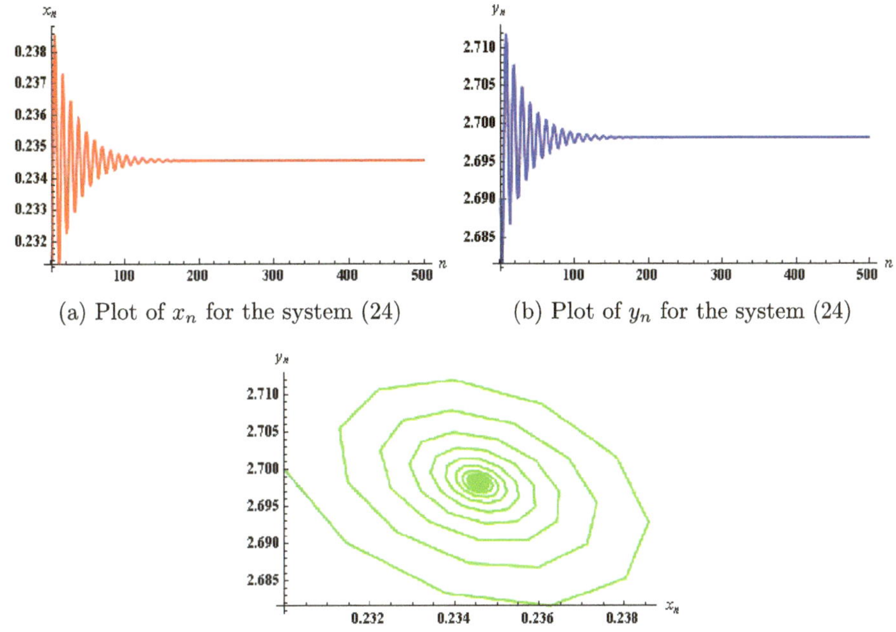

(a) Plot of x_n for the system (24)

(b) Plot of y_n for the system (24)

(c) An attractor of the system (24)

Figure 1 Plots for the system (24).

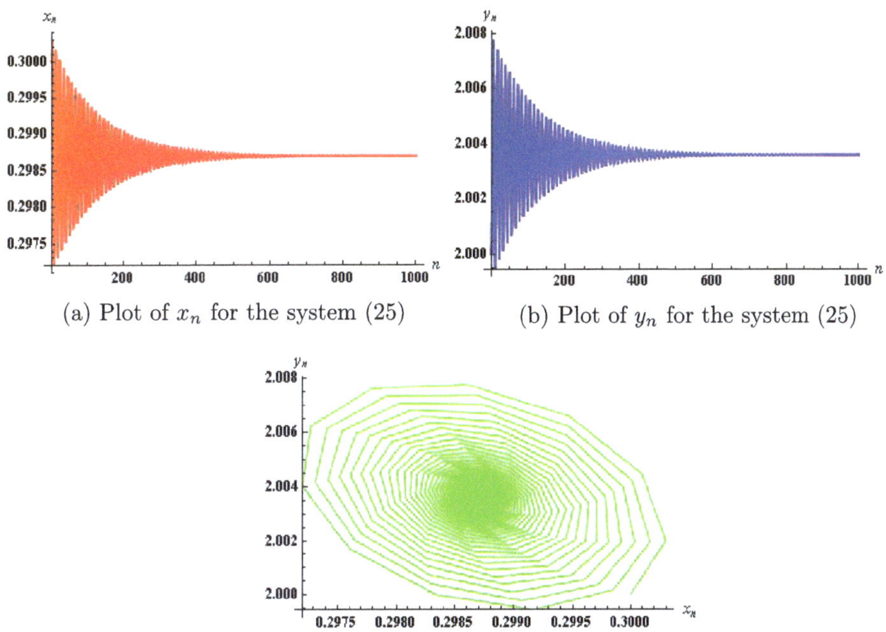

(a) Plot of x_n for the system (25)

(b) Plot of y_n for the system (25)

(c) An attractor of the system (25)

Figure 2 Plots for the system (25).

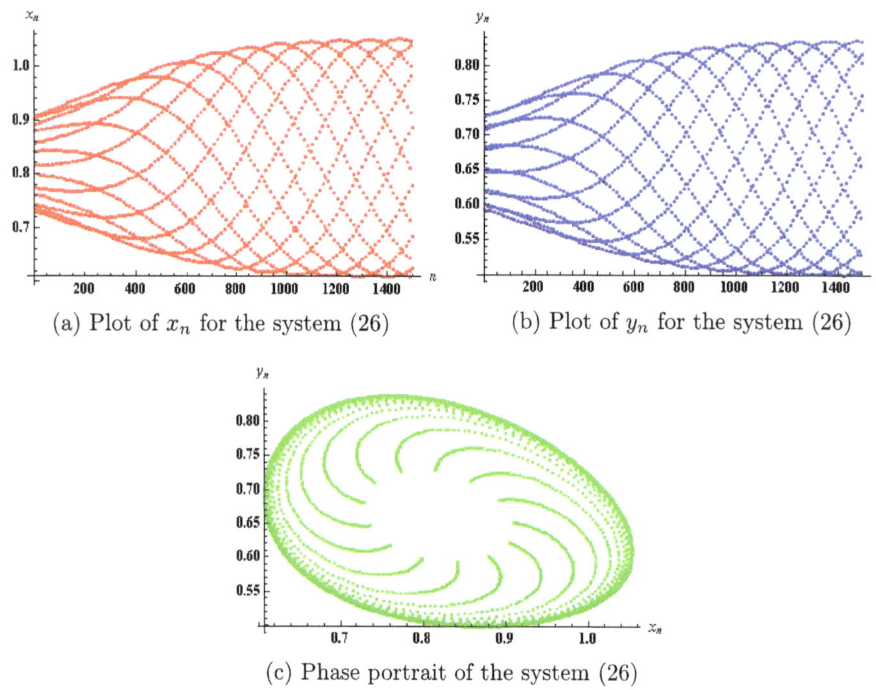

Figure 3 Plots for the system (26).

Example 7.2 Let $\alpha = 11$, $\beta = 12$, $\gamma = 0.77$. Then the system (1) can be written as

$$x_{n+1} = \frac{11x_n}{12x_n + e^{y_n}}, \qquad y_{n+1} = 0.77(x_n + 1)y_n, \quad n = 0, 1, \ldots, \tag{25}$$

with initial conditions $x_0 = 0.3$, $y_0 = 2$. In Figure 2, a plot of x_n is shown in Figure 2(a), a plot of y_n is shown in Figure 2(b), and an attractor of the system (25) is shown in Figure 2(c). In this case, $\ln(\alpha + \beta - \frac{\beta}{\gamma}) = 2.00358$ and $\frac{\beta}{\gamma(\alpha+\beta)-\beta} = 2.10158$. Hence, condition (4) of Theorem 3.2 is satisfied, *i.e.*, $\ln(\alpha + \beta - \frac{\beta}{\gamma}) < \frac{\beta}{\gamma(\alpha+\beta)-\beta}$. The unique positive equilibrium point of the system (25) is given by $(\frac{1-\gamma}{\gamma}, \ln(\frac{\alpha\gamma-\beta+\beta\gamma}{\gamma})) = (0.298701, 2.00358)$.

Example 7.3 Let $\alpha = 2.5$, $\beta = 0.69$, $\gamma = 0.55$. Then the system (1) can be written as

$$x_{n+1} = \frac{2.5x_n}{0.69x_n + e^{y_n}}, \qquad y_{n+1} = 0.55(x_n + 1)y_n, \quad n = 0, 1, \ldots, \tag{26}$$

with initial conditions $x_0 = 0.8$, $y_0 = 0.6$. In Figure 3, a plot of x_n is shown in Figure 3(a), a plot of y_n is shown in Figure 3(b), and a phase portrait of the system (26) is shown in Figure 3(c). In this case, $\ln(\alpha + \beta - \frac{\beta}{\gamma}) = 0.660342$ and $\frac{\beta}{\gamma(\alpha+\beta)-\beta} = 0.648192$. Hence, condition (4) of Theorem 3.2 does not hold, *i.e.*, $\ln(\alpha + \beta - \frac{\beta}{\gamma}) > \frac{\beta}{\gamma(\alpha+\beta)-\beta}$.

Competing interests

The author declares that he has no competing interests.

Author's contributions

The author carried out the proof of the main results and approved the final manuscript.

Acknowledgements
The author thanks the main editor and anonymous referees for their valuable comments and suggestions leading to improvement of this paper. This work was supported by the Higher Education Commission of Pakistan.

References

1. Kulenović, MRS, Ladas, G: Dynamics of Second Order Rational Difference Equations. Chapman & Hall/CRC, Boca Raton (2002)
2. Kocic, VL, Ladas, G: Global Behavior of Nonlinear Difference Equations of Higher Order with Applications. Kluwer Academic, Dordrecht (1993)
3. Allen, LJS, Hannigan, MK, Strauss, MJ: Mathematical analysis of a model for a plant-herbivore system. Bull. Math. Biol. **55**(4), 847-864 (1993)
4. Allen, LJS, Hannigan, MK, Strauss, MJ: Development and analysis of mathematical model for a plant-herbivore system. In: Proceedings of the First World Congress on World Congress of Nonlinear Analysts. WCNA'92, vol. 4, pp. 3723-3732 (1995)
5. Allen, LJS, Strauss, MJ, Tnorvilson, HG, Lipe, WN: A preliminary mathematical model of the apple twig borer (Coleoptera: Bostricidae) and grapes on the Texas High Planes. Ecol. Model. **58**, 369-382 (1991)
6. Din, Q: Dynamics of a discrete Lotka-Volterra model. Adv. Differ. Equ. (2013). doi:10.1186/1687-1847-2013-95
7. Din, Q, Donchev, T: Global character of a host-parasite model. Chaos Solitons Fractals **54**, 1-7 (2013)
8. Din, Q, Elsayed, EM: Stability analysis of a discrete ecological model. Comput. Ecol. Softw. **4**(2), 89-103 (2014)
9. Din, Q, Ibrahim, TF, Khan, KA: Behavior of a competitive system of second-order difference equations. Sci. World J. **2014**, Article ID 283982 (2014)
10. Din, Q: Global stability of a population model. Chaos Solitons Fractals **59**, 119-128 (2014)
11. Din, Q: Stability analysis of a biological network. Netw. Biol. **4**(3), 123-129 (2014)
12. Din, Q, Khan, KA, Nosheen, A: Stability analysis of a system of exponential difference equations. Discrete Dyn. Nat. Soc. **2014**, Article ID 375890 (2014)
13. Din, Q: Asymptotic behavior of an anti-competitive system of second-order difference equations. J. Egypt. Math. Soc. (2014). doi:10.1016/j.joems.2014.08.008
14. Din, Q: Qualitative nature of a discrete predator-prey system. Contemp. Methods Math. Phys. Gravit. (Online) **1**(1), 27-42 (2015)
15. Grove, EA, Ladas, G: Periodicities in Nonlinear Difference Equations. Chapman & Hall/CRC, Boca Raton (2004)
16. Pituk, M: More on Poincaré's and Perron's theorems for difference equations. J. Differ. Equ. Appl. **8**, 201-216 (2002)

9

A stochastic predator-prey model with delays

Bo Du[1,2*], Yamin Wang[3] and Xiuguo Lian[1]

*Correspondence:
dubo7307@163.com
[1]Department of Mathematics, Huaiyin Normal University, Huaian, Jiangsu 223300, China
[2]Department of Mathematics, Yangzhou University, Yangzhou, Jiangsu 225002, China
Full list of author information is available at the end of the article

Abstract
A stochastic delay predator-prey system is considered. Sufficient criteria for global existence, stochastically ultimately bounded in mean and almost surely asymptotic properties are obtained.

Keywords: stochastic perturbation; global existence; ultimately bounded

1 Introduction

One important component of the predator-prey relation is predator's functional response, *i.e.*, the rate of prey consumption by an average predator. However, in many cases, when predators have to search for food and, therefore, have to share or compete for food, the functional response in prey-predator model should be predator-dependent. Skalski and Gilliam [1] pointed out that the predator-dependent model can provide better descriptions of predator feeding over a range of predator-prey abundances by comparing the statistical evidence from some predator-prey systems with the three predator-dependent functional responses (Hassell-Varley [2], Bedding-DeAngelis [3] and Crowley-Martin [4]); furthermore, the Bedding-DeAngelis functional response is performed even better. The classical prey-predator model with Bedding-DeAngelis functional response is

$$\begin{cases} \frac{dx_1}{dt} = x_1[r_1 - a_{11}x_1 - \frac{a_{12}x_2}{1+mx_1+nx_2}], \\ \frac{dx_2}{dt} = x_2[-r_2 + \frac{a_{21}x_1}{1+mx_1+nx_2} - a_{22}x_2]. \end{cases} \tag{1.1}$$

There is extensive literature concerned with the dynamics of this prey-predator model with Bedding-DeAngelis functional response, and we here only mention Liu and Yuan [5], Liu and Zhang [6], Zhao and Lv [7], Fan and Kuang [8], Hwang [9, 10], Guo and Wu [11] among many others.

As was pointed by Kuang [12], any model of species dynamics with delays is an approximation at best. More detailed arguments on the significance of time-delays in realistic models may also be found in the classical books of Macdonald [13] and Gopalsamy [14]. Many authors have studied the delay prey-predator system, see, *e.g.*, [15–21]. For the study of delay population systems with Bedding-DeAngelis functional response, see [22–24]. Particularly, we consider the following population system with delays:

$$\begin{cases} \frac{dx_1}{dt} = x_1[r_1 - a_{11}x_1 - \frac{a_{12}x_2(t-\tau)}{1+mx_1+nx_2}], \\ \frac{dx_2}{dt} = x_2[-r_2 + \frac{a_{21}x_1(t-\tau)}{1+mx_1+nx_2} - a_{22}x_2]. \end{cases} \tag{1.2}$$

In system (1.2), x_1 and x_2 denote the population sizes of the prey and the predator, respectively. We are therefore not only interested in the positive solutions but also require the solutions not to explode at a finite time. To guarantee the positive solutions without explosion (*i.e.*, the global positive solutions), some conditions are in general needed to impose on the coefficients of system (1.2). For example, Kuang [12] discussed the following delay prey-predator system:

$$\begin{cases} \frac{dx_1}{dt} = x_1[r_1 - a_{11}x_1 - a_{12}x_2(t-\tau)], \\ \frac{dx_2}{dt} = x_2[-r_2 + a_{21}x_1(t-\tau) - a_{22}x_2]. \end{cases} \tag{1.3}$$

He claimed that if $\triangle_2 > 0$, then system (1.3) has a positive equilibrium $x^* = (x_1^*, x_2^*) = (\triangle_1/\triangle, \triangle_2/\triangle)$ which is globally asymptotically stable, where $\triangle = a_{11}a_{22} + a_{12}a_{21}$, $\triangle_1 = r_1a_{22} + r_2a_{12}$, $\triangle_2 = r_1a_{21} - r_2a_{11}$.

However, population models are always affected by environmental noises. Therefore stochastic population models have recently been investigated by many authors; see, *e.g.*, [25–33]. Mao *et al.* [34] have recently revealed an important fact that the environmental noise can suppress a potential population explosion. Recently, suppose that the parameter r_i is affected by environmental noises with $r_i \to r_i + \sigma_i\,dB_i$, $i = 1, 2$, then corresponding to system (1.3) the authors [35] obtained the following stochastic mode:

$$\begin{cases} dx_1 = x_1[r_1 - a_{11}x_1 - a_{12}x_2(t-\tau_1)]\,dt + \sigma_1 x_1\,dB_1, \\ dx_2 = x_2[-r_2 + a_{21}x_1(t-\tau_2) - a_{22}x_2]\,dt + \sigma_1 x_2\,dB_2 \end{cases} \tag{1.4}$$

and some sufficient and necessary conditions for stability in the mean and extinction of each population for the above stochastic system (1.4).

Motivated by the above work, we therefore wonder if the explosion problem for system (1.2) can be avoided by taking the environmental noise into account instead of imposing conditions on the coefficients of (1.2). To reveal this interesting fact, we stochastically perturb the delay predator-prey model (1.2) into the Itô stochastic differential delay equation

$$\begin{cases} dx_1(t) = x_1(t)[r_1 - a_{11}x_1 - \frac{a_{12}x_2(t-\tau)}{1+mx_1+nx_2}]\,dt + \sigma_1 x_1^2\,dB_1(t), \\ dx_2(t) = x_2(t)[-r_2 + \frac{a_{21}x_1(t-\tau)}{1+mx_1+nx_2} - a_{22}x_2]\,dt + \sigma_2 x_2^2\,dB_2(t), \end{cases} \tag{SM}$$

where x_1 and x_2 represent predator and prey densities at time t, respectively; r_i, a_{ij}, τ, m, n are positive constants, $i, j = 1, 2$.

In addition, throughout the present paper, let $(\Omega, \mathcal{F}, \{\mathcal{F}_t\}_{t\ge 0}, P)$ be a complete probability space with a filtration $\{\mathcal{F}_t\}_{t\ge 0}$ satisfying the usual conditions (*i.e.*, it is right continuous and $\mathcal{F}_0$ contains all P-null sets). Let $|\cdot|$ denote the Euclidean norm in R^n. For a given constant $\tau > 0$, let $C([-\tau, 0], R_+^n)$ denote the family of all continuous R_+^n-valued functions ξ with its norm $\|\xi\| = \sup\{|\xi(\theta)| : \theta \in [-\tau, 0]\}$, where $R_+ = [0, +\infty)$. Also, denote by $C_{\mathcal{F}_0}^b([-\tau, 0]; R_+^n)$ the family of bounded, $\mathcal{F}_0$-measurable, $C_{\mathcal{F}_0}^b([-\tau, 0]; R_+^n)$-valued random variables.

Remark 1.1 When $m = n = 0$, system (SM) becomes (1.4), so system (SM) is a more general stochastic system. For system (SM), so far as our knowledge is concerned, the work on a predator-prey model with stochastic perturbations seems rare. In this paper, we study system (SM) which is rather general, and some well-known systems may be viewed as its special cases, and obtain some properties of solutions to system (SM).

Remark 1.2 System (SM) is based on assuming that the noise affects parameters a_{11} and a_{22}. In fact, the noise may affect other parameters in (SM), which results in other types of stochastic models, which are the future research topics; for more details, see [36].

Remark 1.3 For the analysis of population dynamical problems of (SM), two difficult issues arise: (i) how to handle the delays in the given model and (ii) how to handle the nonlinear terms in (SM). To deal with these problems, the construction of a Lyapunov functional V is quite crucial, and it is introduced in Sections 3 and 4.

2 Positive and global solutions

In order for a stochastic differential delay equation to have a unique global (*i.e.*, no explosion in a finite time) solution for any given initial data, the coefficients of the equation are generally required to satisfy the linear growth condition and local Lipschitz condition [36]. However, the coefficients of system (SM) do not satisfy the linear growth condition, though they are locally Lipschitz continuous, so the solution of system (SM) may explode at a finite time. In this section we shall show that under simple hypothesis the solution of system (SM) is not only positive but will also not explode to infinity at any finite time.

Theorem 2.1 *For any given initial data* $\{(x_1(\theta), x_2(\theta))^\top : -\tau \le \theta \le 0\} = \xi \in C^b_{\mathcal{F}_0}([-\tau, 0]; R^0_+ \times R^0_+)$, *where* $R^0_+ = (0, +\infty)$. *If* $a_{21} \le 4$, *there is a unique positive local solution* $(x_1(t), x_2(t))$ *to* (SM) *on* $t \ge -\tau$ *with satisfying initial condition* ξ *and the solution will remain in* R^2_+ *with probability* 1.

Proof By the biological meaning, we only focus on the positive solution to system (SM). Thus it is reasonable to make the following change of variables, $x_1(t) = e^{u(t)}$, $x_2(t) = e^{v(t)}$. By using Itô's formula, system (SM) can be reformulated in the following form:

$$\begin{cases} du = [r_1 - a_{11}e^u - a_{12}\frac{e^{v(t-\tau)}}{1+e^{mu}+e^{nv}} - \frac{1}{2}\sigma_1^2]\,dt + \sigma_1 e^u\,dB_1(t), \\ dv = [-r_2(t) + a_{21}\frac{e^{u(t-\tau)}}{1+e^{mu}+e^{nv}} - a_{22}e^v - \frac{1}{2}\sigma_2^2]\,dt + \sigma_2 e^v\,dB_2(t), \quad t \ge 0, \\ u(\theta) = \ln x_1(\theta), v(\theta) = \ln x_2(\theta), \quad \theta \in [-\tau, 0]. \end{cases} \tag{2.1}$$

It is easy to see that the coefficients of (2.1) satisfy the local Lipschitz condition, then for any given initial values $u(\theta) > 0$, $v(\theta) > 0$, $\theta \in [-\tau, 0]$, there is a unique maximal local solution $u(t)$, $v(t)$ on $[-\tau, \tau_e)$, where τ_e is explosion time. To show that this solution is global, we need to show that $\tau_e = \infty$ a.s. Let $n_0 > 0$ be sufficiently large for

$$\frac{1}{n_0} < \min_{-\tau \le t \le 0} |x(t)| \le \max_{-\tau \le t \le 0} |x(t)| < n_0,$$

where $x(t) = (x_1(t), x_2(t))^\top$. For each integer $n > n_0$, define the stopping times:

$$\tau_n = \inf\left\{t \in [0, \tau_e] : x_1(t) \notin \left(\frac{1}{n}, n\right) \text{ or } x_2(t) \notin \left(\frac{1}{n}, n\right)\right\}.$$

Throughout this paper, we set $\inf\emptyset = \infty$. Obviously, τ_n is increasing as $n \to \infty$. Let $\tau_\infty = \lim_{n\to} \tau_n$, whence $\tau_\infty \leq \tau_e$ a.s. If we can show that $\tau_\infty = \infty$ a.s., then $\tau_e = \infty$ a.s. and $x_i(t) \geq 0$ a.s., $i = 1, 2$ for all $t \geq 0$. To show this statement, let us define C^2-function $V : R_+^2 \to R_+$ by

$$V(x_1, x_2) = (\sqrt{x_1} - 1 - 0.5 \ln x_1) + (\sqrt{x_2} - 1 - 0.5 \ln x_2).$$

The nonnegativity of this function can be obtained from $u - 1 - \ln u \geq 0$ on $u > 0$. Let $n \geq n_0$ and $T > 0$ be arbitrary. For $0 \leq t \leq \tau_n \wedge T$, we can use Itô's formula to $\int_{t-\tau}^{t} (x_1^2(s) + x_2^2(s))\, ds + V(x_1, x_2)$ to obtain that

$$\begin{aligned} d\left[\int_{t-\tau}^{t} \left(x_1^2(s) + x_2^2(s)\right) ds + V(x_1, x_2)\right] \\ &= \left[x_1^2 - x_1^2(t-\tau)\right. \\ &\quad \left. + 0.5\left(x_1^{0.5} - 1\right)F(x_1, x_2)\right] dt + \frac{1}{8}\left(-x_1^{0.5} + 2\right)\sigma_1^2 x_1^2\, dt \\ &\quad + 0.5\left(x_1^{0.5} - 1\right)\sigma_1 x_1\, dB_1(t) + \left[x_2^2 - x_2^2(t-\tau) + 0.5\left(x_2^{0.5} - 1\right)G(x_1, x_2)\right] dt \\ &\quad + \frac{1}{8}\left(-x_2^{0.5} + 2\right)\sigma_2^2 x_2^2\, dt + 0.5\left(x_2^{0.5} - 1\right)\sigma_2 x_2\, dB_2(t), \end{aligned} \tag{2.2}$$

where

$$\begin{aligned} F(x_1, x_2) &= r_1 - a_{11}x_1 - \frac{a_{12}x_2(t-\tau)}{1 + mx_1 + nx_2}, \\ G(x_1, x_2) &= -r_2 - a_{22}x_2 + \frac{a_{21}x_1(t-\tau)}{1 + mx_1 + nx_2}. \end{aligned}$$

Compute

$$\begin{aligned} &\left(x_1^{0.5} - 1\right)F(x_1, x_2) \\ &\quad = \left(x_1^{0.5} - 1\right)\left[r_1 - a_{11}x_1 - \frac{a_{12}x_2(t-\tau)}{1 + mx_1 + nx_2}\right] \\ &\quad = r_1\left(x_1^{0.5} - 1\right) - a_{11}x_1^{1.5} + a_{11}x_1 + \frac{a_{12}x_2(t-\tau)}{1 + mx_1 + nx_2} - \frac{a_{12}x_1^{0.5}x_2(t-\tau)}{1 + mx_1 + nx_2} \\ &\quad \leq r_1\left(x_1^{0.5} - 1\right) - a_{11}x_1^{1.5} + a_{11}x_1 + a_{12}x_2(t-\tau) \end{aligned} \tag{2.3}$$

and

$$\begin{aligned} &\left(x_2^{0.5} - 1\right)G(x_1, x_2) \\ &\quad = \left(x_2^{0.5} - 1\right)\left[-r_2 - a_{22}x_2 + \frac{a_{21}x_1(t-\tau)}{1 + mx_1 + nx_2}\right] \\ &\quad = -r_2\left(x_2^{0.5} - 1\right) - a_{22}x_2^{1.5} + a_{22}x_2 + \frac{a_{21}x_2^{0.5}x_1(t-\tau)}{1 + mx_1 + nx_2} - \frac{a_{21}x_1(t-\tau)}{1 + mx_1 + nx_2} \\ &\quad \leq -r_2\left(x_2^{0.5} - 1\right) - a_{22}x_2^{1.5} + a_{22}x_2 + a_{21}x_2^{0.5}x_1(t-\tau) \\ &\quad \leq -r_2\left(x_2^{0.5} - 1\right) - a_{22}x_2^{1.5} + (a_{22} + 0.5a_{21})x_2 + 0.5a_{21}x_1^2(t-\tau). \end{aligned} \tag{2.4}$$

Substituting (2.3) and (2.4) into (2.2) yields

$$d\left[\int_{t-\tau}^{t}\left(x_1^2(s)+x_2^2(s)\right)ds+V(x_1,x_2)\right]$$
$$\leq H(x_1,x_2)\,dt+0.5\left(x_1^{0.5}-1\right)\sigma_1 x_1\,dB_1(t)+0.5\left(x_2^{0.5}-1\right)\sigma_2 x_2\,dB_2(t), \tag{2.5}$$

where

$$\begin{aligned}H(x_1,x_2)&=x_1^2-x_1^2(t-\tau)+0.5r_1\left(x_1^{0.5}-1\right)-0.5a_{11}x_1^{1.5}+0.5a_{11}x_1\\&\quad+0.5a_{12}x_2(t-\tau)+\frac{1}{8}\left(-x_1^{0.5}+2\right)\sigma_1^2x_1^2\\&\quad+0.5x_2^2-0.5x_2^2(t-\tau)-0.5r_2\left(x_2^{0.5}-1\right)\\&\quad-0.5a_{22}x_2^{1.5}+0.5(a_{22}+0.5a_{21})x_2\\&\quad+0.25a_{21}x_1^2(t-\tau)+\frac{1}{8}\left(-x_2^{0.5}+2\right)\sigma_2^2x_2^2\\&=\left(1+0.25\sigma_1^2\right)x_1^2+(0.25a_{21}-1)x_1^2(t-\tau)\\&\quad+0.5r_1\left(x_1^{0.5}-1\right)-0.5a_{11}x_1^{1.5}+0.5a_{11}x_1\\&\quad+\left(0.5+0.25\sigma_2^2\right)x_2^2-0.5x_2^2(t-\tau)-0.5r_2\left(x_2^{0.5}-1\right)\\&\quad-0.5a_{22}x_2^{1.5}+0.5(a_{22}+0.5a_{21})x_2\\&\quad-0.125x_1^{2.5}-0.125x_2^{2.5}.\end{aligned}$$

From $a_{21}\leq 4$, it is easy to see that $H(x_1,x_2)$ is bounded, say by K. By (2.5), we have

$$d\left[\int_{t-\tau}^{t}\left(x_1^2(s)+x_2^2(s)\right)ds+V(x_1,x_2)\right]$$
$$\leq K\,dt+0.5\left(x_1^{0.5}-1\right)\sigma_1 x_1\,dB_1(t)+0.5\left(x_2^{0.5}-1\right)\sigma_2 x_2\,dB_2(t).$$

Integrating both sides from 0 to $\tau_n\wedge T$, and then taking expectations, yields

$$E\left[\int_{\tau_n\wedge T-\tau}^{\tau_n\wedge T}\left(x_1^2(s)+x_2^2(s)\right)ds+V\left(x_1(\tau_n\wedge T),x_2(\tau_n\wedge T)\right)\right]$$
$$\leq\int_{-\tau}^{0}\left(x_1^2(s)+x_2^2(s)\right)ds+V\left(x_1(0),x_2(0)\right)+KT.$$

Consequently,

$$EV\left(x_1(\tau_n\wedge T),x_2(\tau_n\wedge T)\right)\leq\int_{-\tau}^{0}\left(x_1^2(s)+x_2^2(s)\right)ds+V\left(x_1(0),x_2(0)\right)+KT. \tag{2.6}$$

Note that for each $\omega\in\Omega_n=\{\tau_n\leq T\}$, there is some i such that $x_i(\tau_n,\omega)$ equals n or $\frac{1}{n}$ for $i=1,2$. Hence $V(x_1(\tau_n\wedge T),x_2(\tau_n\wedge T))$ is no less than

$$\min\{\sqrt{n}-1-0.5\ln n,\sqrt{1/n}-1-0.5\ln 1/n\}.$$

By (2.6) we have

$$\begin{aligned}&\int_{-\tau}^{0}(x_1^2(s)+x_2^2(s))\,ds+V\big(x_1(0),x_2(0)\big)+KT\\&\quad\ge E\big[1_{\Omega_n}(\omega)V\big(x_1(\tau_n),x_2(\tau_n)\big)\big]\\&\quad\ge P\{\tau_n\le T\}\min\{\sqrt{n}-1-0.5\ln n,\sqrt{1/n}-1-0.5\ln 1/n\},\end{aligned}$$

where 1_{Ω_n} is the indicator function of Ω_n. Letting $n\to\infty$, leads to the contraction

$$\lim_{k\to\infty}P\{\tau_k\le T\}=0.$$

Since $T>0$ is arbitrary, we must have

$$P\{\tau_\infty<\infty\}=0.$$

So $P\{\tau_\infty=\infty\}=1$ as required. The proof is completed. □

Remark 2.1 It is well known that systems (1.2) and (1.3) may explode to infinity at a finite time for some system parameters, see [37]. However, the explosion will no longer happen as long as there is noise. In other words, Theorem 2.1 reveals the important property that the environmental noise suppresses the explosion for the delay equation.

3 Stochastically ultimate boundedness

One of the important issues in the study of population systems is the stochastically ultimate boundedness. System (SM) is said to be stochastically ultimately bounded if for any $\varepsilon\in(0,1)$, there is a positive constant $H=H(\varepsilon)$ such that for any initial data $\xi\in C^b_{\mathcal{F}_0}([-\tau,0];R^0_+\times R^0_+)$, the solution $(x_1,x_2)^\top$ of system (SM) has the property that

$$\lim_{t\to\infty}\sup P\big\{|x(t)|\le H\big\}\ge 1-\varepsilon,$$

where $x=(x_1,x_2)^\top$.

In this section we shall investigate the stochastically ultimate boundedness of system (SM). The following theorem gives a sufficient criterion for the stochastically ultimate boundedness of population.

Lemma 3.1 *Let $\theta\in(0,1)$ and $\theta a_{21}\ge n$. Then there is a positive constant $K=K(\theta)$, which is independent of the initial data $\xi\in C^b_{\mathcal{F}_0}([-\tau,0];R^0_+\times R^0_+)$, such that the solution $x(t)$ of system* (SM) *has the property that*

$$\lim_{t\to\infty}\sup E|x|^\theta\le K.$$

Proof Define

$$V(x_1,x_2)=x_1^\theta+x_2^\theta.$$

Applying Itô's formula to system (SM), we have

$$\begin{aligned} dV(x_1,x_2) &= \left[\theta x_1^{\theta}\left(r_1 - a_{11}x_1 - \frac{a_{12}x_2(t-\tau)}{1+mx_1+nx_2}\right) - \frac{\theta(1-\theta)\sigma_1^2}{2}x_1^{2+\theta}\right]dt \\ &\quad + \theta\sigma_1 x_1^{1+\theta}\,dB_1(t) \\ &\quad + \left[\theta x_2^{\theta}\left(-r_2 + \frac{a_{21}x_1(t-\tau)}{1+mx_1+nx_2} - a_{22}x_2\right) - \frac{\theta(1-\theta)\sigma_2^2}{2}x_2^{2+\theta}\right]dt \\ &\quad + \theta\sigma_2 x_2^{1+\theta}\,dB_2(t). \end{aligned} \tag{3.1}$$

Denote

$$\begin{aligned} LV(x_1,x_2,y_1,y_2) &= \theta x_1^{\theta}\left(r_1 - a_{11}x_1 - \frac{a_{12}y_2}{1+mx_1+nx_2}\right) - \frac{\theta(1-\theta)\sigma_1^2}{2}x_1^{2+\theta} \\ &\quad + \theta x_2^{\theta}\left(-r_2 + \frac{a_{21}y_1}{1+mx_1+nx_2} - a_{22}x_2\right) - \frac{\theta(1-\theta)\sigma_2^2}{2}x_2^{2+\theta} \\ &\le \theta x_1^{\theta} r_1 - \frac{\theta(1-\theta)\sigma_1^2}{2}x_1^{2+\theta} - \frac{\theta(1-\theta)\sigma_2^2}{2}x_2^{2+\theta} + \frac{\theta a_{21}}{n}|y|^2 \\ &= F(x_1,x_2) - V(x_1,x_2) - e^{\tau}|x|^2 + \frac{\theta a_{21}}{n}|y|^2, \end{aligned}$$

where

$$\begin{aligned} &y = (y_1,y_2)^{\top} = \left(x_1(t-\tau), x_2(t-\tau)\right)^{\top}, \qquad x = (x_1,x_2)^{\top}, \\ &F(x_1,x_2) = \theta x_1^{\theta} r_1 - \frac{\theta(1-\theta)\sigma_1^2}{2}x_1^{2+\theta} - \frac{\theta(1-\theta)\sigma_2^2}{2}x_2^{2+\theta} + x_1^{\theta} + x_2^{\theta} + e^{\tau}|x|^2. \end{aligned}$$

Note that $F(x_1,x_2)$ is bounded in R_+^2, namely

$$F(x_1,x_2) \le K_1, \quad \forall x \in R_+^2.$$

Hence we have

$$LV(x_1,x_2,y_1,y_2) \le K_1 - V(x_1,x_2) - e^{\tau}|x|^2 + \frac{\theta a_{21}}{n}|y|^2.$$

Substituting this into (3.1) gives

$$\begin{aligned} dV(x_1,x_2) &\le \left[K_1 - V(x_1,x_2) - e^{\tau}|x|^2 + \frac{\theta a_{21}}{n}|y|^2\right]dt \\ &\quad + \theta\sigma_1 x_1^{1+\theta}\,dB_1(t) + \theta\sigma_2 x_2^{1+\theta}\,dB_2(t). \end{aligned} \tag{3.2}$$

From (3.2) and once again by Itô's formula, we have

$$\begin{aligned} d\left[e^t V(x_1,x_2)\right] &= e^t\left[V(x_1,x_2)\,dt + dV(x_1,x_2)\right] \\ &\le e^t\left[K_1 - e^{\tau}|x|^2 + \frac{\theta a_{21}}{n}\left|x(t-\tau)\right|^2\right]dt \\ &\quad + e^t\theta\sigma_1 x_1^{1+\theta}\,dB_1(t) + e^t\theta\sigma_2 x_2^{1+\theta}\,dB_2(t). \end{aligned}$$

From $\theta a_{21} \geq n$, we hence derive that

$$e^t EV(x_1, x_2) \leq V\big(x_1(0), x_2(0)\big) + K_1 e^t - E\int_0^t e^{s+\tau}\left|x(s)\right|^2 ds$$
$$+ \frac{\theta a_{21}}{n} E\int_0^t e^s \left|x(s-\tau)\right|^2 ds$$
$$= V\big(x_1(0), x_2(0)\big) + K_1 e^t - E\int_0^t e^{s\tau}\left|x(s)\right|^2 ds$$
$$+ \frac{\theta a_{21}}{n} E\int_{-\tau}^{t-\tau} e^s \left|x(s)\right|^2 ds$$
$$\leq V\big(x_1(0), x_2(0)\big) + K_1 e^t + \frac{\theta a_{21}}{n}\int_{-\tau}^{0} e^s \left|x(s)\right|^2 ds.$$

This implies immediately that

$$\lim_{t\to\infty} \sup EV(x_1, x_2) \leq K_1.$$

On the other hand, we have

$$|x|^2 \leq 2\max\{x_1, x_2\}.$$

Thus

$$|x|^\theta \leq 2^{\theta/2}\max\left\{x_1^\theta, x_2^\theta\right\} \leq 2^{\theta/2} V(x_1, x_2).$$

Hence, we have

$$\lim_{t\to\infty} \sup E|x|^\theta \leq 2^{\theta/2} K_1 := K.$$

The proof is completed. □

Theorem 3.1 *Let* $\theta \in (0,1)$ *and* $\theta a_{21} \geq n$. *System* (SM) *is stochastically ultimately bounded.*

Proof From Lemma 3.1, there is $K > 0$ such that

$$\lim_{t\to\infty} \sup E|x|^\theta \leq K.$$

Then, for any $\varepsilon > 0$, let $H = K^2/\varepsilon^2$. Then by Chebyshev's inequality

$$P\left\{\left|x(t)\right| > H\right\} \leq \frac{E(\sqrt{|x(t)|})}{\sqrt{H}}.$$

Thus

$$\lim_{t\to\infty} \sup P\left\{\left|x(t)\right| > H\right\} \leq \frac{K}{\sqrt{H}} = \varepsilon.$$

This implies

$$\limsup_{t\to\infty} P\{|x(t)| \le H\} \ge 1 - \varepsilon.$$

The proof is completed. □

4 Almost surely asymptotic properties

In this section, we study the pathwise properties of system (SM).

Theorem 4.1 *Let the same conditions of Theorem* 2.1 *hold. For any given initial data* $\{(x_1(\theta), x_2(\theta))^\top : -\tau \le \theta \le 0\} = \xi \in C^b_{\mathcal{F}_0}([-\tau, 0]; R^0_+ \times R^0_+)$, *the solution* $x(t)$ *of system* (SM) *has the property that*

$$\limsup_{t\to\infty} \frac{\ln|x(t)|}{\ln t} \le 1 \quad a.s.$$

Proof Define

$$V(x) = x_1 + x_2.$$

It is easy to see that

$$\begin{aligned} dV(x(t)) = {}& x_1(t)\left[r_1 - a_{11}x_1 - \frac{a_{12}x_2(t-\tau)}{1 + mx_1 + nx_2}\right]dt + \sigma_1 x_1^2\, dB_1(t) \\ & + x_2(t)\left[-r_2 + \frac{a_{21}x_1(t-\tau)}{1 + mx_1 + nx_2} - a_{22}x_2\right]dt + \sigma_2 x_2^2\, dB_2(t). \end{aligned}$$

Let $\gamma > 0$ be arbitrary. By Itô's formula, we have

$$\begin{aligned} e^{\gamma t}\ln(V(x(t))) = {}& \ln(V(x(0))) \\ & + \int_0^t e^{\gamma s}\left(\frac{x_1}{V(x(s))}\left[r_1 - a_{11}x_1 - \frac{a_{12}x_2(t-\tau)}{1 + mx_1 + nx_2}\right] - \frac{\sigma_1^2 x_1^4}{2V^2(x(s))}\right)ds \\ & + \int_0^t e^{\gamma s}\left(\frac{x_2}{V(x(s))}\left[-r_2 + \frac{a_{21}x_1(t-\tau)}{1 + mx_1 + nx_2} - a_{22}x_2\right] - \frac{\sigma_2^2 x_2^4}{2V^2(x(s))}\right)ds \\ & + \gamma\int_0^t e^{\gamma s}\ln(V(x(s)))\, ds + M_1(t) + M_2(t), \end{aligned} \tag{4.1}$$

where

$$M_i(t) = \int_0^t \frac{e^{\gamma s}\sigma_i x_i^2}{V(x(s))}\, dB_i(s), \quad i = 1, 2,$$

is a real-valued continuous local martingale vanishing at $t = 0$ and its quadratic form is given by

$$\langle M_i(t), M_i(t)\rangle = \int_0^t \frac{e^{2\gamma s}\sigma_i^2 x_i^4}{V(x(s))}\, ds.$$

Let $\varepsilon \in (0,1)$ and $\theta > 1$ be arbitrary. By the exponential martingale inequality (see, *e.g.*, [36]), for each $k \geq 1$,

$$P\left\{\sup_{0\leq t\leq k}\left[M_i(t) - \frac{\varepsilon}{2}e^{-\gamma k}\langle M_i(t), M_i(t)\rangle\right] > \frac{\theta e^{\gamma k}}{\varepsilon}\ln k\right\} \leq k^{-\theta}.$$

Since the series $\sum_{k=1}^{\infty} k^{-\theta}$ converges, the well-known Borel-Cantelli lemma yields that there exists $\Omega_0 \subset \Omega$ with $P(\Omega_0) = 1$ such that for any $\omega \in \Omega_0$ there exists an integer $k_0 = k_0(\omega)$ such that

$$M_i(t) \leq \frac{\varepsilon}{2}e^{-\gamma k}\langle M_i(t), M_i(t)\rangle + \frac{\theta e^{\gamma k}}{\varepsilon}\ln k \tag{4.2}$$

for all $0 \leq t \leq k$ and $k \geq k_0(\omega)$. Substituting this into (4.1) and then using (4.2), we have

$$\begin{aligned} e^{\gamma t}\ln(V(x(t))) = \ln(V(x(0))) &+ \int_0^t e^{\gamma s}\left(\frac{x_1}{V(x(s))}\left[r_1 - a_{11}x_1 - \frac{a_{12}x_2(t-\tau)}{1+mx_1+nx_2}\right] - \frac{(1-\varepsilon)\sigma_1^2 x_1^4}{2V^2(x(s))}\right)ds \\ &+ \int_0^t e^{\gamma s}\left(\frac{x_2}{V(x(s))}\left[-r_2 + \frac{a_{21}x_1(t-\tau)}{1+mx_1+nx_2} - a_{22}x_2\right] - \frac{(1-\varepsilon)\sigma_2^2 x_2^4}{2V^2(x(s))}\right)ds \\ &+ \gamma\int_0^t e^{\gamma s}\ln(V(x(s)))\,ds + \frac{\theta e^{\gamma k}}{\varepsilon}\ln k \end{aligned} \tag{4.3}$$

for all $0 \leq t \leq k$, $k \geq k_0(\omega)$ and $\omega \in \Omega_0$. Compute

$$\frac{x_1}{V(x(s))}\left[r_1 - a_{11}x_1 - \frac{a_{12}x_2(t-\tau)}{1+mx_1+nx_2}\right] \leq r_1 + a_{12}x_2(t-\tau), \tag{4.4}$$

$$\frac{x_2}{V(x(s))}\left[-r_2 + \frac{a_{21}x_1(t-\tau)}{1+mx_1+nx_2} - a_{22}x_2\right] \leq r_2 + a_{21}x_1(t-\tau), \tag{4.5}$$

$$\frac{\sigma_1^2 x_1^4}{2V^2(x(s))} \geq \frac{\sigma_1^2 k_1^2}{2}x_1^2, \qquad \frac{\sigma_2^2 x_2^4}{2V^2(x(s))} \geq \frac{\sigma_2^2 k_1^2}{2}x_2^2, \tag{4.6}$$

where $k_1, k_2 \in (0,1)$ are positive constants. Substituting (4.4)-(4.6) into (4.3) gives

$$\begin{aligned} &e^{\gamma t}\ln(V(x(t))) \\ &= \ln(V(x(0))) + \gamma\int_0^t e^{\gamma s}\ln(V(x(s)))\,ds + \frac{\theta e^{\gamma k}}{\varepsilon}\ln k \\ &\quad + \int_0^t e^{\gamma s}\left(r_1 + a_{12}x_2(t-\tau) + r_2 + a_{21}x_1(t-\tau)] - \frac{(1-\varepsilon)\sigma_1^2 k_1^2 x_1^2}{2}\right. \\ &\quad \left. - \frac{(1-\varepsilon)\sigma_2^2 k_2^2 x_2^2}{2}\right)ds \end{aligned} \tag{4.7}$$

for all $0 \leq t \leq k$, $k \geq k_0(\omega)$ and $\omega \in \Omega_0$. For $i = 1, 2$, note

$$\int_0^t e^{\gamma s}x_i(s-\tau)\,ds \leq \int_{-\tau}^{t-\tau} e^{\gamma(s+\tau)}x_i(s)\,ds \leq \int_{-\tau}^0 e^{\gamma\tau}x_i(s)\,ds + \int_0^t e^{\gamma(s+\tau)}x_i(s)\,ds.$$

We can rewrite (4.7) as

$$e^{\gamma t}\ln\big(V(x(t))\big) = \ln\big(V(x(0))\big) + a_{12}e^{\gamma\tau}\int_{-\tau}^{0} x_1(s)\,ds + a_{21}e^{\gamma\tau}\int_{-\tau}^{0} x_2(s)\,ds + \frac{\theta e^{\gamma k}}{\varepsilon}\ln k$$
$$+ \int_0^t e^{\gamma s}\Big(r_1 + a_{12}e^{\gamma\tau}x_2 + r_2 + a_{21}e^{\gamma\tau}x_1 - \frac{(1-\varepsilon)\sigma_1^2k_1^2x_1^2}{2}$$
$$- \frac{(1-\varepsilon)\sigma_2^2k_2^2x_2^2}{2}\Big)\,ds + \gamma\int_0^t e^{\gamma s}\ln\big(V(x(s))\big)\,ds. \tag{4.8}$$

Obviously, the following polynomial is bounded by a positive constant, say K_1,

$$\gamma\ln V(x) + r_1 + a_{12}e^{\gamma\tau}x_2 + r_2 + a_{21}e^{\gamma\tau}x_1 - \frac{(1-\varepsilon)\sigma_1^2k_1^2x_1^2}{2} - \frac{(1-\varepsilon)\sigma_2^2k_2^2x_2^2}{2} \le K_1. \tag{4.9}$$

Then by (4.8) and (4.9) we get

$$e^{\gamma t}\ln\big(V(x(t))\big) \le C_1 + \frac{\theta e^{\gamma k}}{\varepsilon}\ln k + \frac{K_1}{\gamma}e^{\gamma t}$$

for all $0 \le t \le k$, $k \ge k_0(\omega)$ and $\omega \in \Omega_0$, where

$$C_1 = \ln\big(V(x(0))\big) + a_{12}e^{\gamma\tau}\int_{-\tau}^{0} x_1(s)\,ds + a_{21}e^{\gamma\tau}\int_{-\tau}^{0} x_2(s)\,ds.$$

Consequently, for any $\omega \in \Omega_0$, if $k-1 \le t \le k$ and $k \ge k_0(\omega)$, we have

$$\frac{\ln(V(x(t)))}{\ln t} \le \frac{1}{\ln(k-1)}\left[e^{-\gamma(k-1)}C_1 + \frac{\theta e^{\gamma}}{\varepsilon}\ln k + \frac{K_1}{\gamma}\right].$$

This implies

$$\limsup_{t\to\infty}\frac{\ln(V(x(t)))}{\ln t} \le \frac{\theta e^{\gamma}}{\varepsilon} \quad \text{a.s.}$$

Letting $\varepsilon \to 1$, $\theta \to 1$ and $\gamma \to 0$, we have

$$\limsup_{t\to\infty}\frac{\ln(V(x(t)))}{\ln t} \le 1 \quad \text{a.s.} \tag{4.10}$$

From $V(x) \le \sqrt{2}|x|$ and (4.10), we get

$$\limsup_{t\to\infty}\frac{\ln|x(t)|}{\ln t} \le 1 \quad \text{a.s.}$$ □

Remark 4.1 Similar to [37], $\forall\varepsilon > 0$, we have

$$|x| \le \kappa + t^{1+\varepsilon} \quad \text{for } t \ge 0,$$

where $\kappa = \sup_{0\le t\le T}|x(t)|$. This means that $|x(t)|$ will grow at most polynomially with order close to 1.

Theorem 4.2 *Let the same conditions of Theorem* 2.1 *hold. For any given initial data* $\{(x_1(\theta), x_2(\theta))^\top : -\tau \le \theta \le 0\} = \xi \in C^b_{\mathcal{F}_0}([-\tau, 0]; R^0_+ \times R^0_+)$, *the solution* $x(t)$ *of system* (SM) *has the property that*

$$\lim_{t\to\infty}\sup\frac{\ln|x(t)|}{\ln t}\le r_1+\frac{\hat{\sigma}^2}{\check{a}^2}\quad a.s.,$$

where $\hat{\sigma}=\min\{\sigma_1 k_1,\sigma_2 k_2\}$, $\check{a}=\max\{a_{12},a_{21}\}$.

Proof Define

$$V(x)=x_1+x_2.$$

By Itô's formula, we have

$$\begin{aligned}\ln\big(V(x(t))\big)&=\ln\big(V(x(0))\big)\\&+\int_0^t\left(\frac{x_1}{V(x(s))}\left[r_1-a_{11}x_1-\frac{a_{12}x_2(s-\tau)}{1+mx_1+nx_2}\right]-\frac{\sigma_1^2x_1^4}{2V^2(x(s))}\right)ds\\&+\int_0^t\left(\frac{x_2}{V(x(s))}\left[-r_2+\frac{a_{21}x_1(s-\tau)}{1+mx_1+nx_2}-a_{22}x_2\right]-\frac{\sigma_2^2x_2^4}{2V^2(x(s))}\right)ds\\&+M_1(t)+M_2(t),\end{aligned}\tag{4.11}$$

where

$$M_i(t)=\int_0^t\frac{\sigma_i x_i^2}{V(x(s))}dB_i(s),\quad i=1,2,$$

is a real-valued continuous local martingale vanishing at $t=0$ and its quadratic form is given by

$$\langle M_i(t),M_i(t)\rangle=\int_0^t\frac{\sigma_i^2x_i^4}{V(x(s))}ds.$$

Let $\varepsilon\in(0,0.5)$ and $\theta=2$ be arbitrary. By the exponential martingale inequality (see, *e.g.*, [36]), for each $k\ge 1$,

$$P\left\{\sup_{0\le t\le k}\left[M_i(t)-\frac{\varepsilon}{2}\langle M_i(t),M_i(t)\rangle\right]>\frac{2\ln k}{\varepsilon}\right\}\le k^{-2}.$$

Since the series $\sum_{k=1}^{\infty}k^{-\theta}$ converges, the well-known Borel-Cantelli lemma yields that there exists $\Omega_0\subset\Omega$ with $P(\Omega_0)=1$ such that for any $\omega\in\Omega_0$ there exists an integer $k_0=k_0(\omega)$ such that

$$M_i(t)\le\frac{\varepsilon}{2}\langle M_i(t),M_i(t)\rangle+\frac{2}{\varepsilon}\ln k$$

for all $0\le<t\le k$ and $k\ge k_0(\omega)$. Substituting this into (4.11) we derive that

$$\begin{aligned}\ln\big(V(x(t))\big)&=\ln\big(V(x(0))\big)\\&+\int_0^t\left(\frac{x_1}{V(x(s))}\left[r_1-a_{11}x_1-\frac{a_{12}x_2(s-\tau)}{1+mx_1+nx_2}\right]-\frac{(1-\varepsilon)\sigma_1^2x_1^4}{2V^2(x(s))}\right)ds\end{aligned}$$

$$+\int_0^t \left(\frac{x_2}{V(x(s))} \left[-r_2 + \frac{a_{21}x_1(s-\tau)}{1+mx_1+nx_2} - a_{22}x_2 \right] - \frac{(1-\varepsilon)\sigma_2^2 x_2^4}{2V^2(x(s))} \right) ds + \frac{4}{\varepsilon}\ln k$$

for all $0 \le t \le k$, $k \ge k_0(\omega)$ and $\omega \in \Omega_0$. Rearranging the above inequality and using (4.4)-(4.6) give

$$\begin{aligned}
\ln(V(x(t))) &= \ln(V(x(0))) + \frac{4}{\varepsilon}\ln k \\
&\quad + \int_0^t \Big(r_1 + a_{12}x_2(s-\tau) + r_2 + a_{21}x_1(s-\tau) \\
&\quad - \frac{(1-\varepsilon)\sigma_1^2 k_1^2 x_1^2}{2} - \frac{(1-\varepsilon)\sigma_2^2 k_2^2 x_2^2}{2} \Big) ds \\
&\le \ln(V(x(0))) + \frac{4}{\varepsilon}\ln k + \int_0^t \Big(r_1 + a_{12}x_2(s-\tau) + r_2 \\
&\quad + a_{21}x_1(s-\tau) - \frac{(1-\varepsilon)\hat{\sigma}^2|x|^2}{2} \Big) ds
\end{aligned} \tag{4.12}$$

for all $0 \le t \le k$, $k \ge k_0(\omega)$ and $\omega \in \Omega_0$. From (4.12), we have

$$\begin{aligned}
&\ln(V(x(t))) + \frac{(1-2\varepsilon)\hat{\sigma}^2}{4}\int_0^t |x(s)|^2 ds \\
&\le \ln(V(x(0))) + \frac{4}{\varepsilon}\ln k \\
&\quad + \int_0^t \left(r_1 + a_{12}x_2(s-\tau) + r_2 + a_{21}x_1(s-\tau) - \frac{\hat{\sigma}^2|x|^2}{4} \right) ds \\
&\le C_2 + \frac{4}{\varepsilon}\ln k + \int_0^t \left(r_1 + a_{12}x_2(s) + r_2 + a_{21}x_1(s) - \frac{\hat{\sigma}^2|x|^2}{4} \right) ds,
\end{aligned}$$

where

$$C_2 = \ln(V(x(0))) + a_{12}\int_{-\tau}^0 x_1(s)\,ds + a_{21}\int_{-\tau}^0 x_2(s)\,ds.$$

It is easy to see that

$$r_1 + a_{12}x_2(t) + r_2 + a_{21}x_1(t) - \frac{\hat{\sigma}^2|x|^2}{4} \le r_1 + \check{a}|x| - \frac{\hat{\sigma}^2|x|^2}{4} \le r_1 + \frac{\check{a}^2}{\hat{\sigma}^2} := K_2.$$

Thus, if $\omega \in \Omega_0$,

$$\ln(V(x(t))) + \frac{(1-2\varepsilon)\hat{\sigma}^2}{4}\int_0^t |x(s)|^2 ds \le C_2 + \frac{4}{\varepsilon}\ln k + K_2 t$$

for all $0 \le t \le k$, $k \ge k_0(\omega)$. Consequently, for any $\omega \in \Omega_0$, if $k-1 \le t \le k$ and $k \ge k_0(\omega)$, we have

$$\frac{1}{t}\left[\ln(V(x(t))) + \frac{(1-2\varepsilon)\hat{\sigma}^2}{4}\int_0^t |x(s)|^2 ds \right] \le \frac{1}{k-1}\left[C_2 + \frac{4}{\varepsilon}\ln k \right] + K_2,$$

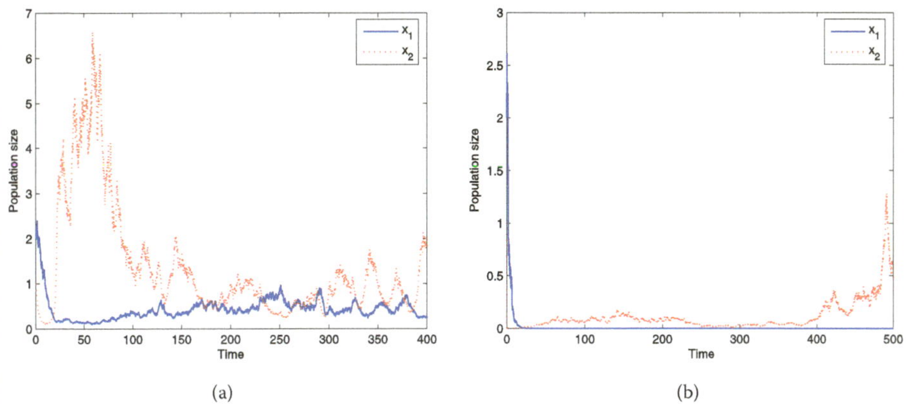

Figure 1 Solutions of system (SM) for $r_1 = 0.02$, $a_{11} = 0.1$, $a_{12} = 0.03$, $r_2 = 3$, $a_{22} = 0.3$, $\sigma_1 = \sigma_2 = 0.02$, $m = n = 1.25$, $\tau = 1$. The horizontal axis represents the time t, the vertical axis represents the population sizes. **(a)** is with $a_{21} = 0.3$; **(b)** is with $a_{21} = 500$.

which implies

$$\limsup_{t\to\infty} \frac{1}{t}\left[\ln\bigl(V(x(t))\bigr) + \frac{(1-2\varepsilon)\hat{\sigma}^2}{4}\int_0^t |x(s)|^2\,ds\right] \le K_2. \tag{4.13}$$

Using (4.13) and letting $\varepsilon \to 0$, we get

$$\limsup_{t\to\infty} \frac{1}{t}\left[\ln\bigl(V(x(t))\bigr) + \frac{\hat{\sigma}^2}{4}\int_0^t |x(s)|^2\,ds\right] \le K_2 \quad \text{a.s.} \tag{4.14}$$

Using $V(x) \ge \frac{|x|}{\sqrt{2}}$ and (4.14), we get

$$\limsup_{t\to\infty} \frac{1}{t}\left[\ln|x(t)| + \frac{\hat{\sigma}^2}{4}\int_0^t |x(s)|^2\,ds\right] \le K_2 \quad \text{a.s.}$$

The proof is completed. □

5 Numerical simulations

Now let us use Milstein's numerical method (see, *e.g.*, [38]) to support our results. In Figure 1, we choose $r_1 = 0.02$, $a_{11} = 0.1$, $a_{12} = 0.03$, $r_2 = 3$, $a_{22} = 0.3$, $\sigma_1 = \sigma_2 = 0.02$, $m = n = 1.25$, $\tau = 1$. The difference between the conditions of Figures 1(a) and (b) is that the values of a_{21} are different. In Figure 1(a) we choose $a_{21} = 0.3 < 4$, then the conditions of Theorem 2.1 hold. Making Theorems 4.1 and 4.2 lead to system (SM) has almost surely asymptotic properties. In Figure 1(b) we choose $a_{21} = 500 > 4$, then the conditions of Theorem 2.1 are not satisfied; furthermore, the conditions of Theorems 4.1 and 4.2 do not hold. Hence, the population x_1 goes to extinction, and x_2 has no almost surely asymptotic properties, see Figure 1(b).

6 Conclusions and future directions

A stochastic delay predator-prey system is considered and system (SM) is more general than the classical predator-prey system with the Beddington-DeAngelis functional response. We have established a sufficient condition under which system (SM) has a global positive solution. We have also discussed the asymptotic properties for the moments as well as the sample paths of the solution. In particular, we have studied a fundamental problem in population systems, namely stochastically ultimate boundedness. Our key contributions in this paper are the following:

- This paper deals with a kind of delay stochastic population system, while most of existing results (see, *e.g.*, [4, 5, 9–11, 22, 31]) are concerned with the non-delayed cases.
- Our stochastic population system is therefore more complicated and the mathematics presented is more difficult.

There are still many interesting and challenging questions that need to be studied. In this paper, we only consider the growth rate a_{11}, a_{22} to be stochastic; other parameters, for example, r_i, $i = 1, 2$, is stochastic, which is not studied. We wish that such questions will be investigated by some authors.

Competing interests
The authors declare that they have no competing interests.

Authors' contributions
All authors contributed equally to the writing of this paper. All authors read and approved the final manuscript.

Author details
[1]Department of Mathematics, Huaiyin Normal University, Huaian, Jiangsu 223300, China. [2]Department of Mathematics, Yangzhou University, Yangzhou, Jiangsu 225002, China. [3]Department of Basis Course, Lianyungang Technical College, Lianyungang, Jiangsu 222006, China.

Acknowledgements
This paper is supported by Postdoctoral Foundation of Jiangsu (1402113C) and Postdoctoral Foundation of China (2014M561716).

References

1. Skalski, GT, Gilliam, JF: Functional responses with predator interference: viable alternatives to the Holling II model. Ecology **82**, 3083-3092 (2001)
2. Hassell, M: Mutual interference between searching insect parasites. J. Anim. Ecol. **40**, 473-486 (1971)
3. Beddington, JR: Mutual interference between parasites or predators and its effect on searching efficiency. J. Anim. Ecol. **44**, 331-341 (1975)
4. Crowley, P, Martin, P: Functional response and interference within and between year classes of dragonfly. J. North Am. Benthol. Soc. **8**, 211-221 (1989)
5. Liu, Z, Yuan, R: Stability and bifurcation in a delayed predator-prey system with Beddington-DeAngelis functional response. J. Math. Anal. Appl. **296**, 521-537 (2004)
6. Liu, S, Zhang, J: Coexistence and stability of predator-prey model with Beddington-DeAngelis functional response and stage structure. J. Math. Anal. Appl. **342**, 446-460 (2008)
7. Zhao, M, Lv, S: Chaos in a three-species food chain model with a Beddington-DeAngelis functional response. Chaos Solitons Fractals **40**, 2305-2316 (2009)
8. Fan, M, Kuang, Y: Dynamics of a non-autonomous predator-prey system with the Beddington-DeAngelis functional response. J. Math. Anal. Appl. **295**, 15-39 (2004)
9. Hwang, TW: Global analysis of the predator-prey system with Beddington-DeAngelis functional response. J. Math. Anal. Appl. **281**, 395-401 (2003)
10. Hwang, TW: Uniqueness of limit cycles of the predator-prey system with Beddington-DeAngelis functional response. J. Math. Anal. Appl. **290**, 113-122 (2004)
11. Guo, G, Wu, J: Multiplicity and uniqueness of positive solutions for a predator-prey model with B-D functional response. Nonlinear Anal. **72**, 1632-1646 (2010)
12. Kuang, Y: Delay Differential Equations with Applications in Population Dynamics. Academic Press, New York (1993)
13. Macdonald, N: Biological Delay Systems: Linear Stability Theory. Cambridge University Press, Cambridge (1989)
14. Gopalsamy, K: Stability and Oscillations in Delay Differential Equations of Population Dynamics. Kluwer Academic, Boston (1992)
15. Fan, M, Wang, K: Existence and global attractivity of positive periodic solutions of periodic *n*-species Lotka-Volterra competition systems with several deviating arguments. Math. Biosci. **160**, 47-61 (1999)
16. Xu, R, Chaplain, M, Davidson, F: Periodic solutions for a delayed predator-prey model of prey dispersal in two-patch environments. Nonlinear Anal., Real World Appl. **5**, 183-206 (2004)
17. Egami, C, Hirano, N: Periodic solutions in a class of periodic delay predator-prey systems. Yokohama Math. J. **51**, 45-61 (2004)
18. Lu, S: On the existence of positive periodic solutions to a Lotka-Volterra cooperative population model with multiple delays. Nonlinear Anal. **68**, 1746-1753 (2008)

19. Lu, S: On the existence of positive periodic solutions for neutral functional differential equation with multiple deviating arguments. J. Math. Anal. Appl. **280**, 321-333 (2003)
20. Lu, S, Chen, L: The problem of existence of periodic solutions for neutral functional differential system with nonlinear difference operator. J. Math. Anal. Appl. **387**, 1127-1136 (2012)
21. Lu, S, Xu, Y, Xia, D: New properties of the *D*-operator and its applications on the problem of periodic solutions to neutral functional differential system. Nonlinear Anal. **74**, 3011-3021 (2011)
22. Ikeda, N, Watanabe, S: Stochastic Differential Equations and Diffusion Processes. Kluwer Academic, Dordrecht (1981)
23. Ding, X, Jiang, J: Positive periodic solutions in delayed Gause-type predator-prey systems. J. Math. Anal. Appl. **339**, 1220-1230 (2008)
24. Kan-on, Y, Mimura, M: Singular perturbation approach to a 3-component reaction-diffusion system arising in population dynamics. SIAM J. Math. Anal. **29**, 1519-1536 (1998)
25. Pang, P, Wang, M: Qualitative analysis of a ratio-dependent predator-prey system with diffusion. Proc. R. Soc. Edinb. A **133**, 919-942 (2003)
26. May, R: Stability and Complexity in Model Ecosystems. Princeton University Press, New York (1992)
27. Liu, M, Wang, K: Extinction and permanence in a stochastic nonautonomous population system. Appl. Math. Lett. **23**, 1464-1467 (2010)
28. Liu, M, Wang, K: Stationary distribution, ergodicity and extinction of a stochastic generalized logistic system. Appl. Math. Lett. **25**, 1980-1985 (2012)
29. Liu, M, Wang, K: Survival analysis of a stochastic cooperation system in a polluted environment. J. Biol. Syst. **19**, 183-204 (2013)
30. Jiang, D, Ji, C, Li, X, O'Regan, D: Analysis of autonomous Lotka-Volterra competition systems with random perturbation. J. Math. Anal. Appl. **390**, 582-595 (2012)
31. Ikeda, N, Watanabe, S: A comparison theorem for solutions of stochastic differential equations and its applications. Osaka J. Math. **14**, 619-633 (1977)
32. Huang, LC: Stochastic delay population systems. Appl. Anal. **88**, 1303-1320 (2009)
33. Li, X, Mao, X: Population dynamical behavior of non-autonomous Lotka-Volterra competitive system with random perturbation. Discrete Contin. Dyn. Syst. **24**, 523-545 (2009)
34. Mao, X, Marion, G, Renshaw, E: Environmental noise suppresses explosion in population dynamics. Stoch. Process. Appl. **97**, 95-110 (2002)
35. Liu, M, Qin, H, Wang, K: A remark on a stochastic predator-prey system with time delays. Appl. Math. Lett. **26**, 318-323 (2011)
36. Mao, X: Stochastic Differential Equations and Applications. Ellis Horwood, Chichester (1997)
37. Bahar, A, Mao, X: Stochastic delay Lotka-Volterra model. J. Math. Anal. Appl. **292**, 364-380 (2004)
38. Kloeden, PE, Shardlow, T: The Milstein scheme for stochastic delay differential equations without using anticipative calculus. Stoch. Anal. Appl. **30**, 181-202 (2012)

10

Approximate controllability result for nonlinear impulsive neutral fuzzy stochastic differential equations with nonlocal conditions

S Narayanamoorthy* and S Sowmiya

*Correspondence:
snm_phd@yahoo.co.in
Department of Applied Mathematics, Bharathiar University, Coimbatore, 641 046, India

Abstract

This paper deals with the approximate controllability of impulsive neutral fuzzy stochastic differential equations with nonlocal conditions in a Banach space by using the concept of fuzzy numbers whose values are normal, convex, upper semicontinuous and compact. The hypotheses are obtained by Schauder's and Banach fixed point theorems. The results are obtained by the evolution operator.

MSC: 34A07; 39A70; 93B05; 49N25

Keywords: approximate controllability; impulsive neutral fuzzy stochastic differential equations; fixed point theorems; nonlocal conditions

1 Introduction

In 1965, Zadeh initiated the development of the modified set theory known as fuzzy set theory, which is a tool that makes possible the description of vague notions and manipulations with them. The basic idea of fuzzy set theory is simple and natural. In [1] the authors studied the basic definitions and properties of fuzzy process, fuzzy number and so on. Stochastic differential equations are used to build more realistic models in economics, social sciences, chemistry, finance, physics and other areas. Therefore, many real world problems can be modeled by stochastic differential equations. The deterministic models often fluctuate due to noise, so we should move from deterministic control to stochastic control problems. In [2, 3] the basic properties and applications of stochastic differential systems and stochastic control problems were studied. The fuzzy stochastic differential equations could be applicable in the investigation of numerous engineering and economic problems where the phenomena are subjected to randomness and fuzziness simultaneously.

Neutral differential equations are one of the most thoroughly studied classes of equations with distributed arguments. They occur naturally in applied problems that contain in their statement some recurrence property and are discussed in [4–6]. Impulsive differential equations have been developed in modeling impulsive problems in physics, population dynamics, ecology, biological systems, biotechnology, industrial robotics, pharmacokinetics, optimal control, and so forth. Again associated with this development, a theory of impulsive differential equations has been given extensive attention. However, in addition to impulsive effects, stochastic effects also exist in real systems. Therefore, impulsive stochastic differential systems describing these dynamical systems are subject to both impulsive and stochastic changes. In [7, 8] these types of impulsive effects and differential systems were studied.

In the literature there are only a few papers that deal with the approximate controllability of fuzzy differential systems, likewise approximate controllability of backward stochastic evolution equations in Hilbert space as discussed in [9]. Dauer *et al.* [10] studied the approximate controllability of semilinear functional equations in a Hilbert space. Sakthivel *et al.* [11, 12] studied the approximate controllability of nonlinear impulsive differential systems and stochastic systems with unbounded delay. Zang and Li [13] discussed the concept of approximate controllability of fraction impulsive neutral stochastic differential equations with nonlocal conditions. Kwun and Park [2] presented the optimal control problem for fuzzy differential equations and nonlocal controllability for the semilinear fuzzy integrodifferential equations in an n-dimensional fuzzy vector space. It should be mentioned that there is no work done on the approximate controllability of fuzzy stochastic differential systems. Motivated by the above considerations, in this paper we investigate the approximate controllability for nonlinear impulsive neutral fuzzy stochastic differential equations with nonlocal conditions described by

$$d\big[x(t)-h\big(t,x(t)\big)\big] = A\big[x(t)-h\big(t,x(t)\big)\big] + Bu(t)\,dt + f\big(t,x(t)\big)\,dt + g\big(t,x(t)\big)\,dw(t), \quad t\in J=[0,a], t\neq t_i, \tag{1}$$

$$\Delta x(t_i) = x(t_i^+) - x(t_i^-)x \in X, \quad i=1,2,\ldots,m, \tag{2}$$

$$x(0)+\mu(x) = x_0. \tag{3}$$

Here, the state variable $x(\cdot)$ takes values in a real line separable Hilbert space X with the inner product $(\cdot,\cdot)$ and the norm $\|\cdot\|$. The control function $u(\cdot)$ takes values in $L^2(J,U)$, a Banach space of admissible control functions for a separable Hilbert space U and $J = [0,b]$, $\Omega = \{(t,s) : 0 \le s \le t \le b\}$. Also, A (*i.e.*, $A(t,x)$) is the infinitesimal generator of a C_0-semigroup in X and B is a bounded linear operator from U into X. Further, f and g are continuous and compact functions and $f : J \to X$ and $g : \Omega \to X$ are measurable mappings in X-norm, and a neutral variable h is continuous and compact; also $h : X \to X$ is a measurable mapping. And here, the function $\mu : PC(J,X) \to X$ is continuous and the impulsive function $I_i : X \to X$ is compact. Furthermore, the fixed time t_i satisfies $0 = t_0 < t_1 < t_2 < \cdots < t_m < a$, $x(t_i^+)$ and $x(t_i^+)$ denotes the right and left limits of $x(t)$ at $t = t_i$, and $\Delta x(t_i) = x(t_i^+) - x(t_i^-)$ represents the jump in the state X at time t_i, where I_i determines the size of the jump.

2 Preliminaries

Here, first we define some properties, theorems and lemmas and also recall some basic definitions which are all used in this paper.

A fuzzy set of R^n is a function $u : R^n \to [0,1]$. For each fuzzy set u, we denote by $[u]^\alpha = x \in R^n; u(x) \ge \alpha$ for any $\alpha \in [0,1]$ its α-level set.

Let E^n denote the collection of all fuzzy sets of R^n that satisfies the conditions, *i.e.*, u is normal, fuzzy convex, upper semicontinuous and $[u]^0$ is compact. We call $u \in E^n$ an n-dimensional fuzzy number.

And we define the abstract phase space B_h, also assume that $h:(-\infty,0]\to(0,\infty)$ is a continuous function with $l=\int_{-\infty}^{0}h(t)\,dt<+\infty=\{\Psi:[-b,0]\to X$ such that $\Psi(t)$ is bounded and measurable$\}$ and equip the space B with the norm

$$\|\Psi\|_{[-b,0]}=\sup_{s\in[-b,0]}|\Psi(s)|\quad\text{for all }\Psi\in B.$$

Let us define

$$B_h=\Big\{\Psi:(-\infty,0]\to X\text{ such that for any }c>0,\Psi|_{[-c,0]}\in B\text{ and}$$
$$\int_{-\infty}^{0}h(s)\|\Psi\|_{[c,0]}\,ds<\infty\Big\}.$$

If B_h is endowed with the norm

$$\|\Psi\|_{B_h}=\int_{-\infty}^{0}h(s)\|\Psi\|_{[s,0]}\,ds\quad\text{for all }\Psi\in B_h,$$

then it is clear that $(B_h,\|\cdot\|_{B_h})$ is a Banach space. Now we consider the space

$$B_h'=\big\{x:(-\infty,b]\to X\text{ such that }x_k\in PC(J_k,X)\text{ and there exist }x(t_k^+)\text{ and}$$
$$x(t_k^-)\text{ with }x(t_k)=x(t_k^-),x_0=\phi\in B_h,k=0,1,\dots,m\big\},$$

where x_k is the restriction of x to $J_k=(t_k,t_k+1]$, $k=0,1,\dots,m$. Set $\|\cdot\|_h'$ to be a seminorm in B_h' defined by

$$\|x\|_h'=\|\phi\|_{B_h}+\sup|x(s)|:s\in[0,b],\quad x\in B_h'.$$

Definition 2.1 The family $A(t):0\le t\le b$ generates a unique linear evolution system $U(t,s):0\le s\le t\le b$ satisfying the following properties:

- $U(t,s)U(s,\tau)=U(t,\tau)$ and $U(t,t)x=x$ for every $s\le\tau\le t$ and all $x\in X$,
- for every $x\in X$, the function for $(t,s)=U(t,s)x$ is continuous and $U(t,s)\in L(X)$ for every $t\ge s$, and
- for $0\le s\le t\le b$, the function $t\to U(t,s)$, $(s,t]\in L(X)$ is differentiable with $\frac{\partial U(t,s)}{\partial t}=A(t)U(t,s)$.

Definition 2.2 Let $x,y\in C(I,E_N)$, here I is a real interval and E_N is the set of all upper semicontinuous convex normal fuzzy numbers with bounded α-level intervals. A mapping $x:I\to E_N$ is called a fuzzy process. We define

$$[x(t)]^\alpha=[x_l^\alpha(t),x_r^\alpha(t)],\quad t\in I,0<\alpha\le1.$$

The derivative $x'(t)$ of a fuzzy process x is defined by

$$[x'(t)]^\alpha=\big[(x_l^\alpha)'(t),(x_r^\alpha(t))'\big],\quad t\in I,0<\alpha\le1,$$

provided that this equation defines fuzzy $x'(t)\in E_N$.

Definition 2.3 A stochastic process x is said to be a mild solution of (1)-(3) if the following conditions are satisfied:

- $X(t,\omega)$ is a measurable function from $J \times \Omega$ to X and $x(t)$ is F_t-adapted,
- $E\|x(t)\|^2 < \infty$ for all $t \in J$,
- $\Delta x(\tau_i) = x(\tau_i^+) - x(\tau_i^-) = I_i(x(\tau_i))$, $x \in X$ and $1 \le i \le m$,
- for each $u \in L_2^F(J,U)$, the process x satisfies the following integral equation:

$$x(t) = U(t,0)\big[x' - \mu(x)\big] + \int_0^t U(t,s)Bu(s)\,ds + \int_0^t U(t,s)f\big(s,x(s)\big)\,ds$$
$$+ \int_0^t U(t,s)g\big(s,x(s)\big)\,dW(s) + \sum_{0 \le t_i \le t} U(t,t_i)I\big(x(t_i^-)\big), \quad t \in J. \tag{4}$$

Lemma 2.4 *Assume that $x \in B_h'$, then for $t \in I$, $x_t \in B_h$. Moreover,*

$$l\big|x(t)\big| \le \|x_t\|_{B_h} \le \|x_0\|_{B_h} + l \sup_{s\in[0,t]} \big\|x(s)\big\|,$$

where $l = \int_{-\infty}^0 h(t)\,dt < \infty$.

Note 2.5 Define $B_h'' = z \in B_h' : z_0 = 0 \in B_h$. For any $z \in B_h''$,

$$\|z\|_{B_h}' = \|z_0\|_{B_h} + \sup\big\{\big|z(s)\big| : s \in [0,b]\big\} = \sup\big\{\big|z(s)\big| : s \in [0,b]\big\},$$

and thus $(B_h'', \|\cdot\|)$ is a Banach space. Set

$$B_r = z \in PC\big((-\infty,b],X\big) : \big\|z(t)\big\|_{B_h}' \le r, \quad 0 \le t \le b.$$

Clearly, B_r is a nonempty, bounded, convex and closed set in B_h'. Then, for any $z \in B_r$, from Lemma 2.4 we have

$$\begin{aligned}
\|z_t + y_t\|_{B_h} &\le \|z_t\|_{B_h} + \|y_t\|_{B_h} \\
&\le \|z_0\|_{B_h} + l \sup_{s\in[0,b]} \big|z(s)\big| + \|y_0\|_{B_h} + l \sup_{s\in[0,b]} \big|y(s)\big| \\
&\le l\big(r + M_1\big|\phi(0)\big|\big) + \|\phi\|_{B_h} = r'.
\end{aligned} \tag{5}$$

For each $t \in J$, $z \in B_r$, we have by the above equation and (A_2)

$$\sup_{t\in J}\big|z(t) + y(t)\big| \le l^{-1}\|z_t + y_t\|_{B_h} \le l^{-1}r',$$
$$\big|I_k\big(z(t_k^-) + z(t_k^-)\big)\big| \le d_k\Big(\sup_{t\in J}\big|z(t) + y(t)\big|\Big) \le d_k\big(l^{-1}r'\big), \quad k = 1,2,\dots,m. \tag{6}$$

Let $x_b(x_0;u)$ be the state value of (1)-(3) at terminal time b corresponding to the control u and the initial value $x_0 = \phi \in B_h$.

Introduce the set $R(b,x_0) = x_b(x_0;u)(0) : u(\cdot) \in L^2(J,U)$, which is called the reachable set at terminal time b, its closure in X is denoted by $\overline{R(b,x_0)}$.

Definition 2.6 If $\overline{R(b,x_0)} = X$, then system (1)-(3) is approximately controllable on the interval J. It is convenient at this point to define operators

$$\Gamma_0^a = \int_0^a U(a,s)BB^*U^*(a,s)\,ds,$$

$$R(\alpha,\Gamma_0^a) = (\alpha I + \Gamma_0^a)^{-1}.$$

(D) $\alpha R(\alpha,\Gamma_0^a) \to 0$ as $\alpha \to 0^+$ is strong operator topology.
It is known that assumption (D) holds iff the linear system $\dot{x}(t) = Ax(t) + Bu(t)$, $x(0) = x_0$ is approximately controllable on J.

The following theorem gives a formula for a control transferring the initial state x_0 to some neighborhood of x_a at time a.

Theorem 2.7 *For arbitrary $x_a \in X$, the control*

$$u(t) = B^*U^*(a,t)R\big(\alpha,\Gamma_0^a\big)p\big(x(\cdot)\big),$$

where

$$\begin{aligned} p\big(x(\cdot)\big) &= x_a - U(a,0)\big[x_0 - \mu(x) - h\big(0,(x_0-\mu(x))\big)\big] - h(a,x_a) \\ &\quad - \int_0^a U(a,s)A(s)h(s,x_s)\,ds - \int_0^a U(a,s)f(s,x_s)\,ds \\ &\quad - \int_0^a U(a,s)g(s,x_s)\,dW(s) - \sum_{k=1}^m U(a,t_k)(I_k)\big(x(t_k^-)\big) \end{aligned} \tag{7}$$

transfers the initial state x_0 to

$$\begin{aligned} x(a) &= x_a - \alpha\big(\alpha I + \Gamma_0^a\big)^{-1}\Bigg[x_a - U(a,0)\big[x_0 - \mu(x) - h\big(0,(x_0-\mu(x))\big)\big] - h(a,x_a) \\ &\quad - \int_0^a U(a,s)A(s)h(s,x_s)\,ds - \int_0^a U(a,s)f(s,x_s)\,ds - \int_0^a U(a,s)g(s,x_s)\,dW(S) \\ &\quad - \sum_{k=1}^m U(a,t_k)(I_k)\big(x(t_k^-)\big)\Bigg]. \end{aligned} \tag{8}$$

3 Approximate controllability

In this section, we state the approximate controllability result for impulsive neutral fuzzy stochastic differential equations under nonlocal conditions by using Schauder's fixed point theorem. Approximate controllability property is the possibility to steer the state of the system from any initial data to a state of arbitrary close to a target by choosing a suitable control.

To establish the result, we introduce the following assumptions on system (1)-(3).

(A_1) When $t > s > 0$, then the strongly continuous semigroup of bounded linear operators $U(t,s)$ generated by $A(t)$ is compact and there exist constants $M_1 > 0$, $M_2 > 0$ such that

$$\|[U(t,s)]^\alpha\| \le M_1^\alpha, \qquad \|[U(t,s)]^\alpha\| \le M_2^\alpha.$$

(A_2) (i) The nonlinear function $f : J \times B_h \times X \to X$ is continuous, uniformly bounded, strongly measurable and there exists $L_f > 0$ such that

$$\|[f(t,\phi,x)]^\alpha\| \le L_f^\alpha \quad \text{for all } (t,\phi,x) \in J \times B_h \times X.$$

(ii) For each positive number r, there exists a function $\delta_r \in L^1(J, R_+)$ such that

$$\sup\{[f(t,\phi,x)]^\alpha : \|\phi\|_{B_h} \le r, \|[x]\| \le r\} \le \delta_r(t) \quad \text{for a.e. } t \in J \text{ and}$$

$$\liminf_{r\to\infty} \frac{1}{r}\int_0^a \delta_r(s)\,ds = \gamma < \infty.$$

(A_3) The nonlinear function $g : J \times B_h \times X \to X$ is continuous, strongly measurable and there exists $L_g > 0$ such that

$$\|[g(t,\phi,x)]^\alpha\| \le L_g^\alpha \quad \text{for all } (t,\phi,x) \in J \times B_h \times X.$$

(A_4) The function $\mu : PC(J,X) \to X$ is a nonlocal condition in the following sense, and there exists a constant $L_\mu > 0$ such that

$$\|[\mu(t,u)]^\alpha - [\mu(s,v)]^\alpha\| \le L_\mu^\alpha[|t-s| - \|u-v\|] \quad \text{for all } u,v \in X.$$

(A_5) $I_k \in C(X,X)$ and there exists a constant d_k^α such that

$$|[I_k(x)]^\alpha| \le d_k^\alpha(|x|), \quad k = 1,2,\ldots,m,$$

for each $x \in X$, and

$$\liminf \frac{d_k(\rho)}{\rho} = \lambda_k < \infty, \quad k = 1,2,\ldots,m.$$

(A_6) The function $h : J \times B_h \to X$ is continuous, and there exists $L_h > 0$ such that

$$\|[h(t,\phi_1)]^\alpha - [h(s,\phi_2)]^\alpha\| < L_h^\alpha[|t-s| + \|\phi_1 - \phi_2\|_{B_h}] \quad \text{for every } t,s \in J \text{ and } \phi_1,\phi_2 \in B_h.$$

The following notations are introduced for convenience:

$$Q' = \frac{L_1^\alpha q'}{M_1^\alpha} + aM_1^\alpha N_1^\alpha q' + M_1^\alpha \int_0^a \lambda_q^\alpha(s)\,ds + M_1^\alpha \int_0^a \lambda_q^\alpha(s)\,dW(s) + M_1^\alpha \sum_{k=1}^m d_k^\alpha(q').$$

Theorem 3.1 *Assume that conditions* (A_1)-(A_6) *are satisfied. Further, suppose that for all* $a > 0$,

$$\left(1 + \frac{1}{\beta}a[M_1^2]^\alpha[M_A^2]^\alpha\right)M_1 Q' < 1, \tag{9}$$

then system (1)-(3) *has a solution on* J.

Proof Now, define the operator $\Re$ on $PC((-\infty,b],X)$, $(\Re x)(t)=\phi(t)$, $t\in(-\infty,0]$.

From the theorem and definition, let

$$\begin{aligned}
[u(t)]^{\alpha} &= [u_l^{\alpha}(t), u_r^{\alpha}(t)] \\
&= \Bigg[B^{*}U^{*}(a,t)(\alpha I+\Gamma_0^{a})^{-1}\Bigg(x_a - U_l^{\alpha}(a,0)[x_0-\mu(x)-h_l^{\alpha}(0,(x_0-\mu(x)))] \\
&\quad - h_l^{\alpha}(a,x(a)) - \int_0^a U_l^{\alpha}(a,s)A(s)h_l^{\alpha}(s,x(s))\,ds - \int_0^a U_l^{\alpha}(a,s)f_l^{\alpha}(s,x(s))\,ds \\
&\quad - \int_0^a U_l^{\alpha}(a,s)g_l^{\alpha}(s,x(s))\,dW(S) - \sum_{k=1}^{m} U_l^{\alpha}(I_k)_l^{\alpha}(x(t_k^{-}))\Bigg), B^{*}U^{*}(a,t) \\
&\quad \times (\alpha I+\Gamma_0^{a})^{-1}\Bigg(x_a - U_r^{\alpha}(a,0)[x_0-\mu(x)-h_r^{\alpha}(0,(x_0-\mu(x)))] - h_r^{\alpha}(a,x(a)) \\
&\quad - \int_0^a U_r^{\alpha}(a,s)A(s)h_r^{\alpha}(s,x(s))\,ds - \int_0^a U_r^{\alpha}(a,s)f_r^{\alpha}(s,x(s))\,ds \\
&\quad - \int_0^a U_r^{\alpha}(a,s)g_r^{\alpha}(s,x(s))\,dW(S) - \sum_{k=1}^{m} U_r^{\alpha}(I_k)_r^{\alpha}(x(t_k^{-}))\Bigg)\Bigg]. \qquad (10)
\end{aligned}$$

Take the function $y(\cdot):(-\infty,a)\to X$ defined by

$$y(t)=\{U(t,0)X(0)t\geq 0, X(0)-\infty<t<0\},$$

then $y_0=\phi$, and the map $t\to y_t$ is continuous. If $[x(t)]^{\alpha}$ is a fuzzy solution of (4), then we can write it as $[x(t)]^{\alpha}=[y(t)]^{\alpha}+[z(t)]^{\alpha}$, $-\infty<t\leq a$.

From the above we describe $Z_0=0$ and the function $[z(t)]^{\alpha}$ that satisfies

$$\begin{aligned}
[z(t)]^{\alpha} &= \Bigg[-U(t,0)h(0,[x_0-\mu(x)]) + h(t,z_t+y_t) + \int_0^t U(t,s)A(s)h(s,z_s+y_s)\,ds \\
&\quad + \int_0^t U(t,\eta)BB^{*}U^{*}(a,s)(\alpha I+\Gamma_0^{a})^{-1} \\
&\quad \times \Bigg(x_a - U(a,0)[x_0-\mu(x)-h(0,(x_0-\mu(x)))] \\
&\quad - h(a,z_a+y_a) - \int_0^a U(a,s)A(s)h(s,z_s+y_s)\,ds - \int_0^a U(a,s)f(s,Z_s+y_s)\,ds \\
&\quad - \int_0^a U(a,s)g(s,z_s+y_s)\,dW(s) - \sum_{k=1}^{m} U(I_k)(z(t_k^{-})+y(t_k^{-}))\Bigg)\eta\,d\eta \\
&\quad + \int_0^t U(t,s)f(s,z_s+y_s)\,ds + \int_0^t U(t,s)g(s,z_s+y_s)\,dW(s) \\
&\quad + \sum_{0<t_k<t} U(t,t_k)I_k(z(t_k^{-})+y(t_k^{-}))\Bigg]^{\alpha}
\end{aligned}$$

iff x satisfies

$$[x(t)]^{\alpha} = \Bigg[-U(t,0)h(0,[x_0-\mu(x)]) + h(t,x_t) + \int_0^t U(t,s)A(s)h(s,x_s)\,ds$$

$$+ \int_0^t U(t,\eta)BB^*U^*(a,s)(\alpha I + \Gamma_0^a)^{-1}$$

$$\times \Bigg(x_a - U(a,0)[x_0 - \mu(x) - h(0,x_0-\mu(x))]$$

$$- h(a,x_a) - \int_0^a U(a,s)A(s)h(s,x_s)\,ds - \int_0^a U(a,s)f(s,x_s)\,ds$$

$$- \int_0^a U(a,s)g(s,x_s)\,dW(s) - \sum_{k=1}^{m} U(I_k)(x(t_k^-)) \Bigg) \eta\, d\eta$$

$$+ \int_0^t U(t,s)f(s,x_s)\,ds + \int_0^t U(t,s)g(s,x_s)\,dW(s) + \sum_{0<t_k<t} U(t,t_k)I_k(x(t_k^-)) \Bigg]^{\alpha}$$

and

$$[x(t)]^{\alpha} = \phi^{\alpha}(t), \quad \text{for all } t \in (-\infty, 0].$$

Let us take the operator $\Psi : B_h'' \to B_h''$ defined by

$$[(\Psi z)(t)]^{\alpha} = 0, \quad t \in (-\infty, 0],$$
$$[(\Psi z)(t)]^{\alpha} = [z(t)]^{\alpha}, \quad t \in J.$$

Obviously, the operator $\Re$ has a fixed point iff Ψ has a fixed point. So we want to prove that Ψ has a fixed point and the proof has the following steps.

Step 1: $\Psi(B_q) \subseteq B_q$ for some $q > 0$.

We claim that there exists a positive constant q such that

$$\Psi(B_q) \subseteq B_q.$$

If this condition is false, then for every positive number q, there exists a function $z^q(\cdot) \in B_q$, but $\Psi(z^q)$ does not belong to B_q, *i.e.*, $|(\psi z^q)(t)| > q$ for some $t \in J$. From conditions (A_1)-(A_6) and

$$q < \left| [(\Psi z^q)(t)]^{\alpha} \right|$$

$$\leq M_1^{\alpha} \left\| h^{\alpha}(0, x_0 - \mu(x)) \right\| + \left\| h^{\alpha}(t, z_t^q + y_t^q) \right\| + M_1^{\alpha} \int_0^t \left\| A(s)h^{\alpha}(s, z_s^q + y_s^q) \right\|$$

$$+ \int_0^t \frac{1}{\beta} [M_1^2]^{\alpha} [M_A^2]^{\alpha} \Bigg[\|x_a\| + M_1^{\alpha} \left\| [x_0 - \mu(x)]^{\alpha} \right\| + M_1^{\alpha} \left\| h^{\alpha}(0, [x_0 - \mu(x)]) \right\|$$

$$+ \left\| h^{\alpha}(a, z_a^q + y_a^q) \right\| + M_1^{\alpha} \int_0^a \left\| A(s)h^{\alpha}(s, z_s^q + y_s^q) \right\| ds + M_1^{\alpha} \int_0^a \left\| f^{\alpha}(s, z_s^q + y_s^q) \right\| ds$$

$$+ M_1^\alpha \int_0^a \|g^\alpha(s, z_s^q + y_s^q)\| dW(s) + M_1^\alpha \sum_{k=1}^m \|I^\alpha - k(z^q(t_k^-) + y^q(t_k^-))\| \Big]$$
$$+ M_1^\alpha \int_0^t \|f^\alpha(s, z_s^q + y_s^q)\| ds + M_1^\alpha \int_0^t \|g^\alpha(s, z_s^q + y_s^q)\| dW(s)$$
$$+ M_1^\alpha \sum_{0<t_k<t} \|I_k^\alpha(z^q(t_k^-) + y^q(t_k^-))\|.$$

Since

$$\int_0^t \|f^\alpha(s, z_s^q + y_s^q)\| ds \le \int_0^a \lambda_q^\alpha(t)\, dt,$$

similarly

$$\int_0^t \|g^\alpha(s, z_s^q + y_s^q)\| dW(s) \le \int_0^a \lambda_q^\alpha(t)\, dW(t).$$

From conditions (A_2), (A_3) and Note 2.5, then

$$q < M_1^\alpha \|h^\alpha(0, x_0 - \mu(x))\| + L_1^\alpha q' + L_2^\alpha + aM_1^\alpha N_1^\alpha q' + N_2^\alpha$$
$$+ \frac{1}{\beta}[M_1^2]^\alpha [M_A^2]^\alpha \Big[\|x_a\| + M_1^\alpha \|[x_0 - \mu(x)]^\alpha\| + M_1^\alpha \|h^\alpha(0, [x_0 - \mu(x)])\|$$
$$+ L_1^\alpha q' + L_2^\alpha + aM_1^\alpha N_1^\alpha q' + N_2^\alpha + M_1^\alpha \int_0^a \lambda_q^\alpha(s)\, ds + M_1^\alpha \int_0^a \lambda_q^\alpha(s)\, dW(s)$$
$$+ M_1^\alpha \sum_{k=1}^m d_k^\alpha(q') \Big] + M_1^\alpha \int_0^a \lambda_q^\alpha(s)\, ds + M_1^\alpha \int_0^a \lambda_q^\alpha(s)\, dW(s) + M_1^\alpha \sum_{k=1}^m d_k^\alpha(q')$$
$$\le Q + \left(1 + \frac{1}{\beta}[M_1^2]^\alpha [M_A^2]^\alpha\right) \Big[L_1^\alpha q' + aM_1^\alpha N_1^\alpha q' + M_1^\alpha \int_0^a \lambda_q^\alpha(s)\, ds$$
$$+ M_1^\alpha \int_0^a \lambda_q^\alpha(s)\, dW(s) + M_1^\alpha \sum_{k=1}^m d^\alpha - k(q') \Big],$$

where Q is independent of q. Dividing both sides by q, we obtain

$$\lim_{q\to+\infty} \inf \sum_{k=1}^m \frac{d^\alpha - k(l^{-1}q')}{q} = \lim_{q\to+\infty} \inf \sum_{k=1}^m \frac{d_k^\alpha(l^{-1}q')}{(l^{-1}q')} \frac{l^{-1}q'}{q},$$
$$\lim_{q\to+\infty} \inf \frac{1}{q} \int_0^a \lambda_q^\alpha(s)\, ds = \lim_{q\to+\infty} \inf \frac{\int_0^a \lambda_q^\alpha(s)\, ds}{q} = (1 + L^0)\Gamma l.$$

Thus we have

$$\left(1 + \frac{1}{\beta} a[M_1^2]^\alpha [M_A^2]^\alpha\right) M_1 Q' \ge 1.$$

This contradicts (9). Hence for $q > 0$, $\Psi(B_q) \subseteq B_q$.

Step 2: Ψ maps B_q into a precompact set in X.

Now we want to prove that $V(t) = (\Psi z)(t) : z \in B_q$ is relatively compact in X. The case $t = 0$ is true since $V(0) = \Psi(0)$. Let $0 < t \leq a$ be fixed, $0 < \epsilon < t$, for $z \in B_q$, we define

$$
\begin{aligned}
\left[(\Psi_\epsilon z)(t)\right]^\alpha = \Bigg[& -U(t,0)h\big(0,[x_0 - \mu(x)]\big) + h(t, z_t + y_t) \\
& + U(t, t-\epsilon)\int_0^{t-\epsilon} U(t-\epsilon, s)A(s)h(s, z_s + y_s)\,ds \\
& + U(t, t-\epsilon)\int_0^{t-\epsilon} U(t-\epsilon, \eta)BB^* U^*(t-\epsilon, \eta)\big(\alpha I + \Gamma_0^a\big)^{-1} \\
& \times \Bigg(x_a - U(a,0)\big[x_0 - \mu(x) - h\big(0, x_0 - \mu(x)\big)\big] - h(a, z_a + y_a) \\
& - \int_0^a U(a,s)A(s)h(s, z_s + y_s)\,ds - \int_0^a U(a,s)f(s, z_s + y_s)\,ds \\
& - \int_0^a U(a,s)g(s, z_s + y_s)\,dW(s) - \sum_{k=1}^m U(I_k)\big(z(t_k^-) + y(t_k^-)\big) \Bigg)\eta\, d\eta \\
& + U(t, t-\epsilon)\int_0^{t-\epsilon} U(t-\epsilon, s)f(s, z_s + y_s)\,ds \\
& + U(t, t-\epsilon)\int_0^{t-\epsilon} U(t-\epsilon, s)g(s, z_s + y_s)\,dW(s) \\
& + \sum_{0<t_k<t} U(t, t_k)I_k\big(z(t_k^-) + y(t_k^-)\big) \Bigg]^\alpha .
\end{aligned}
$$

Set $V_\epsilon(t) = (\Psi_\epsilon z)(t) : z \in B_q$ is relatively compact in X for every ϵ, $0 < \epsilon < t$. That is, a finite set $\tilde{y}_j$, $1 \leq j \leq m$ in X exists such that

$$V_\epsilon(t) \subset \bigcup_{j=0}^m \tilde{L}(\tilde{y}_j, \eta/2),$$

where $\tilde{L}(\tilde{y}_j, \eta/2)$ is an open ball in X with radius $\eta/2$ and center at $\tilde{y}_j$.

$$
\begin{aligned}
& \left\| \left[(\Psi z)(t)\right]^\alpha - \left[(\Psi_\epsilon z)(t)\right]^\alpha \right\| \\
& \leq \int_{t-\epsilon}^t \left\| U(t,s)A(s)h(s, z_s + y_s) \right\| ds \\
& + \int_{t-\epsilon}^t U(t,\eta)BB^* U^*(t,\eta)\big(\alpha I + \Gamma_0^a\big)^{-1}\Big[x_a - U(a,0)\big[\big(x_0 - \mu(x)\big) \\
& - h\big(0, x_0 - \mu(x)\big)\big] - h(a, z_a + y_a) - \int_0^a U(a,s)A(s)h(s, z_s + y_s)\,ds \\
& - \int_0^a U(a,s)f(s, Z_s + y_s)\,ds - \int_0^a U(a,s)g(s, z_s + y_s)\,dW(s) \\
& - \sum_{k=1}^m U(a, t_k)I_k\big(z(t_k^-) + y(t_k^-)\big)\Big]\eta\, d\eta + \int_{t-\epsilon}^t \left\| U(t,s)f(s, z_s + y_s) \right\| ds
\end{aligned}
$$

$$+\int_{t-\epsilon}^{t}\|U(t,s)g(s,z_s+y_s)\|dW(s)$$
$$\leq M_1^\alpha\int_{t-\epsilon}^{t}\left[N_1^\alpha q'+N_2^\alpha\right]ds+\frac{1}{\beta}\left[M_1^2\right]^\alpha\left[M_A^2\right]^\alpha$$
$$\times\int_{t-\epsilon}^{t}\Bigg[\|x_a\|+M_1^\alpha\|x_0-\mu(x)\|+M_1^\alpha\|h^\alpha(0,x_0-\mu(x))\|$$
$$+L_1^\alpha q'+L_2^\alpha+aM_1^\alpha\left[N_1^\alpha q'+N_2^\alpha\right]+M_1^\alpha\int_0^a\lambda_q^\alpha(\tau)\,d\tau+M_1^\alpha\int_0^a\lambda_q^\alpha(\tau)\,dW(\tau)$$
$$+M_1^\alpha\sum_{k=1}^{m}d_k^\alpha\left(l^{-1}q'\right)\Bigg]ds+M_1^\alpha\int_0^a\lambda_q^\alpha(s)\,ds+M_1^\alpha\int_0^a\lambda_q^\alpha(s)\,dW(s)$$
$$\leq\eta/2.$$

Similarly,

$$V(t)\subset\bigcup_{j=0}^{m}\tilde{L}(\tilde{y}_j,\eta).$$

Hence, for every $t\in[0,a]$, the set $V(t)$ is relatively compact in X.

Step 3: Ψ maps B_q into an equicontinuous family.

We claim that $\Psi(B_q)=\Psi z:z\in B_q$ is a family of equicontinuous functions. Let $t_1,t_2\in[0,a]$. By (A_2)-(A_6), we get

$$\left|\left[(\Psi z)(t_1)\right]^\alpha-\left[(\Psi z)(t_2)\right]^\alpha\right|$$
$$\leq\left\|\left[U^\alpha(t_1,0)-U^\alpha(t_2,0)\right]h^\alpha\left(0,\left[x_0-\mu(x)\right]\right)\right\|+\left\|h^\alpha(t_1,z_{t_1}+y_{t_1})\right.$$
$$\left.-h^\alpha(t_2,z_{t_2}+y_{t_2})\right\|+\left\|\int_0^{t_1}\left[U^\alpha(t_1,s)-U^\alpha(t_2,s)\right]A(s)h^\alpha(s,z_s+y_s)\,ds\right\|$$
$$+\left\|\int_{t_1}^{t_2}U^\alpha(t_2,s)A(s)h^\alpha(s,z_s+y_s)\,ds\right\|+\left\|\int_0^{t_1}\left[U^\alpha(t_1,\eta)-U^\alpha(t_2,\eta)\right]BU(s)\,ds\right\|$$
$$+\left\|\int_{t_1}^{t_2}U^\alpha(t_2,\eta)BU(s)\,ds\right\|+\left\|\int_0^{t_1}\left[U^\alpha(t_1,s)-U^\alpha(t_2,s)\right]f^\alpha(s,Z_s+y_s)\,ds\right\|$$
$$+\left\|\int_{t_1}^{t_2}U^\alpha(t_2,s)f^\alpha(s,Z_s+y_s)\,ds\right\|+\left\|\int_0^{t_1}\left[U^\alpha(t_1,s)-U^\alpha(t_2,s)\right]g^\alpha(s,Z_s+y_s)\,ds\right\|$$
$$+\left\|\int_{t_1}^{t_2}U^\alpha(t_2,s)g^\alpha(s,Z_s+y_s)\,dW(s)\right\|$$
$$+\left\|\sum_{0<t_k<t}\left[U^\alpha(t_2,t_k)-U^\alpha(t_1,t_k)\right]I_k\left(z(t_k^-)+y(t_k^-)\right)\right\|$$
$$+\left\|\sum_{t_1<t_k<t_2}U^\alpha(t_2,t_k)I_k\left(z(t_k^-)+y(t_k^-)\right)\right\|$$
$$\leq\left\|\left[U^\alpha(t_1,0)-U^\alpha(t_2,0)\right]\right\|\left|h^\alpha\left(0,\left[x_0-\mu(x)\right]\right)\right|$$
$$+L_h^\alpha\left[|t_1-t_2|+\|z_{t_1}-z_{t_2}\|_{B_h}+\|y_{t_1}-y_{t_2}\|_{B_h}\right]$$
$$+\int_0^{t_1}\left\|U^\alpha(t_1,s)-U^\alpha(t_2,s)\right\|\left[N_1^\alpha\|z_s+y_s\|+N_2^\alpha\right]ds$$

$$+\int_{t_1}^{t_2}\|U^\alpha(t_2,s)\|\left[N_1^\alpha\|z_s+y_s\|_{B_h}+N_2^\alpha\right]ds$$

$$+\frac{1}{\beta}[M_1^2]^\alpha[M_A^2]^\alpha\int_0^{t_1}\|U^\alpha(t_1,\eta)-U^\alpha(t_2,\eta)\|\Bigg[\|x_a\|+M_1^\alpha\big|[x_0-\mu(x)]\big|$$

$$+\big|h^\alpha\big(0,[x_0-\mu(x)]\big)\big|+\left[L_1^\alpha\|z_a+y_a\|_{B_h}+L_2^\alpha\right]+M_1^\alpha\int_0^a\left[N_1^\alpha\|z_s+y_s\|_{B_h}+N_2^\alpha\right]ds$$

$$+M_1^\alpha\left(\int_0^a\lambda_q^\alpha(s)\,ds+\int_0^a\lambda_q^\alpha(s)\,dW(s)+\sum_{k=1}^m d_k^\alpha(l^{-1}q')\right)\Bigg]d\eta$$

$$+\frac{1}{\beta}[M_1^2]^\alpha[M_A^2]^\alpha\int_{t_1}^{t_2}\|U^\alpha(t_2,\eta)\|$$

$$\times\Bigg[\|x_a\|+M_1^\alpha\big|[x_0-\mu(x)]\big|+\big|h^\alpha\big(0,[x_0-\mu(x)]\big)\big|+\left[L_1^\alpha\|z_a+y_a\|_{B_h}+L_2^\alpha\right]$$

$$+M_1^\alpha\int_0^a\left[N_1^\alpha\|z_s+y_s\|_{B_h}+N_2^\alpha\right]ds+M_1^\alpha\int_0^a\lambda_q^\alpha(s)\,ds+M_1^\alpha\int_0^a\lambda_q^\alpha(s)\,dW(s)$$

$$+M_1^\alpha\sum_{k=1}^m d_k^\alpha(l^{-1}q')\Bigg]d\eta+\int_0^{t_1}\|U^\alpha(t_1,s)-U^\alpha(t_2,s)\|\lambda_q^\alpha(s)\,ds$$

$$+\int_{t_1}^{t_2}\|U^\alpha(t_2,s)\|\lambda_q^\alpha(s)\,ds$$

$$+\int_0^{t_1}\|U^\alpha(t_1,s)-U^\alpha(t_2,s)\|\lambda_q^\alpha(s)\,dW(s)+\int_{t_1}^{t_2}\|U^\alpha(t_2,s)\|\lambda_q^\alpha(s)\,dW(s)$$

$$+\sum_{0<t_k<t}\|U^\alpha(t_1,t_k)-U^\alpha(t_2,t_k)\|d_k^\alpha(l^{-1}q')+M_1^\alpha\sum_{t_1<t_k<t_2}d_k^\alpha(l^{-1}q').$$

This implies that $|[(\Psi z)(t_1)]^\alpha-[(\Psi z)(t_2)]^\alpha|\to 0$ as $t_2-t_1\to 0$, since the compactness of $U^\alpha(t,s)$ is the continuity of $U^\alpha(t,s)$ in the uniform operator topology. Thus Ψ maps B_q into an equicontinuous family of functions.

Step 4: $B_h''\to B_h''$ is continuous.

First we show that Ψ is continuous in B_h''. Let $[z_t^{(n)}(t)]^\alpha{}_0^\infty\subseteq B_h''$ with $z_n\to z$ in B_h''. Then there exists a number $q>0$ such that $|[z^{(n)}(t)]^\alpha|\le r$ for all n and a.e. $t\in J$, so $z^{(n)}\in B_q$ and $z\in B_q$. By note we have

$$\left\|z_t^{(n)}+y_t\right\|\le q',\quad t\in J.$$

From conditions (A_2)-(A_6), we have

(1) $I_k,k=1,2,\ldots,m$, is continuous.

(2) $h^\alpha(t,z_t^{(n)}+y_t)\to h^\alpha(t,z_t+y_t)$ for each $t\in J$ and since $\|h^\alpha(t,z_t^{(n)}+y_t)-h^\alpha(t,z_t+y_t)\|<2[L_1^\alpha\|y\|+L_2^\alpha]$.

(3) $f^\alpha(s,z_s^{(n)}+y_s)\to f^\alpha(s,z_s+y_s)$ for each $s\in J$ and since $\|f^\alpha(s,z_s^{(n)}+y_s)-f^\alpha(s,z_s+y_s)\|<2\lambda_q^\alpha(s)$.

From the dominated convergence theorem, we obtain

$$\left|[(\Psi z)^{(n)}]^\alpha-[(\Psi z)]^\alpha\right|_{B_h}$$

$$\le\left\|h^\alpha\big(t,z_t^{(n)}+y_t\big)-h^\alpha(t,z_t+y_t)\right\|+M_1^\alpha\int_0^t\|A(s)h^\alpha\big(s,z_s^{(n)}+y_s\big)$$

$$
\begin{aligned}
&- A(s)h^{\alpha}(s, z_s + y_s)\| \, ds + \frac{1}{\beta}[M_1^2]^{\alpha}[M_A^2]^{\alpha} \int_0^a \Bigg[\|h^{\alpha}(a, z_a^{(n)} + y_a) \\
&- h^{\alpha}(a, z_a + y_a) \, ds\| + M_1^{\alpha} \int_0^a \|A(s)h^{\alpha}(a, z_a^{(n)} + y_a) \\
&- A(s)h^{\alpha}(a, z_a + y_a)\| \, ds + M_1^{\alpha} \int_0^a \|f^{\alpha}(s, z_s^{(n)} + y_s) - f^{\alpha}(s, z_s + y_s)\| \, ds \\
&+ M_1^{\alpha} \int_0^a \|g^{\alpha}(s, z_s^{(n)} + y_s) - g^{\alpha}(s, z_s + y_s)\| dW(s) \\
&+ M_1^{\alpha} \sum_{k=1}^{m} \|I_k^{\alpha}(z^{(n)}(t_k^-) + y(t_k^-) - I_k^{\alpha}(z(t_k^-) + y(t_k^-))\| \Bigg] d\eta \\
&+ M_1^{\alpha} \int_0^t \|f^{\alpha}(s, z_s^{(n)} + y_s) - f^{\alpha}(s, z_s + y_s)\| \, ds \\
&+ M_1^{\alpha} \int_0^t \|g^{\alpha}(s, z_s^{(n)} + y_s) - g^{\alpha}(s, z_s + y_s)\| dW(s) \\
&+ M_1^{\alpha} \sum_{0<t_k<t} \|I_k^{\alpha}(z^{(n)}(t_k^-) + y(t_k^-) - I_k^{\alpha}(z(t_k^-) + y(t_k^-))\| \to 0 \quad \text{as } n \to \infty,
\end{aligned}
$$

which shows that Ψ is continuous in B_h''. Hence the conditions of Schauder's fixed point theorem are all satisfied, and also the operator Ψ has a fixed point in B_q. Thus system (1)-(3) has a solution on J. □

Theorem 3.2 *Conditions* (A_1), (A_3), (A_5) *and* (D) *are satisfied. Then system* (1)-(3) *is approximately controllable on* J.

Proof Let $\tilde{x}^{\gamma}(\cdot)$ be a fixed point of Ψ in B_q, and any fixed point of Ψ is a mild solution of (1)-(3) on $[0, b]$.

By Theorem 2.7, the control

$$
\tilde{u}^{\gamma}(t) = B^* U^*(a, t)\tilde{R}(\gamma, \Gamma_0^a)[P(\tilde{x}^{\gamma})]^{\alpha},
$$

which satisfies

$$
\begin{aligned}
[\tilde{x}^{\gamma}(a)]^{\alpha} = x_a - \gamma \tilde{R}(\gamma, \Gamma_0^a) \Bigg[& x_a - U^{\alpha}(a, 0)[x_0 - \mu(x) - h^{\alpha}(0, (x_0 - \mu(x)))] \\
& - h^{\alpha}(a, x_a) - \int_0^a U^{\alpha}(a, s)A(s)h^{\alpha}(s, \tilde{x}_s^{\gamma}) \, ds - \int_0^a U^{\alpha}(a, s)f^{\alpha}(s, \tilde{x}_s^{\gamma}) \, ds \\
& - \int_0^a U^{\alpha}(a, s)g^{\alpha}(s, \tilde{x}_s^{\gamma}) \, dW(S) - \sum_{k=1}^{m} U^{\alpha}(a, t_k)(I_k)(x(t_k^-)) \Bigg].
\end{aligned} \tag{11}
$$

By (A_3)-(A_5)

$$
\int_0^a \|f^{\alpha}(s, \tilde{x}_s^{\gamma})\| \, ds \le L_f^{\alpha} a, \qquad \int_0^a \|g^{\alpha}(s, \tilde{x}_s^{\gamma})\| dW(s) \le L_g^{\alpha} a,
$$

$$
\int_0^a \|A(s)h^{\alpha}(s, \tilde{x}_s^{\gamma})\| \, ds \le L_h^{\alpha} a.
$$

Here, the sequences $f^\alpha(s,\tilde{x}_s^\gamma)$, $g^\alpha(s,\tilde{x}_s^\gamma)$ and $A(s)h^\alpha(s,\tilde{x}_s^\gamma)$ are bounded in $L^2(J,X)$. Therefore, there is a subsequence $f^\alpha(s,\tilde{x}_s^\gamma)$, $g^\alpha(s,\tilde{x}_s^\gamma)$ and $A(s)h^\alpha(s,\tilde{x}_s^\gamma)$ are all weakly convergent to say $f^\alpha(s)$, $g^\alpha(s)$ and $A(s)h^\alpha(s)$ are bounded in $L^2(J,X)$, respectively. Take

$$\begin{aligned}\tilde{\omega} &= x_a - U^\alpha(a,0)\big[x_0 - \mu(x) - h^\alpha\big(0,(x_0-\mu(x))\big)\big] - h^\alpha\big(a,x(a)\big)\\ &\quad - \int_0^a U^\alpha(a,s)A(s)h^\alpha(s)\,ds\\ &\quad - \int_0^a U^\alpha(a,s)f^\alpha(s)\,ds - \int_0^a U^\alpha(a,s)g\alpha(s)\,dW(S) - \sum_{k=1}^m U^\alpha(a,t_k)I_k^\alpha\big(x(t_k^-)\big).\end{aligned}$$

It follows by the compactness of the operators $q(\cdot) \to \int_0^a U(\cdot,s)q(s)\,ds : L^2(J,X) \to C(J,X)$, we obtain that

$$\begin{aligned}\big\|[P(\tilde{x}^\gamma)]^\alpha - \tilde{\omega}\big\| &\le \left\|\int_0^a U^\alpha(a,s)\big[A(s)h^\alpha(s,\tilde{x}_s^\gamma) - A(s)h^\alpha(s)\big]\,ds\right\|\\ &\quad + \left\|\int_0^a U^\alpha(a,s)\big[f^\alpha(s,\tilde{x}_s^\gamma) - f^\alpha(s)\big]\,ds\right\|\\ &\quad + \left\|\int_0^a U^\alpha(a,s)\big[g^\alpha(s,\tilde{x}_s^\gamma) - g^\alpha(s)\big]\,dW(S)\right\| \to 0 \quad \text{as } \gamma \to 0^+.\end{aligned}$$

By (11) and definition

$$\begin{aligned}\big\|[\tilde{x}^\gamma(a)]^\alpha - x_a\big\| &= \big\|\gamma\tilde{R}(\gamma,\Gamma_0^a)[P(\tilde{x}^\gamma)]^\alpha\big\|\\ &= \big\|\gamma\tilde{R}(\gamma,\Gamma_0^a)[P(\tilde{x}^\gamma)]^\alpha - \tilde{\omega} + \tilde{\omega}\big\|\\ &\le \big\|\gamma\tilde{R}(\gamma,\Gamma_0^a)\tilde{\omega}\big\| + \big\|\gamma\tilde{R}(\gamma,\Gamma_0^a)[P(\tilde{x}^\gamma)]^\alpha - \tilde{\omega}\big\|\\ &\le \big\|\gamma\tilde{R}(\gamma,\Gamma_0^a)\tilde{\omega}\big\| + \big\|[P(\tilde{x}^\gamma)]^\alpha - \tilde{\omega}\big\| \to 0 \quad \text{as } \gamma \to 0^+,\end{aligned}$$

i.e., $\|[\tilde{x}^\gamma(a)]^\alpha - x_a\| \to 0$.

This proves the approximate controllability of (1)-(3). □

4 Example

We consider an impulsive neutral fuzzy stochastic differential equation with nonlocal condition:

$$\begin{aligned}&{}^cD_t^\alpha\left[z(t,x) - \int_{-\infty}^t e^{(s-t)}z(s,x)\,ds\right]\\ &= \frac{\partial^2}{\partial z^2}\left[z(t,x) - \int_{-\infty}^t e^{(s-t)}z(s,x)\,ds\right] + \eta(t,x)\\ &\quad + \int_{-\infty}^t \hat{a}(s)\sin z(t+s,x)\,ds + \int_{-\infty}^t \alpha(t-s)z(s,x)\,dW(s), \quad x \in [0,\Pi],\end{aligned} \tag{12}$$

$$z(t_k^+,x) - z(t_k^-,x) = I_k\big(z(\tau_k^-,x)\big), \quad k = 1,2,\ldots,n, \tag{13}$$

$$z(t,0) = z(t,\Pi) = 0, \quad t \in J = [0,b], \tag{14}$$

$$z(0,x) + \int_0^\Pi k_1(x,y)z(t,y)\,dy = \phi(t,x), \quad t \in (-\infty,0]. \tag{15}$$

Let $U = L^2([0,\Pi])$ and $v(t) = e^{2t}$, $t > 0$, where ${}^cD_t^\alpha$ is a Caputo fractional partial derivative of order $0 < \alpha < 1$, $b > 0$, $k_1(x,y) \in L_2([0,\Pi] \times [0,\Pi])$ and $\int_{-\infty}^0 |\hat{a}(s)|\, ds < \infty$. In our formulation we take U defined on a fuzzy stochastic space (Ω, F, P). To study the approximate controllability of (12)-(15), assume that $\alpha(t)$ is measurable and continuous with finite $L_\alpha^2 = \int_{-\infty}^0 \frac{\alpha^2(-s)}{v(s)}\, ds$. We define the operator A by $Az = \frac{\partial^2 z}{\partial x^2}$ with domain

$$D(A) = \left\{ z \in U \text{ and } \frac{\partial z}{\partial x}, \frac{\partial^2 z}{\partial x^2} \text{ are absolutely continuous, } \frac{\partial^2 z}{\partial x^2} \in U \text{ and } z(0) = z(\Pi) = 0 \right\}.$$

From this we know that A generates a strongly continuous semigroup $T(t)$, $t > 0$, given by $T(t)z = \sum_{n=1}^\infty e^{-n^2 t} \langle z, e_n \rangle e_n$, which is compact, analytic, and also spectrum of A consists of the eigenvalues $-n^2$ for $n \in N$ with corresponding eigenvectors $z_n(x) = \sqrt{\frac{2}{\Pi}} \sin(nx)$, and z_n is an orthonormal basis of H.

Define the bounded linear operator B by

$$Bu(t)(x) = \eta(t,x), \quad 0 \leq x \leq \Pi, u \in U.$$

Now, we present a special space C_v. For $(t,\varphi) \in J \times C_v$, where $z(s,x) = \varphi(s,x) \in (-\infty, 0] \times [0,\Pi]$, and the operator $h,f : J \times C_v \to U$, $g : J \times C_v \to L_2^0(U,U)$ by

$$h(t,\varphi)(x) = \int_{-\infty}^0 e^{-s}\varphi(s,x)\, ds, \qquad f(t,\varphi)(x) = \int_{-\infty}^0 \hat{a}(s) \sin\big(\varphi(s,x)\big)\, ds,$$

$$g(t,\varphi)(x) = \int_{-\infty}^0 \alpha(-s)\varphi(s,x)\, dW(s)$$

with the choice of A, h, f, g of system (12)-(15) can be rewritten as the abstract form of system (1)-(3). Thus, under the appropriate conditions on the function h, f, g and I_k as those in (A_1)-(A_6), system (12)-(15) is approximately controllable.

5 Conclusion

In this paper, we focused on approximate controllability results for impulsive neutral fuzzy stochastic differential equations with nonlocal conditions in a Banach space by using Schauder's and Banach fixed point theorems. According to the hypotheses, it is proved that the system is approximately controllable under nonlocal conditions. Then, the result showed that these fixed point theorems can strongly be used in control problems to obtain the sufficient conditions. Upon making some approximate assumptions, by the ideas and techniques as suggested in this paper, one can establish the approximate controllability results for a wide class of linear fuzzy stochastic evolution equations.

Competing interests

The authors declare that they have no competing interests.

Authors' contributions

All authors contributed equally to the writing of this paper. All authors read and approved the final manuscript.

References

1. Negoita, CV, Ralescu, DA: Applications of Fuzzy Sets to System Analysis. Wiley, New York (1975)
2. Kwun, YC, Park, DG: Optimal control problem for fuzzy differential equations. In: Proceedings of the Korea-Vietnam Joint Seminar, pp. 103-114 (1998)
3. Mao, XR: Stochastic Differential Equations and Their Application (1997)
4. Ruan, S: Oscillation of second order neutral differential equations. Can. Math. Bull. **36**, 485-496 (1991)
5. Zhang, X, Feng, M: Transformation technique, fixed point theorem and positive solutions for second-order impulsive differential equations with deviating arguments. Adv. Differ. Equ. **2014**, 312 (2014)
6. Zhang, MR: Periodic solutions of linear and quasilinear neutral functional differential equations. J. Math. Anal. Appl. **189**(2), 378-392 (1995)
7. Benchohra, M, Henderson, J, Ntouyas, S: Impulsive Differential Equations and Inclusions. Contemporary Mathematics and Its Applications, vol. 2 (2006)
8. Lakshmikantham, V, Bainov, DD, Simeonov, PS: Theory of Impulsive Differential Equations. World Scientific, Singapore (1989)
9. Dauer, JP, Mahmudov, NI, Matar, MM: Approximate controllability of backward stochastic evolution equations in Hilbert spaces. J. Math. Anal. Appl. **323**, 42-56 (2006)
10. Dauer, JP, Mahmudov, NI: Approximate controllability of semilinear functional equations in Hilbert spaces. J. Math. Anal. Appl. **273**, 310-327 (2002)
11. Sakthivel, R, Mahmudov, NI, Kim, JH: Approximate controllability of nonlinear impulsive differential systems. Rep. Math. Phys. **60**, 85-96 (2007)
12. Sakthivel, R, Nieto, J, Mahmudov, NI: Approximate controllability of nonlinear deterministic and stochastic systems with unbounded delay. Taiwan. J. Math. **14**(5), 1777-1797 (2010)
13. Zang, Y, Li, J: Approximate controllability of fractional impulsive neutral stochastic differential equations with nonlocal conditions. Bound. Value Probl. **2013**, 193 (2013)

11

Solvability for some boundary value problems with discrete ϕ-Laplacian operators

Tianlan Chen* and Ruyun Ma

*Correspondence:
chentianlan511@126.com
Department of Mathematics,
Northwest Normal University,
Lanzhou, 730070, P.R. China

Abstract

In this paper, we study the existence of solutions of the discrete ϕ-Laplacian equation $\nabla[\phi(\Delta u_k)] = \lambda f(k, u_k, \Delta u_k)$, $k \in [2, n-1]_{\mathbb{Z}}$, with Dirichlet or mixed boundary conditions. Under general conditions, an explicit estimate of λ_0 is given such that the problem possesses a solution for any $|\lambda| < \lambda_0$.

Keywords: ϕ-Laplacian; existence; Schauder's fixed point theorem; mixed boundary value problem; Dirichlet boundary value problem

1 Introduction

The study of difference equations represents a very important field in mathematical research. Different mathematical models coupled with the basic theory of this type of equation can be found in the classical monograph by Goldberg [1] and in the book by Lakshmikantham and Trigiante [2]. Besides, they are also natural consequences of the discretization of differential problems.

In this paper, we consider the existence of solutions of discrete ϕ-Laplacian problems

$$\nabla\big[\phi(\Delta u_k)\big] = \lambda f(k, u_k, \Delta u_k), \quad k \in [2, n-1]_{\mathbb{Z}}, \tag{1}$$

with Dirichlet boundary conditions

$$u_1 = 0 = u_n \tag{2}$$

or mixed boundary conditions

$$u_1 = 0 = \Delta u_{n-1} \tag{3}$$

under the following assumptions:

(H1) $\phi : (-a, a) \to (-b, b)$ is an increasing homeomorphism with $\phi(0) = 0$ and $0 < a, b \leq \infty$;

(H2) for all $k \in [2, n-1]_{\mathbb{Z}}$, $f(k, \cdot, \cdot) : \mathbb{R} \times (-a, a) \to \mathbb{R}$ with $\mathbf{f} = (f(2, u, v), \ldots, f(n-1, u, v))$, and for each compact set $A \subset \mathbb{R} \times (-a, a)$, there exists a bounded function $\mathbf{h}_A = (h_A(2), \ldots, h_A(n-1))$ defined from $\mathbb{R}$ to $\mathbb{R}^{n-2}$ such that

$$\big|f(k, u, v)\big| \leq h_A(k), \quad \text{for } k \in [2, n-1]_{\mathbb{Z}} \text{ and all } (u, v) \in A.$$

The study of the ϕ-Laplacian equation is a classical topic that has attracted the attention of many researchers because of its interest for applications. Usually, a ϕ-Laplacian operator is said to be *singular* when the domain of ϕ is finite (that is, $a < +\infty$); in the contrary case, the operator is said to be *regular*. On the other hand we say that ϕ is *bounded* if its range is finite (that is, $b < +\infty$) and *unbounded* in the other case. There are three paradigmatic models in this context:

(i) $a = b = +\infty$ (regular unbounded): we have the p-Laplacian operator

$$\phi_1(x) = |x|^{p-2}x \quad \text{with } p > 1.$$

(ii) $a < +\infty$, $b = +\infty$ (singular unbounded): we have the relativistic operator

$$\phi_2(x) = \frac{x}{\sqrt{1-x^2}}.$$

(iii) $a = +\infty$, $b < +\infty$ (regular bounded): we have the one-dimensional mean curvature operator

$$\phi_3(x) = \frac{x}{\sqrt{1+x^2}}.$$

Among them, the p-Laplacian operator has received a lot of attention and the number of related references is huge (we only mention [3–6] and references therein). For the relativistic operator, it has recently been proved in [7–9] that the Dirichlet problem is always solvable. This is a striking result closely related to the '*a priori*' bound of the derivatives of the solutions. For the curvature operator, this is no longer true, but other results as regards the existence of solutions of differential problems can be found in [10, 11]. To the best of our knowledge, the discrete problem has received almost no attention. In this article, we will discuss it in detail.

The purpose of this paper is to show that analogs of the existence results of solutions for differential problems proved in [10] hold for the corresponding difference equations. However, some basic ideas from differential calculus are not necessarily available in the field of difference equations, such as Rolle's theorem and symmetry of the domain of solutions. Thus, new challenges are faced and innovation is required. In addition, we extend some results of Bereanu and Mawhin in [7]; see Remark 1. The proof is elementary and relies on Schauder's fixed point theorem after a suitable reduction of the problem to a first-order summing-difference equation.

We end this section with some notations. For $\mathbf{x} \in \mathbb{R}^p$ set

$$|\mathbf{x}|_\infty = \max_{1\le k\le p} |x_k|, \qquad |\mathbf{x}|_1 = \sum_{k=1}^{p} |x_k|.$$

For every $l, m \in \mathbb{N}$ with $m > l$, we set $\sum_{k=m}^{l} x_k = 0$, and $[l, m]_{\mathbb{Z}} := \{l, l+1, \dots, m\}$.

Let $n \in \mathbb{N}$, $n \ge 4$ be fixed and $\mathbf{x} = (x_1, \dots, x_n) \in \mathbb{R}^n$. Define

$$\Delta\mathbf{x} = (\Delta x_1, \dots, \Delta x_{n-1}) \in \mathbb{R}^{n-1},$$

as

$$\Delta x_k = x_{k+1} - x_k \quad (1 \le k \le n-1),$$

and if $|\Delta \mathbf{x}|_\infty < a$, define

$$\nabla\big[\phi(\Delta \mathbf{x})\big] = \big(\nabla\big[\phi(\Delta x_2)\big], \dots, \nabla\big[\phi(\Delta x_{n-1})\big]\big) \in \mathbb{R}^{n-2},$$

as

$$\nabla\big[\phi(\Delta x_k)\big] = \phi(\Delta x_k) - \phi(\Delta x_{k-1}) \quad (2 \le k \le n-1).$$

For convenience, for each $0 < r < b$, let M_r be defined as

$$M_r := |\mathbf{h}_{A_r}|_1, \tag{4}$$

where $A_r = [(n-1)\phi^{-1}(-r), (n-1)\phi^{-1}(r)] \times [\phi^{-1}(-r), \phi^{-1}(r)]$.

2 The Dirichlet boundary value problem

Let us consider the boundary value problem

$$\nabla\big[\phi(\Delta u_k)\big] = \lambda f(k, u_k, \Delta u_k), \quad k \in [2, n-1]_{\mathbb{Z}}, \qquad u_1 = 0 = u_n, \tag{5}$$

where $\lambda \in \mathbb{R}$ is a parameter and (H1) and (H2) hold.

Let us define the space

$$H = \left\{ \mathbf{y} \in \mathbb{R}^{n-1} \,\middle|\, \mathbf{y} = (y_1, \dots, y_{n-1}) \text{ with } |\mathbf{y}|_\infty < \frac{b}{2} \right\},$$

of course, $\frac{b}{2}$ must be understood as $+\infty$ when $b = +\infty$. The following result is a slight modification of Lemma 2 in [7], but we include the proof for the sake of completeness.

Lemma 2.1 *For any $\mathbf{y} \in H$, there exists a unique constant $\gamma := P_\phi[\mathbf{y}]$ such that*

$$\sum_{k=1}^{n-1} \phi^{-1}(\gamma + y_k) = 0. \tag{6}$$

Besides, $|P_\phi[\mathbf{y}]| \le |\mathbf{y}|_\infty$, and the function $P_\phi : H \to (-\frac{b}{2}, \frac{b}{2})$ is continuous.

Proof From the properties of ϕ, it is clear that

$$\sum_{k=1}^{n-1} \phi^{-1}\big(-|\mathbf{y}|_\infty + y_k\big) \le 0 \le \sum_{k=1}^{n-1} \phi^{-1}\big(|\mathbf{y}|_\infty + y_k\big). \tag{7}$$

Because ϕ is a homeomorphism such that $\phi(0) = 0$, it follows that ϕ^{-1} is strictly monotone and satisfies (7), so there exists a unique $\gamma := P_\phi[\mathbf{y}]$ verifying (6) with $\gamma \le |\mathbf{y}|_\infty$. Note that P_ϕ takes bounded sets into bounded sets. So the continuity of P_ϕ follows easily. □

By means of a suitable change of variable, we relate the problem (5) with the nonlocal first-order equation

$$\nabla y_k = \lambda f\left(k, \sum_{i=1}^{k-1} \phi^{-1}\big(P_\phi[\mathbf{y}] + y_i\big), \phi^{-1}\big(P_\phi[\mathbf{y}] + y_k\big)\right), \quad k \in [2, n-1]_{\mathbb{Z}}. \tag{8}$$

By a simple computation, we can get the following result.

Lemma 2.2 *If* $\mathbf{y} = (y_1, \ldots, y_{n-1})$ *is a solution of the problem* (8) *with* $|\mathbf{y}|_\infty < \frac{b}{2}$, *then*

$$u_k = \sum_{i=1}^{k-1} \phi^{-1}\big(P_\phi[\mathbf{y}] + y_i\big)$$

is a solution of (5).

Now we are in a position to prove the main result of this section: the solvability of problem (5) for small λ.

Theorem 2.3 *For each* $0 < r < \frac{b}{2}$, *let* M_r *be defined by* (4). *If*

$$|\lambda| < \lambda_0 := \sup_{0<r<\frac{b}{2}} \frac{r}{M_{2r}},$$

then problem (5) *has a solution.*

Proof Let $0 < r_1 < \frac{b}{2}$ be such that $|\lambda| \le \frac{r_1}{M_{2r_1}}$ and consider the ball

$$B_{r_1} = \big\{\mathbf{y} \in \mathbb{R}^n \mid \mathbf{y} = (y_1, \ldots, y_n) \text{ with } |\mathbf{y}|_\infty \le r_1\big\}.$$

For each $\mathbf{y} \in B_{r_1}$, we define the operator

$$Ty_k = \lambda \sum_{j=2}^{k} f\left(j, \sum_{i=1}^{j-1} \phi^{-1}\big(P_\phi[\mathbf{y}] + y_i\big), \phi^{-1}\big(P_\phi[\mathbf{y}] + y_j\big)\right).$$

It is easy to show that T is completely continuous. Moreover, by our assumptions and the choice of r_1 we have

$$|T\mathbf{y}|_\infty \le |\lambda| M_{2r_1} \le r_1,$$

which implies that $T(B_{r_1}) \subset B_{r_1}$. Thus Schauder's fixed point theorem yields a fixed point for T which is a solution of (8), and therefore by Lemma 2.2 it is also a solution of problem (5). □

Now we are going to apply Theorem 2.3 to study the solvability of the Dirichlet problem

$$\nabla\big[\phi(\Delta u_k)\big] = f(k, u_k, \Delta u_k), \quad k \in [2, n-1]_{\mathbb{Z}}, \qquad u_1 = 0 = u_n. \tag{9}$$

We point out that problem (9) presents interesting different features depending on the bounded or unbounded behavior of ϕ.

2.1 Unbounded ϕ-Laplacian ($b = +\infty$)

A consequence of Theorem 2.3 is that, whenever ϕ is unbounded, then (9) is solvable.

Corollary 1 *Assume that ϕ is unbounded (that is, $b = +\infty$) and there exists a bounded function $\mathbf{h} = (h(2),\dots,h(n-1))$ such that*

$$|f(k,u,v)| \leq h(k), \quad k \in [2,n-1]_{\mathbb{Z}}, u \in \big(-(n-1)a,(n-1)a\big), v \in (-a,a). \tag{10}$$

Then the Dirichlet problem (9) has at least one solution.

Proof By conditions (10) it is clear that

$$M_r = |\mathbf{h}|_1, \quad \text{for each } r > 0.$$

Therefore

$$\lambda_0 = \sup_{0<r<+\infty} \frac{r}{M_{2r}} = +\infty,$$

and then Theorem 2.3 ensures us that the problem (9) has a solution for each $\lambda \in \mathbb{R}$, and in particular for $\lambda = 1$. □

Remark 1 Corollary 1 applies in particular if ϕ is also singular $a < +\infty$ and for each $k \in [2,n-1]_{\mathbb{Z}}, f(k,\cdot,\cdot)$ is continuous on $\mathbb{R}^2$. In this way Corollary 1 improves Theorem 1 in [7].

Example 1 The assumptions of Corollary 1 are satisfied for $\phi(v) = \frac{v}{\sqrt{1-v^2}}$ and for all $k \in [2,n-1]_{\mathbb{Z}}$,

$$f(k,u,v) = \begin{cases} 1, & u \in [0,(n-1)a), v \in (-a,a), \\ -1, & u \in (-(n-1)a,0), v \in (-a,a). \end{cases}$$

Thus, by Corollary 1, the problem (9) has a solution.

2.2 Bounded ϕ-Laplacian ($b < +\infty$)

In the case of a bounded ϕ-Laplacian, the 'universal' solvability of (9) is no longer true even for a constant nonlinearity $f(t,u,v) \equiv M$ as we show in the following result.

Proposition 1 *Assume that ϕ is bounded (that is, $b < +\infty$), let $M \in \mathbb{R}$ and consider the Dirichlet problem*

$$\nabla\big[\phi(\Delta u_k)\big] = M, \quad k \in [2,n-1]_{\mathbb{Z}}, \qquad u_1 = 0 = u_n. \tag{11}$$

Then the following claims hold:

(i) *If $|M| \geq \frac{b}{n-2}$, then (11) has no solution.*
(ii) *If $|M| < \frac{b}{2(n-2)}$, then (11) has a solution.*

Before proving the Proposition 1, we first give the following result, which is a slight modification of Lemma 2 in [7].

Lemma 2.4 *For any $\mathbf{y} \in E$,*

$$E := \left\{\mathbf{y} \in \mathbb{R}^{n-1} \,\middle|\, \mathbf{y} = (y_1,\dots,y_{n-1}) \text{ with } |\mathbf{y}|_1 < \frac{b}{2}\right\},$$

there exists a unique constant $\gamma := Q_\phi[\mathbf{y}]$ *such that*

$$\phi^{-1}(\gamma) + \sum_{k=2}^{n-1} \phi^{-1}\left(\gamma + \sum_{j=2}^{k} y_j\right) = 0.$$

Besides, $|Q_\phi[\mathbf{y}]| \leq |\mathbf{y}|_1$, *and the function* $Q_\phi : E \to (-\frac{b}{2}, \frac{b}{2})$ *is continuous.*

Proof The proof of Lemma 2.4 is similar to Lemma 2.1 and we omit it. □

Proof of the Proposition 1 (i) If $\mathbf{u} = (u_1, \ldots, u_n)$ is a solution of (11), then there exists some $\tau_0 \in [2, n]_\mathbb{Z}$ such that $\Delta u_{\tau_0 - 1} \geq 0$, and therefore from (11) we have

$$\phi(\Delta u_k) = \phi(\Delta u_{\tau_0 - 1}) + \sum_{j=\tau_0}^{k} M$$

and

$$u_k = \sum_{j=1}^{k-1} \phi^{-1}\big[M(j - \tau_0 + 1) + \phi(\Delta u_{\tau_0 - 1})\big]. \tag{12}$$

Observe that $\phi(\Delta u_{\tau_0 - 1}) \geq 0$ and suppose that $|M| \geq \frac{b}{n-2}$, then

$$\max_{j \in [1, n-1]_\mathbb{Z}, \tau_0 \in [2, n]_\mathbb{Z}} \left|M(j - \tau_0 + 1) + \phi(\Delta u_{\tau_0 - 1})\right| \geq b.$$

In this case $\mathbf{u}$ given by (12) would not be well defined since the domain of ϕ^{-1} is the interval $(-b, b)$, and thus a solution of (11) cannot exist.

(ii) If $|M| < \frac{b}{2(n-2)}$, then the function $\mathbf{u}$ given by (12) is well defined for $\tau_0 = 2$, and satisfies

$$\nabla\big[\phi(\Delta u_k)\big] = M, \quad k \in [2, n-1]_\mathbb{Z} \quad \text{and} \quad u_1 = 0,$$

and hence

$$u_k = \phi^{-1}(\gamma) + \sum_{l=2}^{k-1} \phi^{-1}\left(\gamma + \sum_{j=2}^{l} M\right), \quad k = 2, \ldots, n.$$

Using Lemma 2.4, we can show that $u_n = 0$ (in fact $\gamma = \phi(\Delta u_1)$). Thus $\mathbf{u}$ is a solution of (11). □

Remark 2 Regrettably our method does not determine whether or not there exists a solution of (11) when $\frac{b}{2(n-2)} \leq |M| < \frac{b}{(n-2)}$, since we cannot guarantee $u_n = 0$ (in fact, the difference equation loses the symmetric property about the domain of solution). However, in [10], Proposition 1, the authors prove that the differential problem corresponding to (11) has the complete result.

As a consequence of Theorem 2.3 we obtain the following sufficient condition for the solvability of the Dirichlet problem.

Corollary 2 *Assume that ϕ is bounded (that is, $b < +\infty$) and there exists a bounded function $\mathbf{h} = (h(2), \ldots, h(n-1))$ such that*

$$|f(k,u,v)| \leq h(k), \quad k \in [2, n-1]_{\mathbb{Z}}, u \in \big(-(n-1)a, (n-1)a\big), v \in (-a, a),$$

with $|\mathbf{h}|_1 < \frac{b}{2}$. Then the Dirichlet problem (9) *has at least one solution.*

Proof Now, for each $0 < r < \frac{b}{2}$ we have $M_{2r} = |\mathbf{h}|_1 < \frac{b}{2}$. Therefore

$$\lambda_0 = \sup_{0<r<\frac{b}{2}} \frac{r}{M_{2r}} > 1,$$

and thus Theorem 2.3 implies the existence of a solution for problem (9) with $\lambda = 1$. □

Example 2 Let $n \in \mathbb{N}$, $n \geq 4$ be fixed. We take $\phi(v) = \frac{v}{\sqrt{1+v^2}}$ and $f(k,u,v) = \frac{1}{4n}\sin(k+u-v)\pi$. Thus, by a simple computation, we show by Corollary 2 that problem (9) has a solution.

3 The mixed boundary value problem

Compared with the Dirichlet problem, the mixed boundary value problem

$$\nabla\big[\phi(\Delta u_k)\big] = \lambda f(k, u_k, \Delta u_k), \quad k \in [2, n-1]_{\mathbb{Z}}, \qquad u_1 = 0 = \Delta u_{n-1} \tag{13}$$

is less studied in the related literature. In this case, by means of the change of variable $y_k = \phi(\Delta u_k)$, $k = 1, \ldots, n-1$, we see that a solution $\mathbf{u} = (u_1, \ldots, u_n)$ of (13) is equivalent to a solution $\mathbf{y} = (y_1, \ldots, y_{n-1})$ of the following nonlocal first-order problem:

$$\nabla y_k = \lambda f\left(k, \sum_{j=1}^{k-1} \phi^{-1}(y_j), \phi^{-1}(y_k)\right), \quad k \in [2, n-1]_{\mathbb{Z}}, \qquad y_{n-1} = 0. \tag{14}$$

By using the same idea as in Theorem 2.3, we can prove the following result.

Theorem 3.1 *If*

$$|\lambda| < \tilde{\lambda}_0 := \sup_{0<r<b} \frac{r}{M_r},$$

then problem (13) *has a solution.*

Proof Let $0 < r_1 < b$ be such that $\lambda \leq \frac{r_1}{M_{r_1}}$, define

$$B_{r_1} = \big\{\mathbf{y} \mid \mathbf{y} = (y_1, \ldots, y_{n-1}), \text{for all } k \in [1, n-1]_{\mathbb{Z}}, y_k \in (-b, b) \text{ with } |\mathbf{y}|_\infty \leq r_1\big\}$$

and apply Schauder's fixed point theorem to the completely continuous operator $T : B_{r_1} \to B_{r_1}$ defined as

$$Ty_k = \lambda \sum_{j=k+1}^{n-1} f\left(j, \sum_{l=1}^{j-1} \phi^{-1}(y_l), \phi^{-1}(y_j)\right) \leq \lambda \sum_{j=k+1}^{n-1} h(j) \leq \lambda M_{r_1} \leq r_1.$$

Thus (14) has a solution $\mathbf{y}$, which implies that $\mathbf{u}$ is a solution of (13). □

Remark 3 The preceding theorem is sharp in the following sense: when $\phi(x) = \frac{x}{\sqrt{1+x^2}}$ and $f(k,u,v) \equiv M$, then it is easy to show that problem (13) has a solution if and only if $|\lambda| < \tilde{\lambda}_0 = \sup_{0<r<1} \frac{r}{M_r} = \frac{1}{M}$.

Competing interests
The authors declare that they have no competing interests.

Authors' contributions
The authors have achieved equal contributions to each part of this paper. All the authors read and approved the final manuscript.

Acknowledgements
The authors are very grateful to the anonymous referees for their valuable suggestions. This work was supported by the NSFC (No. 11361054), SRFDP (No. 20126203110004) and Gansu Provincial National Science Foundation of China (No. 1208RJZA258).

References

1. Goldberg, S: Introduction to Difference Equations. Wiley, New York (1960)
2. Lakshmikantham, V, Trigiante, D: Theory of Difference Equations: Numerical Methods and Applications. Mathematics in Science and Engineering, vol. 181. Academic Press, Boston (1988)
3. Cabada, A, Otero-Espinar, V: Existence and comparison results for difference ϕ-Laplacian boundary value problems with lower and upper solutions in reverse order. J. Math. Anal. Appl. **267**, 501-521 (2002)
4. Wang, D, Guan, G: Three positive solutions of boundary value problems for p-Laplacian difference equations. Comput. Math. Appl. **55**, 1943-1949 (2008)
5. Coster, C: Pairs of positive solutions for the one dimensional p-Laplacian. Nonlinear Anal. **23**, 669-681 (1994)
6. Jiang, L, Zhou, Z: Three solutions to Dirichlet boundary value problems for p-Laplacian difference equations. Adv. Differ. Equ. **2008**, Article ID 345916 (2008)
7. Bereanu, C, Mawhin, J: Boundary value problem for second-order nonlinear difference equations with discrete ϕ-Laplacian and singular ϕ. J. Differ. Equ. Appl. **14**, 1099-1118 (2008)
8. Bereanu, C, Thompson, H: Periodic solutions of second order nonlinear difference equation with discrete ϕ-Laplacian. J. Math. Anal. Appl. **330**, 1002-1015 (2007)
9. Bereanu, C, Mawhin, J: Existence and multiplicity results for some nonlinear problems with singular ϕ-Laplacian. J. Differ. Equ. **243**, 536-557 (2007)
10. Cid, J, Torres, P: Solvability for some boundary value problems with ϕ-Laplacian operators. Discrete Contin. Dyn. Syst. **23**, 727-732 (2009)
11. Torres, P: Nondegeneracy of the periodically forced Liénard differential equation with ϕ-Laplacian. Commun. Contemp. Math. **13**, 283-292 (2011)

12

Dynamical analysis of an SIRS network model with direct immunization and infective vector

Rongzhong Yu[1*], Kezan Li[2], Baidi Chen[1] and Dingqin Shi[1]

*Correspondence: yurongzhong@163.com
[1]College of Sciences, Jiujiang University, Qianjin East Road, Jiujiang, 332005, China
Full list of author information is available at the end of the article

Abstract

With the awareness of risk in infective disease spreading, healthy individuals (the susceptible ones) will take some measures to acquire temporary immunity. This paper addresses an SIRS model with direct immunization and an infective vector in complex networks and performs the dynamical analysis for this model. By theoretical analysis, we obtain the epidemic threshold λ_c and prove that if infection rate $\lambda < \lambda_c$, the disease-free equilibrium is globally asymptotically stable; if $\lambda > \lambda_c$, there exists a unique endemic equilibrium, and it is globally attractive. These theoretical results are confirmed by numerical simulations.

Keywords: epidemic threshold; dynamical analysis; epidemic equilibrium; SIRS model; global attraction

1 Introduction

In recent years, many epidemic models on complex networks, such as SIS (susceptible-infected-susceptible) [1–8] and SIR (susceptible-infected-removed) [9–16] and so on, have been widely studied by researchers from different subjects. Classical studies have revealed that there is an epidemic threshold λ_c for an epidemic model on homogeneous networks, below which the disease will die out; otherwise there will exist a persistence. However, Pastor-Satorras and Vespignani further showed a striking result that the epidemic threshold λ_c will vanish for a heterogenous network with sufficiently large sizes [1, 2, 4].

In fact, apart from the human behavior [17, 18] and the external environment [19], the infection vector (*e.g.*, mosquitoes) may also play an important role in epidemic transmission [3, 20, 21]. The infective vector generally acts as a carrier of an infective disease and can transmit it to a human. By considering the disease spreading on a human network caused by an infection vector, Cooke and Busenberg [22, 23] have addressed some epidemic compartment models. As we known, some diseases spread not only by contacts between people and infected vectors but also by blood contacts within human. By noting this fact, Shi *et al.* [21] proposed a new SIS model with an infective medium on complex networks, which models the spread of a class of infectious diseases. Then a modified SIS model is proposed in [3] by assuming that the human contacts can be considered as a scale-free network, but the infective media may contact a person without any selectivity. The epidemic threshold and the stability of endemic equilibrium are investigated theoretically. A more general modified SIS model with an infective medium on complex networks was introduced in [20], and the authors investigated the global attraction of endemic equilibrium by the basic reproduction number R_0. However, the direct relation between the epidemic threshold λ_c in [3] and the basic reproduction number R_0 in [20] has not been revealed.

In the real world, with the awareness of an infectious disease spreading, some healthy individuals will usually take some protective measures (*e.g.*, vaccine inoculation) to acquire temporary immunity. By investigating an SIRS epidemic model [24] with direct immunization on complex networks, the result shows that the direct immunization can increase the epidemic threshold and reduce the prevalence of infectious disease.

In this paper, we propose a new SIRS model with direct immunization and an infective vector on complex networks. We get the epidemic threshold λ_c, below which the disease-free equilibrium is globally stable; otherwise, the disease-free equilibrium is unstable and a unique endemic equilibrium exists, and it is also globally attractive. More importantly, according to our method, one can directly determine the relation between the epidemic threshold and the basic reproduction number.

The rest of the paper is organized as follows. In Section 2, we propose a new SIRS model with direct immunization and an infective vector on complex networks. Section 3 analyzes the dynamics of the model and shows some theoretical results. Some numerical simulations are performed to confirm our theoretical predictions in Section 4.

2 The model

Based on the real mechanism of some relevant epidemic networks, our model will be constructed with the following context:

- Two infection mechanisms: the disease spreads not only by contacts between individuals, but also by contacts between individuals and infective vectors.
- Two removed ways: (i) with awareness of risk infective disease spreading, some health individuals may make some protective measures (vaccination) to acquire temporary immunity; (ii) an infected individual becoming a removed individual after cure may acquire temporary immunity.

In addition, we further suppose that the individuals' contacts can be treated as heterogeneous, but the contacts between individuals and vectors can be considered as homogeneous. This assumption results from the selective contacts in a human network and non-selective contacts between people and vectors [3, 20].

Let $S_k(t)$, $I_k(t)$ and $R_k(t)$ be the densities of susceptible, infected and removed nodes with degree k at time t respectively, and let $V(t)$ be the density of the infective medium at time t. Let $\rho(t) = \sum p(k)I_k(t)$ denote the density of infected individuals on the network, and Θ represents the probability that a randomly chosen link emanating from a node of degree k leads to infected nodes. In this paper, we consider the situation of uncorrelated networks, then Θ can be written as $\Theta = \frac{1}{\langle k \rangle} \sum_{k'} k'p(k')I_{k'}(t)$ [1, 2], where $p(k)$ denotes the degree distribution of the network, $\langle k \rangle = \sum kp(k)$ is its average degree.

We assume that the susceptible nodes become the removed nodes with rate α for acquiring temporary immunity. At the same time, each susceptible (health) node is infected with rates λ and γ_1 if it is contacted to infected nodes and infective vectors, respectively. Infected nodes are cured with rate β and removed nodes again become susceptible with rate δ for immunization-lost. Health vectors are infected with rate γ_2 if they are contacted to infected individuals and infected vectors recover with rate ξ.

In the real world, an epidemic always occurs on a finite network [4] even though the size of the network is very large. Hence, we consider disease transmission in a finite population in this paper, and let n be the maximum degree. Then, neglecting of contact duration, the proposed SIRS propagation model can be described by the following differential equations:

$$\begin{cases} \frac{dS_k(t)}{dt} = -\lambda k S_k(t)\Theta + \delta R_k(t) - \gamma_1 V(t) S_k(t) - \alpha S_k(t), \\ \frac{dI_k(t)}{dt} = \lambda k S_k(t)\Theta - \beta I_k(t) + \gamma_1 V(t) S_k(t), \\ \frac{dR_k(t)}{dt} = \beta I_k(t) - \delta R_k(t) + \alpha S_k(t), \\ \frac{dV(t)}{dt} = -\xi V(t) + \gamma_2 (1 - V(t))\rho(t). \end{cases} \tag{1}$$

Without loss of generality, assume that the infective vector has unit recovery, *i.e.*, $\xi = 1$. In addition, the variables $S_k(t), I_k(t), R_k(t)$ satisfy the normalization condition $S_k(t) + I_k(t) + R_k(t) = 1$. Then Eq. (1) can be rewritten as

$$\begin{cases} \frac{dI_k(t)}{dt} = [\lambda k\Theta + \gamma_1 V(t)][1 - I_k(t) - R_k(t)] - \beta I_k(t), \\ \frac{dR_k(t)}{dt} = (\beta - \alpha) I_k(t) - (\alpha + \delta) R_k(t) + \alpha, \\ \frac{dV(t)}{dt} = -V(t) + \gamma_2 (1 - V(t))\rho(t). \end{cases} \tag{2}$$

3 Epidemic threshold and global analysis

3.1 Epidemic threshold

Theorem 1 *Let* $\lambda_c = \frac{[\beta(\alpha+\delta) - \delta\gamma_1\gamma_2]\beta(\alpha+\delta)\langle k\rangle}{\delta^2\gamma_1\gamma_2(\langle k\rangle^2 - \langle k^2\rangle) + \delta\beta(\delta+\alpha)\langle k^2\rangle}$ *and* $\beta - \frac{\delta}{\alpha+\delta}\gamma_1\gamma_2 > 0$, *if* $\lambda > \lambda_c$, *then one and only one endemic equilibrium solution of system* (2) *exists, i.e., the epidemic propagation may outbreak on complex networks.*

Proof By letting the right-hand side of system (2) be zero, we have

$$\beta I_k(t) = \left[\lambda k\Theta + \gamma_1 V(t)\right]\left[1 - R_k(t) - I_k(t)\right], \tag{3}$$

$$(\beta - \alpha) I_k(t) - (\alpha + \delta) R_k(t) + \alpha = 0 \tag{4}$$

and

$$V(t) = \frac{\gamma_2 \rho(t)}{1 + \gamma_2 \rho(t)}. \tag{5}$$

Substituting (4) into (3), one obtains

$$\beta I_k(t) = \left[\lambda k\Theta + \gamma_1 V(t)\right]\left[\frac{\delta}{\delta + \alpha} - \frac{\beta + \delta}{\delta + \alpha} I_k(t)\right]. \tag{6}$$

Then substituting (5) into (6), we have

$$I_k(t) = \frac{\delta[\lambda k\Theta + \lambda k\gamma_2\rho(t)\Theta + \gamma_1\gamma_2\rho(t)]}{\beta(\delta + \alpha)[1 + \gamma_2\rho(t)] + (\beta + \delta)[\lambda k\Theta + \lambda k\gamma_2\rho(t)\Theta + \gamma_1\gamma_2\rho(t)]}. \tag{7}$$

Let

$$\mathcal{F}_1 = \delta\left[\lambda k\Theta + \lambda k\gamma_2\rho(t)\Theta + \gamma_1\gamma_2\rho(t)\right]$$

and

$$\mathcal{F}_2 = \beta(\delta + \alpha)\left[1 + \gamma_2\rho(t)\right] + (\beta + \delta)\left[\lambda k\Theta + \lambda k\gamma_2\rho(t)\Theta + \gamma_1\gamma_2\rho(t)\right].$$

Then one has a self-consistency equation as follows:

$$\Theta(t) = \frac{1}{\langle k\rangle}\sum_k kp(k)I_k(t) = \frac{1}{\langle k\rangle}\left\langle k\frac{\mathcal{F}_1}{\mathcal{F}_2}\right\rangle \equiv \mathcal{F}(\Theta). \tag{8}$$

It is obvious that $\Theta(t) = 0$ is a trivial solution to Eq. (8). What we are interested in is the condition under which the epidemic propagation outbreaks. Since $\Theta \in [0,1]$, $\mathcal{F}(0) = 0$ and $\mathcal{F}(\Theta) \in [0,1)$, then Eq. (8) must have a non-trivial solution if $\frac{d\mathcal{F}}{d\Theta}|_{\Theta=0} > 1$.

By computing the following expression

$$\left.\frac{d\mathcal{F}}{d\Theta}\right|_{\Theta=0} = \frac{1}{\langle k\rangle}\left\langle k\frac{(\frac{\partial\mathcal{F}_1}{\partial\rho}\cdot\frac{d\rho}{d\Theta} + \frac{\partial\mathcal{F}_1}{\partial\Theta})\mathcal{F}_2 - (\frac{\partial\mathcal{F}_2}{\partial\rho}\cdot\frac{d\rho}{d\Theta} + \frac{\partial\mathcal{F}_2}{\partial\Theta})\mathcal{F}_1}{\mathcal{F}_2^2}\right\rangle\right|_{\Theta=0} > 1, \tag{9}$$

we have the epidemic threshold λ_c taking the following expression in the case where $\beta - \frac{\delta}{\alpha+\delta}\gamma_1\gamma_2 > 0$,

$$\lambda_c = \frac{[\beta(\alpha+\delta) - \delta\gamma_1\gamma_2]\beta(\alpha+\delta)\langle k\rangle}{\delta^2\gamma_1\gamma_2(\langle k\rangle^2 - \langle k^2\rangle) + \delta\beta(\delta+\alpha)\langle k^2\rangle}.$$

Inequality (9) holds if and only if $\lambda > \lambda_c$. Furthermore, we will ascertain the uniqueness of endemic equilibrium in a similar way as paper [20]. Assume that $I = (I_1, I_2, \ldots, I_n, V)$ and $I^* = (I_1^*, I_2^*, \ldots, I_n^*, V^*)$ are two different roots to Eqs. (5) and (6). Let

$$\eta = \max\left\{\max_{k=1,\ldots,n}\left\{\frac{I_k}{I_k^*}\right\}, \frac{V}{V^*}\right\}.$$

Moreover, assume that $\eta > 1$ without loss of generality. We will complete the proof in two cases as follows.

Case 1: If there exists a natural number $k_0 \in \{1, 2, \ldots, n\}$ such that $\eta = \frac{I_{k_0}}{I_{k_0}^*}$, then we, by (6), have

$$\beta I_{k_0} = \left[\lambda k_0\Theta(I) + \gamma_1 V\right]\left[\frac{\delta}{\delta+\alpha} - \frac{\beta+\delta}{\delta+\alpha}I_{k_0}\right], \tag{10}$$

$$\beta I_{k_0}^* = \left[\lambda k_0\Theta(I^*) + \gamma_1 V^*\right]\left[\frac{\delta}{\delta+\alpha} - \frac{\beta+\delta}{\delta+\alpha}I_{k_0}^*\right], \tag{11}$$

where $\Theta(I) = \frac{1}{\langle k\rangle}\sum kp(k)I_k$ and $\Theta(I^*) = \frac{1}{\langle k\rangle}\sum kp(k)I_k^*$. From (10) and (11), we can obtain

$$\left[\lambda k_0\Theta(I) + \gamma_1 V\right]\left[\frac{\delta}{\delta+\alpha} - \frac{\beta+\delta}{\delta+\alpha}I_{k_0}\right] = \eta\cdot\left[\lambda k_0\Theta(I^*) + \gamma_1 V^*\right]\left[\frac{\delta}{\delta+\alpha} - \frac{\beta+\delta}{\delta+\alpha}I_{k_0}^*\right], \tag{12}$$

while we have the following inequalities according to the definition of η:

$$\eta\cdot\left[\lambda k_0\Theta(I^*) + \gamma_1 V^*\right] = \lambda k_0\Theta(I^*)\cdot\eta + \gamma_1 V^*\cdot\eta > \lambda k_0\Theta(I) + \gamma_1 V, \tag{13}$$

$$\frac{\delta}{\delta+\alpha} - \frac{\beta+\delta}{\delta+\alpha}I_{k_0}^* > \frac{\delta}{\delta+\alpha} - \frac{\beta+\delta}{\delta+\alpha}I_{k_0}. \tag{14}$$

It is obvious that (13) and (14) contradict (12).

Case 2: If $\eta = \frac{V}{V*}$, according to (5), then we have

$$V = (1 - V)\gamma_2 \rho(I), \tag{15}$$

$$V^* = \left(1 - V^*\right)\gamma_2 \rho\left(I^*\right), \tag{16}$$

where $\rho(I) = \sum p(k)I_k$ and $\rho(I^*) = \sum p(k)I_k^*$. From (15), (16) and the definition of η, one obtains

$$(1 - V)\gamma_2 \rho(I) = \left(1 - V^*\right)\gamma_2 \rho\left(I^*\right) \cdot \eta > (1 - V)\gamma_2 \sum p(k)\left(I_k^* \cdot \eta\right) > (1 - V)\gamma_2 \rho(I). \tag{17}$$

A contradiction appears. In conclusion, when $\lambda > \lambda_c$, one and only one endemic equilibrium solution exists for system (1). □

Letting $\tau = \frac{\delta}{\delta + \alpha}$, the epidemic threshold λ_c can be rewritten as

$$\lambda_c = \frac{[\beta - \tau\gamma_1\gamma_2]\beta\langle k\rangle}{\tau^2\gamma_1\gamma_2(\langle k\rangle^2 - \langle k^2\rangle) + \tau\beta\langle k^2\rangle}. \tag{18}$$

Note that the assumption that $\beta - \tau\gamma_1\gamma_2 > 0$, which is a default condition below, ensures that the threshold λ_c is larger than zero for a finite size network. One can see that the critical threshold vanishes for a scale-free network with sufficiently large sizes, which agrees with the previous papers [1, 2, 4].

The reader should find that the model parameters are general in this model. If γ_1 or γ_2 vanishes, the proposed model may become an SIRS model with direct immunization via one infection mechanism (contacts between individuals), and the epidemic $\lambda_c = \frac{\beta\langle k\rangle}{\tau\langle k^2\rangle}$ which agrees with the one of paper [24]. In addition, the proposed model may become one SIRS model with an infection vector via two infection mechanisms when $\alpha = 0$. And the epidemic threshold $\lambda_c = \frac{[\beta - \gamma_1\gamma_2]\beta\langle k\rangle}{\gamma_1\gamma_2(\langle k\rangle^2 - \langle k^2\rangle) + \beta\langle k^2\rangle}$, especially, $\lambda_c = \frac{[1 - \gamma_1\gamma_2]\langle k\rangle}{\gamma_1\gamma_2(\langle k\rangle^2 - \langle k^2\rangle) + \langle k^2\rangle}$ if $\beta = 1$, which is in accordance with the one of paper [3].

3.2 Global stability of disease-free equilibrium

Theorem 2 *For system* (1), *let* λ_c *be the epidemic threshold defined as* (18). *If* $\lambda < \lambda_c$, *then the disease-free equilibrium is globally asymptotically stable. Otherwise, there exists a unique endemic equilibrium.*

Proof For convenience, system (1) can be rewritten as follows:

$$\begin{cases} \frac{dS_k(t)}{dt} = -[\lambda k\Theta + \gamma_1 V(t) + \delta + \alpha]S_k(t) - \delta I_k(t) + \delta, \\ \frac{dI_k(t)}{dt} = [\lambda k\Theta + \gamma_1 V(t)]S_k(t) - \beta I_k(t), \\ \frac{dV(t)}{dt} = -V(t) + \gamma_2(1 - V(t))\rho(t). \end{cases} \tag{19}$$

The disease-free equilibrium of Eq. (19) is $S_k = \frac{\delta}{\delta + \alpha} = \tau$, $I_k = 0$, $V = 0$, $k = 1, 2, \ldots, n$. Moreover, the Jacobian matrix at disease-free equilibrium can be represented as

$$J = \begin{bmatrix} D & F \\ 0 & M \end{bmatrix}, \tag{20}$$

where $D = -(\delta + \alpha)E_n$, and E_n is an nth identity matrix, F and M are $n \times (n+1)$ and $(n+1) \times (n+1)$ matrices taking the following forms, respectively:

$$F = \begin{bmatrix} -\frac{\tau\lambda}{\langle k\rangle}p(1) - \delta & -\frac{\tau\lambda}{\langle k\rangle}2\cdot p(2) & \cdots & -\frac{\tau\lambda}{\langle k\rangle}n\cdot p(n) & -\gamma_1\tau \\ -\frac{\tau\lambda}{\langle k\rangle}2\cdot p(1) & -\frac{\tau\lambda}{\langle k\rangle}2^2\cdot p(2) - \delta & \cdots & -\frac{\tau\lambda}{\langle k\rangle}2n\cdot p(n) & -\gamma_1\tau \\ \vdots & \vdots & \ddots & \vdots & \vdots \\ -\frac{\tau\lambda}{\langle k\rangle}n\cdot p(1) & -\frac{\tau\lambda}{\langle k\rangle}n\cdot 2\cdot p(2) & \cdots & -\frac{\tau\lambda}{\langle k\rangle}n^2\cdot p(n) - \delta & -\gamma_1\tau \end{bmatrix},$$

$$M = \begin{bmatrix} \frac{\tau\lambda}{\langle k\rangle}p(1) - \beta & \frac{\tau\lambda}{\langle k\rangle}2\cdot p(2) & \cdots & \frac{\tau\lambda}{\langle k\rangle}n\cdot p(n) & \gamma_1\tau \\ \frac{\tau\lambda}{\langle k\rangle}2\cdot p(1) & \frac{\tau\lambda}{\langle k\rangle}2^2\cdot p(2) - \beta & \cdots & \frac{\tau\lambda}{\langle k\rangle}2n\cdot p(n) & \gamma_1\tau \\ \vdots & \vdots & \ddots & \vdots & \vdots \\ \frac{\tau\lambda}{\langle k\rangle}n\cdot p(1) & \frac{\tau\lambda}{\langle k\rangle}n\cdot 2\cdot p(2) & \cdots & \frac{\tau\lambda}{\langle k\rangle}n^2\cdot p(n) - \beta & \gamma_1\tau \\ \gamma_2 p(1) & \gamma_2 p(2) & \cdots & \gamma_2 p(n) & -1 \end{bmatrix}.$$

It is obvious that all of eigenvalues of J have negative part if and only if all of eigenvalues of M have negative part. Denote $A = M - \mu\cdot E$ and $\nu = \mu + \beta$, where E is an $(n+1)$-order unit matrix, and μ is eigenvalue, then

$$\det A = \begin{vmatrix} \frac{\tau\lambda}{\langle k\rangle}p(1) - \nu & \frac{\tau\lambda}{\langle k\rangle}2\cdot p(2) & \cdots & \frac{\tau\lambda}{\langle k\rangle}n\cdot p(n) & \gamma_1\tau \\ \frac{\tau\lambda}{\langle k\rangle}2\cdot p(1) & \frac{\tau\lambda}{\langle k\rangle}2^2\cdot p(2) - \nu & \cdots & \frac{\tau\lambda}{\langle k\rangle}2n\cdot p(n) & \gamma_1\tau \\ \vdots & \vdots & \ddots & \vdots & \vdots \\ \frac{\tau\lambda}{\langle k\rangle}n\cdot p(1) & \frac{\tau\lambda}{\langle k\rangle}n\cdot 2\cdot p(2) & \cdots & \frac{\tau\lambda}{\langle k\rangle}n^2\cdot p(n) - \nu & \gamma_1\tau \\ \gamma_2 p(1) & \gamma_2 p(2) & \cdots & \gamma_2 p(n) & -\nu - (1-\beta) \end{vmatrix}.$$

It follows that all of eigenvalues of M have negative real part if and only if all of roots of the characteristic polynomial $\det A = 0$ have real part of at most β. By some elementary transformations, one obtains that the issue above holds if and only all of roots of the following cubic polynomial have real part of at most β:

$$g(\nu) = \nu^3 - a_2\nu^2 - a_1\nu + a_0 = 0, \tag{21}$$

where

$$a_2 = \frac{\tau\lambda\langle k^2\rangle - (1-\beta)\langle k\rangle}{\langle k\rangle},$$

$$a_1 = \frac{(1-\beta)\tau\lambda\langle k^2\rangle + \gamma_1\gamma_2\tau\langle k\rangle}{\langle k\rangle},$$

$$a_0 = \frac{\gamma_1\gamma_2\tau^2\lambda[\langle hk^2\rangle - \langle k\rangle^2]}{\langle k\rangle}.$$

It is easy to obtain that

$$\lim_{\nu\to-\infty} g(\nu) = -\infty, \qquad \lim_{\nu\to+\infty} g(\nu) = +\infty,$$

$$g(0) = \frac{\gamma_1\gamma_2\tau^2\lambda[\langle k^2\rangle - \langle k\rangle^2]}{\langle k\rangle} > 0, \qquad g\left(\lambda\tau\frac{\langle k^2\rangle}{\langle k\rangle}\right) = -\gamma_1\gamma_2\tau^2\lambda\langle k\rangle < 0,$$

since $\langle k^2\rangle \gg \langle k\rangle^2$ for a sufficiently large network. We will investigate the roots to the cubic equation (21) below:

$$g(\beta) = \beta^2 - \beta\tau\lambda\frac{\langle k^2\rangle}{\langle k\rangle} - \gamma_1\gamma_2\tau\beta + \gamma_1\gamma_2\tau^2\lambda\frac{\langle k^2\rangle - \langle k\rangle^2}{\langle k\rangle}\begin{cases} > 0, & \text{if } \lambda < \lambda_c, \\ < 0, & \text{if } \lambda > \lambda_c. \end{cases} \tag{22}$$

If $\lambda < \lambda_c$, then $\lambda\tau\frac{\langle k^2\rangle}{\langle k\rangle} < \beta$ since $\lambda_c < \frac{\beta\langle k\rangle}{\tau\langle k^2\rangle}$. It implies that the cubic equation (21) has three roots ν_1, ν_2 and ν_3 satisfying

$$-\infty < \nu_1 < 0 < \nu_2 < \lambda\tau\frac{\langle k^2\rangle}{\langle k\rangle} < \nu_3 < \beta.$$

If $\lambda > \lambda_c$, then the cubic equation (21) has three roots satisfying

$$-\infty < \nu_1 < 0 < \nu_2 < \beta < \nu_3 < +\infty.$$

In a word, there exists a unique positive eigenvalue of J if and only if $\lambda > \lambda_c$, below which the unique epidemic equilibrium exists. Otherwise all real-valued eigenvalues of J are negative, this implies that the disease-free equilibrium is globally stable according to Lemma 1 in paper [14]. □

In paper [20], the authors follow the concepts of next-generation matrix (NGM) to give a threshold - the basic reproduction number R_0, by which the global stability of a modified SIS model is studied. The NGM is a matrix that relates the numbers of newly infected individuals in various categories in consecutive generation, and the basic reproduction number R_0 is the spectral radius of the NGM (refer to the papers [25, 26] for details). However, the direct relationship between the epidemic threshold and the basic reproduction number is not clearly revealed. In fact, we can reveal that $\lambda = \lambda_c$ if and only if $R_0 = 1$ by the same way as above.

3.3 Global attraction of endemic equilibrium

In this part, we show a proposition and prove the global attraction of the endemic equilibrium by the same way as the one in [3, 8]. Inequalities (27) and (28) in Proposition 2 are helpful to prove the main result (Theorem 3).

Proposition 1 *Suppose that* $\alpha \geq \beta$ *and the solution* $I_k(t)$ *of system* (2) *satisfies* $\limsup_{t\to\infty} I_k(t) \leq U_k$, $\liminf_{t\to\infty} I_k(t) \geq L_k$, *then*

$$\limsup_{t\to\infty} V(t) \leq \frac{\gamma_2\langle U_k\rangle}{1+\gamma_2\langle U_k\rangle}, \qquad \liminf_{t\to\infty} V(t) \geq \frac{\gamma_2\langle L_k\rangle}{1+\gamma_2\langle L_k\rangle}$$

and

$$\limsup_{t\to\infty}(1 - R_k(t)) \leq \frac{(\alpha-\beta)U_k+\delta}{\alpha+\delta}, \qquad \liminf_{t\to\infty}(1 - R_k(t)) \geq \frac{(\alpha-\beta)L_k+\delta}{\alpha+\delta}.$$

Proof Without loss of generality, we only verify the upper limit inequalities. From $\limsup_{t\to\infty} I_k(t) \leq U_k$, we obtain that for any $\epsilon > 0$, there exists $\tau_0 > 0$ such that $I_k(t) \leq$

$U_k + \epsilon$ for $t > \tau_0$. It follows that

$$\frac{dV(t)}{dt} = -V(t) + \gamma_2\big(1 - V(t)\big)\rho(t) \le -V(t) + \gamma_2\big(1 - V(t)\big)\langle U_k + \epsilon\rangle$$
$$= -\big(1 + \gamma_2\langle U_k + \epsilon\rangle\big)V(t) + \gamma_2\langle U_k + \epsilon\rangle \tag{23}$$

for $t > \tau_0$. Therefore, the following inequality holds since $\epsilon > 0$ is arbitrarily small:

$$\limsup_{t\to\infty} V(t) \le \frac{\gamma_2\langle U_k\rangle}{1 + \gamma_2\langle U_k\rangle}. \tag{24}$$

At the same time, it follows from the second equation of system (2) and $\alpha \ge \beta$ that

$$\frac{d(1 - R_k(t))}{dt} = -(\alpha + \delta)\big(1 - R_k(t)\big) + (\alpha - \beta)I_k(t) + \delta$$
$$\le -(\alpha + \delta)\big(1 - R_k(t)\big) + (\alpha - \beta)(U_k + \epsilon) + \delta. \tag{25}$$

Consequently,

$$\limsup_{t\to\infty}\big(1 - R_k(t)\big) \le \frac{(\alpha - \beta)U_k + \delta}{\alpha + \delta}. \tag{26}$$

□

Proposition 2 *Suppose that* $\alpha \ge \beta$ *and the solution* $I_k(t)$ *of system* (2) *satisfies* $\limsup_{t\to\infty} I_k(t) \le U_k$, $\liminf_{t\to\infty} I_k(t) \ge L_k$, *then*

$$\limsup_{t\to\infty} I_k(t) \le \frac{[\lambda\frac{k}{\langle k\rangle}\langle kU_k\rangle + \frac{\gamma_1\gamma_2\langle U_k\rangle}{1+\gamma_2\langle U_k\rangle}][\frac{\delta+(\alpha-\beta)U_k}{\alpha+\delta}]}{\beta + \lambda\frac{k}{\langle k\rangle}\langle kU_k\rangle + \frac{\gamma_1\gamma_2\langle U_k\rangle}{1+\gamma_2\langle U_k\rangle}}, \tag{27}$$

$$\liminf_{t\to\infty} I_k(t) \ge \frac{[\lambda\frac{k}{\langle k\rangle}\langle kL_k\rangle + \frac{\gamma_1\gamma_2\langle L_k\rangle}{1+\gamma_2\langle L_k\rangle}][\frac{\delta+(\alpha-\beta)L_k}{\alpha+\delta}]}{\beta + \lambda\frac{k}{\langle k\rangle}\langle kL_k\rangle + \frac{\gamma_1\gamma_2\langle L_k\rangle}{1+\gamma_2\langle L_k\rangle}}. \tag{28}$$

Proof From $\limsup_{t\to\infty} I_k(t) \le U_k$ and Proposition 1, one obtains that for any $\epsilon > 0$, there exists large enough τ such that the following inequalities hold for $t > \tau$:

$$\Theta(t,k) \le \frac{\langle k(U_k + \epsilon)\rangle}{\langle k\rangle}, \tag{29}$$

$$V(t) \le \frac{\gamma_2 U_k}{1 + \gamma_2 U_k} + \epsilon, \tag{30}$$

$$1 - R_k(t) \le \frac{(\alpha - \beta)U_k + \delta}{\alpha + \delta} + \epsilon. \tag{31}$$

Considering the first equation of system (2), for $t > \tau$, it follows from (29), (30) and (31) that

$$\frac{dI_k(t)}{dt} = \big[\lambda k\Theta(t,k) + \gamma_1 V(t)\big]\big[1 - I_k(t) - R_k(t)\big] - \beta I_k(t)$$
$$\le \left[\lambda\frac{k}{\langle k\rangle}\langle k(U_k + \epsilon)\rangle + \gamma_1\left(\frac{\gamma_2 U_k}{1 + \gamma_2 U_k} + \epsilon\right)\right]$$
$$\times\left[\frac{(\alpha - \beta)U_k + \delta}{\alpha + \delta} + \epsilon - I_k(t)\right] - \beta I_k(t)$$

$$= -\left[\beta + \lambda\frac{k}{\langle k\rangle}\langle k(U_k+\epsilon)\rangle + \gamma_1\left(\frac{\gamma_2 U_k}{1+\gamma_2 U_k}+\epsilon\right)\right]I_k$$
$$+\left[\lambda\frac{k}{\langle k\rangle}\langle k(U_k+\epsilon)\rangle + \gamma_1\left(\frac{\gamma_2 U_k}{1+\gamma_2 U_k}+\epsilon\right)\right]\left[\frac{(\alpha-\beta)U_k+\delta}{\alpha+\delta}+\epsilon\right]. \qquad (32)$$

Since $\epsilon > 0$ is arbitrary small, we get

$$\limsup_{t\to\infty} I_k(t) \le \frac{[\lambda\frac{k}{\langle k\rangle}\langle kU_k\rangle + \frac{\gamma_1\gamma_2\langle U_k\rangle}{1+\gamma_2\langle U_k\rangle}][\frac{\delta+(\alpha-\beta)U_k}{\alpha+\delta}]}{\beta+\lambda\frac{k}{\langle k\rangle}\langle kU_k\rangle + \frac{\gamma_1\gamma_2\langle U_k\rangle}{1+\gamma_2\langle U_k\rangle}}.$$

Similarly, we can prove inequality (28). □

Denote $\triangle_k = \{(I_k, R_k)|0 \le I_k + R_k \le 1, I_k \ge 0, R_k \ge 0\}$, $k = 1, 2, \ldots, n$, and $\triangle = \prod_{k=1}^{n} \triangle_k \times [0, 1]$.

Theorem 3 *If $\lambda > \lambda_c$ and $\alpha \ge \beta$, then system* (2) *has a unique endemic equilibrium $E^* = \{I_1^*, R_1^*, \ldots, I_n^*, R_n^*, V^*\}$ which is of global attraction in $\triangle - \{F^*\}$, where $F^* = \{0, \frac{\alpha}{\alpha+\delta}, 0, \frac{\alpha}{\alpha+\delta}, \ldots, 0, \frac{\alpha}{\alpha+\delta}, 0\}$ is the disease-free equilibrium of system* (2).

Proof Define a map $\mathcal{G} = \{\mathcal{G}_1, \mathcal{G}_2, \ldots, \mathcal{G}_n\} : \mathbb{R}^n \mapsto \mathbb{R}^n$ as follows:

$$\mathcal{G}_k(x_1, x_2, \ldots, x_n) = \frac{[\lambda\frac{k}{\langle k\rangle}\langle kx_k\rangle + \frac{\gamma_1\gamma_2\langle x_k\rangle}{1+\gamma_2\langle x_k\rangle}][\frac{\delta+(\alpha-\beta)x_k}{\alpha+\delta}]}{\beta+\lambda\frac{k}{\langle k\rangle}\langle kx_k\rangle + \frac{\gamma_1\gamma_2\langle x_k\rangle}{1+\gamma_2\langle x_k\rangle}} \quad (k = 1, 2, \ldots, n). \qquad (33)$$

Let $U_k^{(1)} = 1$ for all $k = 1, 2, \ldots, n$ and $U_k^{(m+1)} = \mathcal{G}_k(U_1^{(m)}, U_2^{(m)}, \ldots, U_n^{(m)})$, it is obvious that $\limsup_{t\to\infty} I_k(t) \le U_k^{(1)} = 1$. According to Proposition 2, we then obtain

$$\limsup_{t\to\infty} I_k(t) \le U_k^{(m)}, \quad k = 1, 2, \ldots, n, m = 1, 2, \ldots.$$

Moreover, for all $k = 1, 2, \ldots, n$, we can testify the convergence of the sequences $\{U_k^{(m)}\}_{m=1}^{+\infty}$ by induction. First, it is obvious that $U_k^{(2)} < U_k^{(1)} = 1$. Secondly, if $U_k^{(m+1)} \le U_k^{(m)}$, then the reader can easily verify that $U_k^{(m+2)} \le U_k^{(m+1)}$. It implies that the sequence $\{U_k^{(m)}\}$ is convergent. Denoted by $U_k = \lim_{m\to\infty} U_k^{(m)}$, we then have $\limsup_{t\to\infty} I_k(t) \le U_k$, $k = 1, 2, \ldots, n$.

Let $\mathcal{H}(\Theta) = \frac{1}{\langle k\rangle}\langle k\mathcal{G}_k\rangle = \frac{1}{\langle k\rangle}\langle k\frac{[\lambda\frac{k}{\langle k\rangle}\langle kx_k\rangle + \frac{\gamma_1\gamma_2\langle x_k\rangle}{1+\gamma_2\langle x_k\rangle}][\frac{\delta+(\alpha-\beta)x_k}{\alpha+\delta}]}{\beta+\lambda\frac{k}{\langle k\rangle}\langle kx_k\rangle + \frac{\gamma_1\gamma_2\langle x_k\rangle}{1+\gamma_2\langle x_k\rangle}}\rangle$, where $\Theta = \frac{1}{\langle k\rangle}\langle kx_k\rangle$. One obtains that $\mathcal{H}'(\Theta)|_{\Theta=0} > 1$ if $\lambda > \lambda_c$. It implies that $\frac{1}{\langle k\rangle}\langle k\mathcal{G}_k\rangle > \frac{1}{\langle k\rangle}\langle kx_k\rangle$ when $x_k > 0$ $(k = 1, 2, \ldots, n)$ is small enough.

Denote $L_k^{(m+1)} = \mathcal{G}_k(L_1^{(m)}, L_2^{(m)}, \ldots, L_n^{(m)})$ for each $k = 1, 2, \ldots, n$. According to Lemma 1 in paper [14], we can take $L_k^{(1)}$ small enough such that $\forall k$, $0 < L_k^{(1)} < \liminf_{t\to\infty} I_k(t)$ and $L_k^{(2)} > L_k^{(1)}$. If $L_k^{(m)} \ge L_k^{(m-1)}$, it is easy to testify that $L_k^{(m+1)} \ge L_k^{(m)}$. In result, the sequences $\{L_k^{(m)}\}_{m=1}^{+\infty}$ are convergent for all k, and denote $L_k = \lim_{m\to\infty} L_k^{(m)}$.

Both $\{L_k\}$ and $\{U_k\}$ satisfy the following equation:

$$I_k = \frac{[\lambda\frac{k}{\langle k\rangle}\langle kI_k\rangle + \gamma_1\frac{\gamma_2\langle I_k\rangle}{1+\gamma_2\langle I_k\rangle}][\frac{\delta+(\alpha-\beta)I_k}{\alpha+\delta}]}{\beta+\lambda\frac{k}{\langle k\rangle}\langle kI_k\rangle + \gamma_1\frac{\gamma_2\langle I_k\rangle}{1+\gamma_2\langle I_k\rangle}}. \qquad (34)$$

After some transformations, we can find that both $\{L_k\}$ and $\{U_k\}$ satisfy Eq. (7). Thus one obtains that $U_k = L_k = I_k^*$ and $\lim_{t\to\infty} I_k(t) = I_k^*$ according to the uniqueness of endemic equilibrium of system (1). The proof of Theorem 3 is completed. □

4 Simulations

In the section above, some theoretical results for the proposed mean-field equations are revealed. For a finite size network, there exists an epidemic threshold λ_c, if the infection rate $\lambda < \lambda_c$, then the disease-free equilibrium is globally asymptotically stable. Otherwise, the disease-free equilibrium is unstable and a unique globally attracting endemic equilibrium exists.

In this section, we will perform some numerical simulations to confirm the theoretical results over BA (Barabási-Albert) scale-free networks which are generated by the preferential attachment algorithm [27]. All the networks used in the simulations were built using $N = 10^4$ nodes.

The epidemics are seeded with randomly chosen fraction of nodes to avoid stochastic extinction. The probability of a susceptible node with n ($\leq$ degree k) infectious neighbors being infected in small interval of time h is $1 - (1 - \lambda h)^n(1 - \gamma_1 hV(t))$ at step $t + 1$. The term $(1 - \lambda h)^n$ represents the probability that a susceptible node cannot be infected by his (her) infectious neighbors, and the term $1 - \gamma_1 hV(t)$ represents the probability that a susceptible node cannot be infected by infectious vectors. In this paper, we give the numerical simulations from different angles:

- The epidemic threshold λ_c that changes as functions of different model parameters.
- The final density of the infected nodes that changes as functions of different model parameters.

In stochastic simulations, the dynamics are totally evolved for 2,000 time steps, we set the time interval $h = 0.1$ and let $\rho = \frac{1}{T}\sum_{t=t_0}^{t_0-1+T} \rho(t)$ (here, $T = 100$, $t_0 = 1{,}901$) be the time average to reduce the fluctuation of $\rho(t)$. At the same time, to minimize random fluctuation caused by the initial conditions, we make average of ρ over 100 realizations of different initial infectious nodes. Let λ increase systematically by $\Delta\lambda$ beginning with $\lambda = 0$, if $\rho > 0.0005$ as $\lambda = \lambda^*$ and $\rho < 0.0005$ as $\lambda < \lambda^*$, we set $\lambda_c = \lambda^* - \Delta\lambda$.

In Figure 1, we illustrate the variation of the epidemic threshold λ_c with respect to the parameters γ_1 and γ_2 both for stochastic simulations (SS) and also for mean-field (MF)

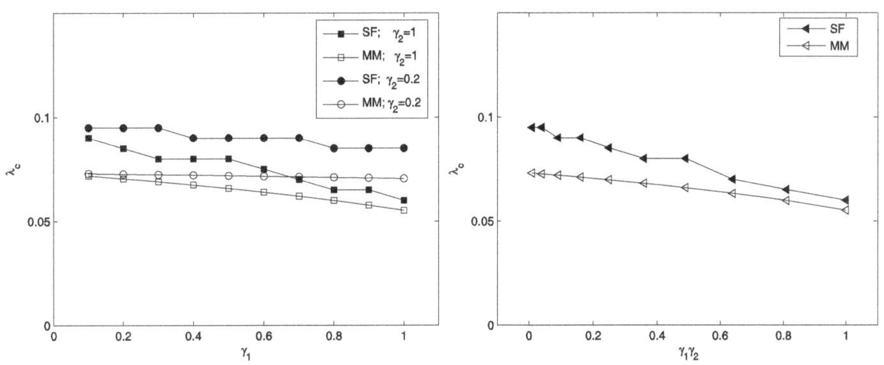

Figure 1 Illustration of the variation of the epidemic threshold with respect to the parameters γ_1 and γ_2. Left: Plot of λ_c *versus* γ_1 for different γ_2. Right: Plot of λ_c *versus* $\gamma_1\gamma_2$. Other parameters are chosen as $\alpha = 1.05$, $\beta = 0.9$, $\delta = 0.8$ and $\langle k\rangle = 10$. 'SF' means stochastic simulations and 'MM' means mean-field predictions.

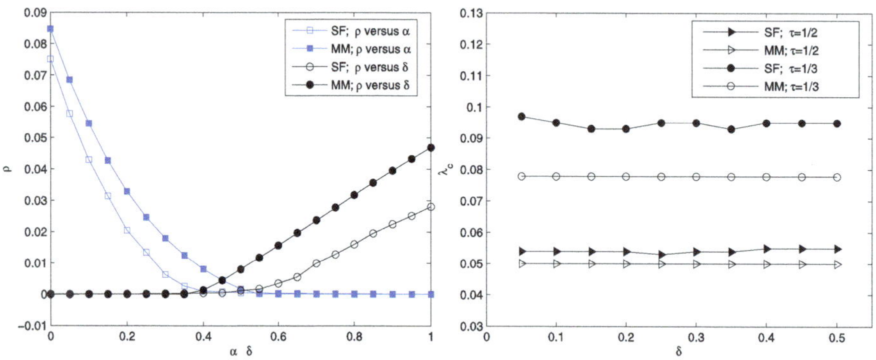

Figure 2 Illustration of the variations of the final infected density and the epidemic threshold with respect to the parameters α and δ. Left: Plot of ρ *versus* α and δ. When considering ρ *versus* α, we set $\delta = 0.5$; when considering ρ *versus* δ, we set $\alpha = 0.4$. Right: The variation of the infection threshold λ_c with respect to δ over BA scale-free network for different $\tau = \frac{\delta}{\alpha+\delta}$. Other parameters are chosen as $\gamma_1 = \gamma_2 = 0.6$, $\beta = 0.8$ and $\langle k \rangle = 10$.

predictions formula (18). It is clear that the epidemic threshold λ_c decreases as γ_1 (or γ_2) increases. We thus conclude that reducing the contacts between individuals and vectors can effectively control the spread of the disease over networks. From this angle, the stochastic simulations agree well with the mean-field predictions. The discrepancy between these is also shown in our simulations, we can see the stochastic simulations are slightly larger than the theoretical predictions for threshold λ_c, which is likely to be due to a distribution cutoff effect on a finite size network [28] and other neglected factors (for instance, network is static and has degree-correlations). Moreover, one can find that for larger infection rates γ_1 and γ_2, the error between stochastic simulation and theoretical predictions is smaller, which is likely to be attributed to the contacts homogeneity between individuals and vectors.

Figure 2 illustrates the variations of the final infected density ρ (left) and the epidemic threshold λ_c (right) with respect to parameters α and δ, respectively. It is clear that the discrepancy remains for the effect of a finite size network. The theoretical prediction of formula (18) shows that for different parameters α and δ, the epidemic threshold λ_c is unchanged as long as $\tau = \frac{\delta}{\alpha+\delta}$ is unchanged. From Figure 2 we can see that the stochastic simulations are in accordance with mean-field predictions disregarding the slight fluctuations for the effect of the stochastic factor. Also, one can see that the larger direct immunization rate α, the larger the epidemic threshold is from Figure 2 (right). Namely, the direct immunization can increase the critical threshold of epidemic spreading on complex networks and reduce the prevalence of infectious disease, which agrees with the results of paper [24]. Consequently, we can prevent the disease spreading by improving immunization strength.

Figure 3 reflects the relation between the final densities of infected nodes and infection rate λ for different average degree. One can see that the larger average degree, the lower the discrepancy rate is. We simulate the time series of total densities of infected nodes on the BA network with $\langle k \rangle = 20$ in Figure 4. We can obtain that the epidemic threshold $\lambda_c = 0.032$ from formula (18). It is clear that if $\lambda < \lambda_c$, the disease will disappear quickly; otherwise $\lambda > \lambda_c$ the disease will persist in this system. Considering the factor that the epidemic threshold λ_c of stochastic simulations is larger than the one of the mean-field

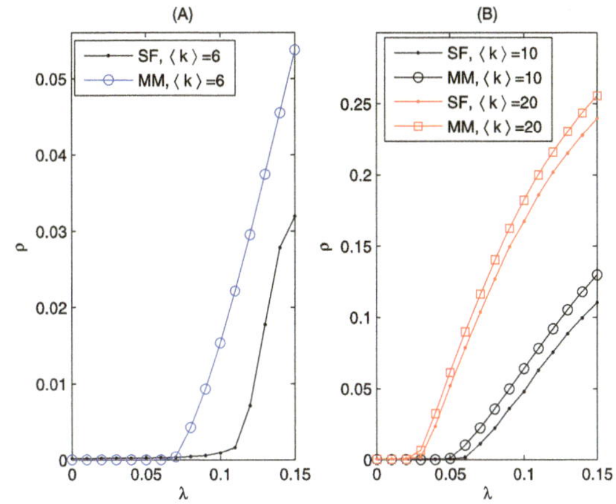

Figure 3 The variation of the stable densities of infected nodes with respect to infection rate λ for BA scale-free network, where $\alpha = 0.83$, $\beta = 0.7$, $\delta = 0.71$, $\gamma_1 = 0.45$, $\gamma_2 = 0.54$.

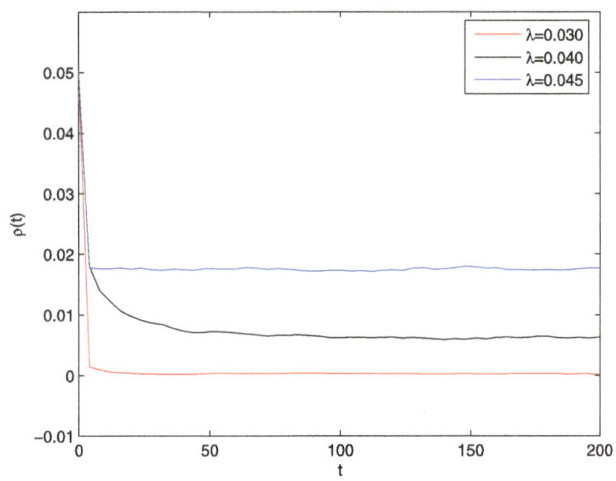

Figure 4 The total densities of infected nodes that change as time series on BA scale-free network with $\langle k \rangle = 20$, where $\alpha = 0.83$, $\beta = 0.8$, $\delta = 0.7$, $\gamma_1 = 0.60$, $\gamma_2 = 0.60$.

predictions, we think the stochastic simulations are reasonably consistent with the theoretic results.

5 Conclusions

In order to better explain the mechanism of spreading of epidemics, we have investigated a novel SIRS model with direct immunization via two infection mechanisms in this paper. The model is approximately described by the mean-field method neglecting contact duration.

The disease-free equilibrium and endemic equilibrium and their dynamics are discussed in this paper. Our theoretic results show that for finite size networks, there exists epidemic threshold λ_c, below which the disease-free equilibrium is globally asymptotically stable; otherwise a unique endemic equilibrium exists. We prove theoretically that the endemic equilibrium is of attraction when $\lambda > \lambda_c$ under the assumption that $\alpha \geq \beta$. To go a step further, we perform some numerical simulations to test and verify our theoretical results.

From simulations as above, we can find that the discrepancies between stochastic simulations and theoretical predictions remain for the effect of a finite size network [28] and stochastic factors. Disregarding these slight errors, we think that the numerical simulations confirm to the theoretical results, and the mean-field approach is of effectiveness. Especially, one can see that the larger transmission rates from infected vectors to susceptible individuals γ_1 or from infected individuals to susceptible vectors γ_2, the better the simulations accord with the mean-field predictions. Just as mentioned above, this may attribute to the homogeneity of contacts between individuals and vectors.

Competing interests
The authors declare that they have no competing interests.

Authors' contributions
All authors contributed equally to the writing of this paper. All authors read and approved the final manuscript.

Author details
[1]College of Sciences, Jiujiang University, Qianjin East Road, Jiujiang, 332005, China. [2]School of Mathematics and Computing Science, Guilin University of Electronic Technology, Guilin, 541004, China.

Acknowledgements
This research was jointly supported by NSFC Grant 11401274 and the Natural Science Foundation of Jiangxi Province (Grants 20142BDH80027 and 20132BAB201012). RZ Yu was also supported by the Foundations of Jiangxi Education Bureau (Grants GJJ12617 and GJJ13714).

References

1. Pastor-Satorras, R, Vespignani, A: Epidemic spreading in scale-free networks. Phys. Rev. Lett. **86**, 3200-3203 (2001)
2. Pastor-Satorras, R, Vespignani, A: Epidemic dynamics and endemic states in complex networks. Phys. Rev. E **63**, 066117 (2001)
3. Yang, M, Chen, GR, Fu, XC: A modified SIS model with an infective medium on complex networks and its global stability. Physica A **390**, 2408-2413 (2011)
4. Pastor-Satorras, R, Vespignani, A: Epidemic dynamics in finite size scale-free networks. Phys. Rev. E **65**, 035108 (2002)
5. d'Onofrio, A: A note on the global behavior of the network-based SIS epidemic model. Nonlinear Anal., Real World Appl. **9**, 1567-1572 (2008)
6. Fu, X, Small, M, Walker, DM, Zhang, H: Epidemic dynamics on scale-free networks with piecewise linear infectivity and immunizations. Phys. Rev. E **77**, 036113 (2008)
7. Xia, C, Wang, Z, Sanz, J, Meloni, S, Morenoc, Y: Effects of delayed recovery and nonuniform transmission on the spreading of diseases in complex networks. Physica A **392**, 1577-1585 (2013)
8. Wang, L, Dai, GZ: Global stability of virus spreading in complex heterogeneous networks. SIAM J. Appl. Math. **68**, 1495-1502 (2008)
9. Yang, R, Wang, BH, Ren, J, Bai, WJ, Shi, ZW, Xu, W, Zhou, T: Epidemic spreading on heterogeneous networks with identical infectivity. Phys. Lett. A **364**, 189-193 (2007)
10. Li, K, Small, M, Zhang, H, Fu, X: Epidemic outbreaks on networks with effective contacts. Nonlinear Anal., Real World Appl. **11**, 1017-1025 (2010)
11. Ball, F, Neal, P: Network epidemic models with two types of mixing. Math. Biosci. **212**, 69-87 (2008)
12. Volz, E: SIR dynamics in random networks with heterogeneous connectivity. J. Math. Biol. **56**, 293-310 (2008)
13. Kiss, IZ, Green, DM, Kao, RR: The effect of contact heterogeneity and multiple routes of transmission on final epidemic size. Math. Biosci. **203**, 124-136 (2006)
14. Lou, J, Ruggeri, T: The dynamics of spreading and immune strategies of sexually transmitted diseases on scale-free network. J. Math. Anal. Appl. **365**, 210-219 (2010)
15. Miller, JC: A note on a paper by Erik Volz: SIR dynamics in random networks. J. Math. Biol. **62**, 349-358 (2011)
16. Xia, C, Wang, L, Sun, S, Wang, J: An SIR model with infection delay and propagation vector in complex networks. Nonlinear Dyn. **69**, 927-934 (2012)
17. Valdez, LD, Macri, PA, Braunstein, LA: Intermittent social distancing strategy for epidemic control. Phys. Rev. E **85**, 036108 (2012)
18. Wu, Q, Fu, X, Small, M, Xu, X-J: The impact of awareness on epidemic spreading in networks. Chaos **22**, 013101 (2012)
19. Molesworth, AM, Cuevas, LE, Connor, SJ, Morse, AP, Thomson, MC: Environmental risk and meningitis epidemics in Africa. Emerg. Infect. Dis. **9**(10), 1287-1293 (2003)
20. Wang, Y, Jin, Z, Yang, Z, Zhang, Z-K, Zhou, T, Sun, G-Q: Global analysis of an SIS model with an infective vector on complex networks. Nonlinear Anal., Real World Appl. **13**, 543-557 (2012)
21. Shi, HJ, Duan, ZS, Chen, GR: An SIS model with infective medium on complex networks. Physica A **387**, 2133-2144 (2008)
22. Cooke, KL: Stability analysis for a vector disease model. Rocky Mt. J. Math. **9**, 31-42 (1979)

23. Busenberg, S, Cooke, KL: Periodic solutions of a periodic nonlinear delay differential equation. SIAM J. Appl. Math. **35**, 704-721 (1978)
24. Xia, C, Liu, Z, Chen, Z, Yuan, Z: SIRS epidemic model with direct immunization on complex networks. Control Decis. **23**, 468-472 (2008) (in Chinese)
25. Diekmann, O, Heesterbeek, JAP, Roberts, MG: The construction of next-generation matrices for compartmental epidemic models. J. R. Soc. Interface **7**(47), 873-885 (2010)
26. Diekmann, O, Heesterbeek, JAP, Roberts, MG: On the definition and the computation of the basic reproduction ratio R_0 in models for infectious diseases in heterogeneous populations. J. Math. Biol. **28**, 365-382 (1990)
27. Barabási, AL, Albert, R: Emergence of scaling in random networks. Science **286**, 509-512 (1999)
28. Olinky, R, Stone, L: Unexpected epidemic thresholds in heterogeneous networks: the role of disease transmission. Phys. Rev. E **70**, 030902(R) (2004)

13

Complex projective synchronization in drive-response stochastic networks with switching topology and complex-variable systems

Xuefei Wu*

*Correspondence:
wuxuefei@szpt.edu.cn
School of Computer and Software Engineering, Shenzhen Polytechnic, Shenzhen, 518055, P.R. China

Abstract

In this paper, the complex projective synchronization in drive-response stochastic switching networks with complex-variable systems is considered. The pinning control scheme and the adaptive feedback algorithms are adopted to achieve complex projective synchronization, and the structure of stochastic switching networks with complex-variable systems makes our research more universal and practical. Using a suitable Lyapunov function, we obtain some simple and practical sufficient conditions which guarantee the complex projective synchronization in drive-response stochastic switching networks with complex-variable systems. Illustrated examples have been given to show the effectiveness and feasibility of the proposed methods.

Keywords: complex projective synchronization; stochastic switching networks; drive-response; time delay

1 Introduction

Closely related with people's life, network ranges from the Internet, the communication network to the biological networks, neural networks in nature, *etc.* It can be manifested in the form of complex networks. Generally speaking, it exists in nature and society. The studies on complex networks have become one of the hottest topics in the scientific research, and they have attracted wide attention of researchers working in the fields of information science, mathematics, physics, biology, system control, engineering, economics, society, military and so on [1–5]. In the studies on a variety of dynamical behaviors of complex networks, synchronization, as a typical form of describing collective motion of networks is one of the most important group dynamic behaviors of the network. Because of scientific importance and universality of real networks as well as a wealth of theoretical basis and challenge, it occupies a very important position in the studies of complex networks, and fruitful research results are achieved. In the literature, there are many widely studied synchronization patterns, for example, complete synchronization [6–8], lag synchronization [9–11], anti-synchronization [12–14], phase synchronization [15, 16], projective synchronization [17–29], and so on. Projective synchronization refers to the state variable response network gradually tending to a percentage value of the drive network state variables under some control. Due to the flexibility of the synchronization state scaling factor in projective synchronization, it is popular in the field of security digital communication.

Recently, projective synchronization under various cases of complex dynamical networks has been studied [17–29]. In Ref. [25], Li studied the generalized projective synchronization between two different chaotic systems: Lorenz system and Chen's system. The proposed method combines backstepping methods and active control without having to calculate the Lyapunov exponents and the eigenvalues of the Jacobian matrix, which makes it simple and convenient. In Ref. [26], Lü *et al.* proposed a method of the lag projective synchronization of a class of complex networks containing nodes with chaotic behavior. Discrete chaotic systems are taken as nodes to constitute a complex network, and the topological structure of the network can be arbitrary. Considering the lag effect between a network node and a chaos signal of the target system, the control input of the network and the identification law of adjustment parameters are designed based on the Lyapunov theorem. In Ref. [27], Zhu *et al.* explored the mode-dependent projective synchronization problem of a couple of stochastic neutral-type neural networks with distributed time-delays. By using the Lyapunov stability theory and the adaptive control method, a sufficient projective synchronization criterion for this neutral-type neural network model is derived.

In the existing research, the most complex networks are usually described by the real variable differential system. The two complex networks (the so-called drive-response networks) based on real number, real matrix, or even real function evolve along the same or inverse directions [17–27]. However, the drive-response networks based on complex number can often evolve in different directions with a constant intersection angle, for example, $y = \rho e^{j\theta} x$, where x denotes the drive system, y denotes the response system, $\rho > 0$ denotes the zoom rate, $\theta \in [0, 2\pi)$ denotes the rotation angle and $j = \sqrt{-1}$. The state variables of the system defined in the complex field can describe a lot of practical problems. For example, in Ref. [30], authors use the state variables of a complex Lorenz system to describe the physical properties of parameters of atom polarization amplitude in the laser harmonic, electric field and population inversion, has realized the synchronization between two chaotic attractors, which fully illustrates that the complex system has the important application in the engineering. At the same time, if the projective synchronization methods combined with complex system are applied in the field of secure communications, this will further enhance the security of secret communication. Recently, some related works have come out, such as [28, 29]. In Refs. [28, 29], Wu and Fu introduced the concept of complex projective synchronization based on Lyapunov stability theory, several typical chaotic complex dynamical systems are considered and the corresponding controllers are designed to achieve the complex projective synchronization. This synchronization scheme has a large number of real-life examples. For instance, in a social network or games of economic activities, behaviors of individuals (response systems) will be affected not only by powerful one (drive system), but also by those with a similar role as themselves [28]. Another example, in a distributed computers collaboration, each distributed computer (response system) not only receives unified command from a server (drive system), but they also share resources between each other for collaboration [31].

Besides, in a lot of real network systems, nodes coupling way (the so-called topology) is not fixed but changing over time. For example, in the urban traffic network, the city partial obstruction caused by a traffic accident in the main street can result in the change of urban traffic network structure; in an interpersonal network, along with the development of society, economy *etc.*, the relationships between different people are also changing. These factors also lead to changes in the whole network topology, they all belong to the structure of switch network topology, and the existing research on the static dynamic network synchronization methods is no longer suitable for this switch topology with time-varying dynamic network, which requires us to design a new network synchronization method [32–35].

Furthermore, a signal transmitted between the network nodes usually is affected by the network bandwidth, transmission medium and measuring noise factors, which result in time delays [21, 23, 25, 36], randomly missing or incomplete information [14–18]. Therefore, it is important to study the effect of time delay and stochastic noise in complex projective synchronization of drive-response networks.

Based on the above, the complex projective synchronization in drive-response stochastic switching networks with complex-variable systems is considered in this paper. The complex projective synchronization is achieved via a pinning control scheme and an adaptive coupling strength method. Several simple and practical criteria for complex projective synchronization are obtained by using the Lyapunov functional method, stochastic differential theory and linear matrix inequality (LMI) approaches.

Notation: Throughout this paper, $\mathbb{C}^n$ and $\mathbb{C}^{m\times n}$ denote n-dimensional complex vectors and the set of $m \times n$ complex matrices, respectively. For the Hermite matrix H, the notation $H > 0$ ($H < 0$) means that the matrix H is positive definite (negative definite). For any complex (real) matrix M, $M^s = M^T + \overline{M}$. For any complex number (or complex vector) x, the notations x^r and x^i denote its real and imaginary parts, respectively, and $\bar{x}$ denotes the complex conjugate of x. $\lambda_{\min}(A)$ ($\lambda_{\max}(A)$) represents the smallest (largest) eigenvalue of a symmetric matrix A. $\otimes$ is the Kronecker product. The superscript T of x^T or A^T denotes the transpose of the vector $x \in \mathbb{R}^n$ or the matrix $A \in \mathbb{C}^{m\times n}$. I_n is an identity matrix with n nodes.

2 Model description and preliminaries

Consider a drive-response network coupled with $1 + N$ identical partially linear stochastic complex network with coupling time delay, which is described as follows:

$$\begin{cases} \dot{x}(t) = M(z(t))x(t), \\ \dot{z}(t) = f(x(t), z(t)), \end{cases} \tag{1}$$

$$\begin{aligned} dy_i(t) = &\left\{ M\big(z(t)\big)y_i(t) + \varepsilon_1 \sum_{k=1}^{N} a_{ik}\big(r(t)\big)\Gamma y_k(t) + \varepsilon_2 \sum_{k=1}^{N} b_{ik}\big(r(t)\big)\Gamma y_k(t-\tau) \right\} dt \\ &+ \sigma_i\big(y(t), y(t-\tau)\big)\, dw_i(t), \quad i = 1, 2, \dots, N, \end{aligned} \tag{2}$$

where $x(t) = (x_1, x_2, \dots, x_m)^T \in \mathbb{C}^m$, and $z(t) \in \mathbb{R}$ is the drive system variable, $y_i(t) = (y_{i1}, y_{i2}, \dots, y_{im})^T \in \mathbb{C}^m$ is the state variable of a node i in the response network. $M(z(t)) \in \mathbb{R}^{m\times m}$ is a complex matrix function, $\varepsilon_1 > 0$, $\varepsilon_2 > 0$ is the coupling strength and $\Gamma \in \mathbb{R}^{m\times m}$ is the inner coupling matrix. τ is the coupling time delay; $r(t) = r : [0, +\infty) \to (1, 2, \dots, M)$ is a switching signal. Matrices $A(r(t)) = (a_{ij}(r(t)))_{N\times N}$ and $B(r(t)) = (b_{ij}(r(t)))_{N\times N}$ are the zero-row-sum outer coupling matrices, which denote the network switching topology and are defined as follows: if there is a connection (information transmission) from node j to node i $(i \neq j)$, then $a_{ij}(r(t)) \neq 0$ and $b_{ij}(r(t)) \neq 0$; otherwise, $a_{ij}(r(t)) = 0$ and $b_{ij}(r(t)) = 0$; and $w_i(t) = (w_{i1}(t), w_{i2}(t), \ldots, w_{in}(t))^T \in \mathbb{R}^n$ is a bounded vector-form Weiner process satisfying

$$\mathbb{E}w_{ij}(t) = 0, \qquad \mathbb{E}w_{ij}^2(t) = 1, \qquad \mathbb{E}w_{ij}(t)w_{ij}(s) = 0 \quad (s \neq t).$$

Now, two mathematical definitions for the generalized projective synchronization are introduced as follows.

Definition 1 If there is a complex α such that

$$\sum_{i=1}^{N} \mathbb{E}\left\| y_i(t) - \alpha x(t) \right\|^2 \leq Ke^{-\kappa t}$$

for some $K > 0$ and some $\kappa > 0$, then the drive-response network (1) and (2) is said to achieve complex projective synchronization in the mean-square, and the parameter α is called a scaling factor.

Without loss of generality, let $\alpha = \rho(\cos\theta + j\sin\theta)$, where $\rho = |\alpha|$ is the module of α and $\theta \in [0, 2\pi)$ is the phase of α. Therefore, the projective synchronization is achieved when $\theta = 0$ or π. Furthermore, the complete synchronization is achieved when $\rho = 1$ and $\theta = 0$, the anti-synchronization is achieved when $\rho = 1$ and $\theta = \pi$ [28].

Definition 2 [28] Matrix $A = (a_{ik})_{i,k}^{N}$ is said to belong to class A_1, denoted as $A \in A_1$, if

(1) $a_{ik} \geq 0$, $i \neq k$, $a_{ii} = -\sum_{k=1,k\neq i}^{N} a_{ik} = -\sum_{k=1,k\neq i}^{N} a_{ki}$, $i = 1, 2, \ldots, N$;

(2) A is irreducible.

The following lemmas and assumption are used throughout the paper.

Lemma 1 [28] *Let $m \times m$ be a complex matrix, H be Hermitian, then*

(1) *$x^T H\bar{x}$ is real for all $x \in C^m$;*

(2) *all the eigenvalues of H are real.*

Lemma 2 [36] *If $A = (a_{ij})_{m\times m}$ is irreducible, $a_{ij} = a_{ji} \geq 0$ for $i \neq j$, and $\sum_{j=1}^{m} a_{ij} = 0$ for all $i = 1, 2, \ldots, m$, then all eigenvalues of the matrix*

$$\begin{pmatrix} a_{11} - \varepsilon & a_{12} & \cdots & a_{1m} \\ a_{21} & a_{22} & \cdots & a_{2m} \\ \vdots & \vdots & \ddots & \vdots \\ a_{m1} & a_{m2} & \cdots & a_{mm} \end{pmatrix}$$

are negative for any positive constant ε.

Lemma 3 [31, 37] *Consider an n-dimensional stochastic differential equation*

$$dx(t) = f\left(t, x(t), x(t-\tau)\right) dt + \sigma\left(t, x(t), x(t-\tau)\right) d\omega(t). \tag{3}$$

Let $C^{2,1}(\mathbb{C}_+ \times \mathbb{C}^n; \mathbb{R}_+)$ denote the family of all nonnegative functions $V(t,x)$ on $\mathbb{R}_+ \times \mathbb{C}^n$, which are twice continuously differentiable in x and once differentiable in t. If $V \in C^{2,1}(\mathbb{R}_+ \times \mathbb{C}^n; \mathbb{R}_+)$, define an operator $\mathcal{L}V$ from $\mathbb{R}_+ \times \mathbb{C}^n$ to $\mathbb{R}$ by

$$\mathcal{L}V(t,x) = V_t(t,x) + V_x(t,x)f(t,x,y) + \frac{1}{2}\mathrm{Tr}\big[\sigma(t,x,y)^T V_{xx}\sigma(t,x,y)\big],$$

where $V_t(t,x) = \partial V(t,x)/\partial t$, $V_x(t,x) = (\partial V(t,x)/\partial x_1, \ldots, \partial V(t,x)/\partial x_n)$, $V_{xx}(t,x) = (\frac{\partial^2 V(t,x)}{\partial x_i x_j})_{n\times n}$. If $V \in C^{2,1}(\mathbb{R}_+ \times \mathbb{C}^n; \mathbb{R}_+)$, then for any $\infty > t > t_0 \geq 0$,

$$\mathbb{E}V\big(t,x(t)\big) = \mathbb{E}V\big(t_0,x(t_0)\big) + \mathbb{E}\int_{t_0}^{t} \mathcal{L}V\big(s,x(s)\big)\,ds$$

as long as the expectations of the integrals exist.

Assumption 1 [28] Suppose that there exists a constant L such that the largest eigenvalue of $M^s(z(t))$ satisfies

$$\lambda_{\max}\big(M^s\big(z(t)\big)\big) \leq L.$$

Remark 1 All the chaotic systems satisfy Assumption 1 due to $z(t)$ is bounded [28].

Assumption 2 Denote $e_i(t) = y_i(t) - \alpha x(t)$, suppose $\sigma_i(e(t), e(t-\tau)) = \sigma_i(y(t), y(t-\tau))$. Then there exist positive definite constant matrices Υ_{i1}, Υ_{i2} for $i = 1, 2, \ldots, N$ such that

$$\begin{aligned}&\mathrm{Tr}\big[\sigma_i^T\big(e(t), e(t-\tau)\big)\overline{\sigma_i\big(e(t), e(t-\tau)\big)}\big]\\ &\quad\leq \sum_{j=1}^{N} e_j^T(t)\Upsilon_{i1}\overline{e_j(t)} + \sum_{j=1}^{N} e_j^T(t-\tau)\Upsilon_{i2}\overline{e_j(t-\tau)}.\end{aligned}$$

Remark 2 Assumption 2 is easily satisfied, for instance, because of existing noise in the process of information transmission, the noise strength $\sigma_i(y(t), y(t-\tau)) = |\bar{\sigma}_i \times \sum_{k=1}^{N} a_{ik}(r(t))(y_k(t) - y_i(t)) + \tilde{\sigma}_i \sum_{k=1}^{N} b_{ik}(r(t))(y_k(t-\tau) - y_i(t-\tau))|$, which depends on the states of the nodes, where $\bar{\sigma}_i$ and $\tilde{\sigma}_i$ are constants, $i = 1, 2, \ldots, N$, so that $\Upsilon_{i1} = \bar{\sigma}_i N\,\mathrm{diag}\{a_{i1}^2, a_{i2}^2, \ldots, a_{iN}^2\}$, $\Upsilon_{i2} = \tilde{\sigma}_i N\,\mathrm{diag}\{b_{i1}^2, b_{i2}^2, \ldots, b_{iN}^2\}$.

3 Main results

Our objective here is to achieve complex projective synchronization in the drive-response network (1) and (2) by adopting different control schemes. Firstly, several sufficient conditions for achieving complex projective synchronization in the drive-response network (1) and (2) by applying proper controllers $u_i(t)$ on the response network are obtained. Then the controlled response network is

$$\begin{aligned}dy_i(t) = &\left\{M\big(z(t)\big)y_i(t) + \varepsilon_1\sum_{k=1}^{N} a_{ik}\big(r(t)\big)\Gamma y_k(t) + \varepsilon_2\sum_{k=1}^{N} b_{ik}\big(r(t)\big)\Gamma y_k(t-\tau) + u_i(t)\right\}dt\\ &+ \sigma_i\big(y(t), y(t-\tau)\big)\,dw_i(t), \quad i = 1, 2, \ldots, N. \qquad (4)\end{aligned}$$

Define the synchronization errors between the drive network (1) and the response network (2) as $e_i(t) = y_i(t) - \alpha x(t)$ because of $\sigma_i(e(t), e(t-\tau)) = \sigma_i(y(t), y(t-\tau))$, then we have the following error system:

$$de_i(t) = \left\{ M(z(t))e_i(t) + \varepsilon_1 \sum_{k=1}^{N} a_{ik}(r(t))\Gamma e_k(t) + \varepsilon_2 \sum_{k=1}^{N} b_{ik}(r(t))\Gamma e_k(t-\tau) + u_i(t) \right\} dt$$
$$+ \sigma_i(e(t), e(t-\tau))\, dw_i(t), \quad i = 1, 2, \ldots, N. \tag{5}$$

Next, we consider complex projective synchronization between (1) and (3) via pinning control under the assumption $A \in A_1$, $B \in A_1$ and $\Gamma > 0$. Especially, only one node is pinning for achieving complex projective synchronization.

Theorem 1 *Suppose that Assumption* 1 *holds,* $A(r) \in A_1$, $B(r) \in A_1$ *for* $r = 1, 2, \ldots, M$ *and* $\Gamma > 0$. *The complex projective synchronization in the drive-response network* (1) *and* (3) *with the following single controller*

$$u_1(t) = -\varepsilon_1 d_1 \Gamma e_1(t),$$
$$u_i(t) = 0, \quad i = 2, \ldots, N, \tag{6}$$

can be achieved if the following condition is satisfied:

$$\gamma + a + be^{\gamma\tau} < 0, \tag{7}$$

where $\gamma > 0$, $a = \min_r \lambda_{\min}((L+\alpha)I_{Nm} + \varepsilon_1((A(r))^s - D_1) \otimes \Gamma) + \Upsilon_{i1}$, $b = \max_r \lambda_{\max}(\frac{1}{\alpha}\varepsilon_2^2((B(r)) \otimes \Gamma)^{sT}((B(r)) \otimes \Gamma) + \Upsilon_{i2})$, $d_1 > 0$ *and* $D_1 = \mathrm{diag}(2d_1, 0, \ldots, 0)$.

Proof Consider the Lyapunov functional candidate

$$V(t) = \sum_{i=1}^{N} e_i^T(t)\overline{e_i(t)}.$$

Calculating $\mathcal{L}V(t)$ with respect to t along the solution of (4) and noticing the adaptive feedback controllers (5), for $r(t) = r$, one has

$$\mathcal{L}V(t) = \sum_{i=1}^{N} \left\{ \left[e_i^T(t)M^T(z(t)) + \varepsilon_1 \sum_{k=1}^{N} a_{ik}(r)e_k^T(t)\Gamma^T + \varepsilon_2 \sum_{k=1}^{N} b_{ik}(r)e_k^T(t-\tau)\Gamma^T \right] \overline{e_i(t)} \right.$$
$$\left. + e_i^T(t) \left[\overline{M(z(t))e_i(t)} + \varepsilon_1 \sum_{k=1}^{N} a_{ik}(r)\Gamma\overline{e_k(t)} + \varepsilon_2 \sum_{k=1}^{N} b_{ik}(r)\Gamma\overline{e_k(t-\tau)} \right] \right\}$$
$$- 2\varepsilon_1 d_1 e_1^T(t)\Gamma\overline{e_1(t)} + \frac{1}{2}\sum_{i=1}^{N} \mathrm{Tr}\left[\sigma_i(e(t), e(t-\tau))^T \overline{\sigma_i(e(t), e(t-\tau))}\right]$$
$$= \sum_{i=1}^{N} \left[e_i^T(t)M^T(z(t))\overline{e_i(t)} + e_i^T(t)\overline{M(z(t))e_i(t)} \right]$$
$$+ \varepsilon_1 \sum_{i=1}^{N}\sum_{k=1}^{N} a_{ik}(r)\left[e_k^T(t)\Gamma^T + e_i^T(t)\Gamma\right]\overline{e_k(t)}$$

$$+\varepsilon_2\sum_{i=1}^{N}\sum_{k=1}^{N}b_{ik}(r)\big[e_k^T(t-\tau)\Gamma^T\overline{e_k(t)}+e_i^T(t)\Gamma\overline{e_k(t-\tau)}\big]-2\varepsilon_1 d_1 e_1^T(t)\Gamma\overline{e_1(t)}$$

$$+\frac{1}{2}\sum_{i=1}^{N}\mathrm{Tr}\big[\sigma_i\big(e(t),e(t-\tau)\big)^T\overline{\sigma_i\big(e(t),e(t-\tau)\big)}\big]$$

$$\le\sum_{i=1}^{N}Le_i^T(t)\overline{e_i(t)}+\varepsilon_1\sum_{i=1}^{N}\sum_{k=1}^{N}a_{ik}(r)\big[e_k^T(t)\Gamma^T+e_i^T(t)\Gamma\big]\overline{e_k(t)}-2\varepsilon_1 d_1 e_1^T(t)\Gamma\overline{e_1(t)}$$

$$+\varepsilon_2\sum_{i=1}^{N}\sum_{k=1}^{N}b_{ik}(r)\big[e_k^T(t)\Gamma^T+e_i^T(t)\Gamma\big]\overline{e_k(t-\tau)}$$

$$+\sum_{i=1}^{N}e_i^T(t)\Upsilon_{i1}\overline{e_i(t)}+\sum_{i=1}^{N}e_i^T(t-\tau)\Upsilon_{i2}\overline{e_i(t-\tau)}.$$

Let $e(t)=(e_1^T(t),e_2^T(t),\dots,e_N^T(t))^T$, then one has

$$\mathcal{L}V(t)\le e^T(t)\big(\big((L+\alpha)I_{Nm}+\varepsilon_1\big(\big(A(r)\big)^s-D_1\big)\otimes\Gamma\big)+\Upsilon_{i1}\big)\overline{e(t)}$$
$$+e^T(t-\tau)\left(\frac{1}{\alpha}\varepsilon_2^2\big(B(r)\otimes\Gamma\big)^{sT}\big(B(r)\otimes\Gamma\big)+\Upsilon_{i2}\right)\overline{e(t-\tau)}.$$

In view of condition (6), we have

$$\mathcal{L}V(t)\le aV(t)+bV(t-\tau). \tag{8}$$

Define

$$W(t)=e^{\gamma t}V(t)$$

and use equation (6) to compute the operator

$$\mathcal{L}W(t)=e^{\gamma t}\big[\gamma V(t)+\mathcal{L}V(t)\big]$$
$$\le e^{\gamma t}\big[\gamma V(t)+aV(t)+bV(t-\tau)\big],$$

which, after applying the generalized Itô formula, gives

$$e^{\gamma t}\mathbb{E}V(t)=\mathbb{E}V(0)+\mathbb{E}\int_0^t\mathcal{L}W(s)\,ds \tag{9}$$

for any $t\ge 0$. Hence we have

$$e^{\gamma t}\mathbb{E}V(t)\le\mathbb{E}V(0)+\mathbb{E}\int_0^t e^{\gamma s}\big[\gamma V(s)+aV(s)+bV(s-\tau)\big]\,ds$$
$$\le\mathbb{E}V(0)+(\gamma+a)\int_0^t e^{\gamma s}\mathbb{E}V(s)\,ds+be^{\gamma\tau}\int_0^t e^{\gamma(s-\tau)}\mathbb{E}V(s-\tau)\,ds. \tag{10}$$

By changing variable $s-\tau=u$, we have

$$\int_0^t e^{\gamma(s-\tau)}\mathbb{E}V(s-\tau)\,ds=\int_{-\tau}^{t-\tau}e^{\gamma u}\mathbb{E}V(u)\,du\le\int_{-\tau}^{t}e^{\gamma u}\mathbb{E}V(u)\,du. \tag{11}$$

Substituting equation (11) into equation (10), we get

$$e^{\gamma t}\mathbb{E}V(t) \leq \mathbb{E}V(0) + \left(\gamma + a + be^{\gamma\tau}\right)\int_{-\tau}^{t} e^{\gamma u}\mathbb{E}V(u)\,du.$$

By using Gronwall's inequality, we get

$$e^{\gamma t}\mathbb{E}V(t) \leq Ke^{-\kappa t},$$

where $K = \mathbb{E}V(0)e^{(\gamma+a+be^{\gamma\tau})\tau}$ and $\kappa = -(\gamma + a + be^{\gamma\tau})$. In light of condition (7), the proof is completed. □

Remark 3 In Theorem 1, the coupling matrix A must be strongly connected and the coupling matrix B is not necessarily a symmetrical or irreducible matrix. From condition (7) of Theorem 1, we can determine the control strength ε_1 and ε_2 to reach complex project synchronization.

If considering the system without delay, that is, $\tau = 0$, we can derive the following controlled response network and the error system:

$$\begin{aligned} dy_i(t) = {} & \left\{M(z(t))y_i(t) + \varepsilon_1\sum_{k=1}^{N} a_{ik}(r(t))\Gamma y_k(t) + u_i(t)\right\}dt \\ & + \sigma_i(y(t))\,dw_i(t), \quad i = 1,2,\ldots,N \end{aligned} \tag{12}$$

and

$$\begin{aligned} de_i(t) = {} & \left\{M(z(t))e_i(t) + \varepsilon_1\sum_{k=1}^{N} a_{ik}(r(t))\Gamma e_k(t) + u_i(t)\right\}dt \\ & + \sigma_i(e(t))\,dw_i(t), \quad i = 1,2,\ldots,N, \end{aligned} \tag{13}$$

then, without loss of generality, one has the following corollary.

Corollary 1 *Suppose that Assumption* 1 *holds,* $A(r) \in A_1$, $\Gamma > 0$ *for* $r = 1,2,\ldots,M$. *The complex projective synchronization in the drive-response network* (1) *and* (3) *with the following single controller*:

$$\begin{cases} u_1(t) = -\varepsilon_1 d_1 \Gamma e_1(t), \\ u_i(t) = 0, \quad i = 2,\ldots,N, \end{cases}$$

can be achieved if $a < 0$ *is satisfied where* $a = \min_r \lambda_{\min}(LI_{Nm} + \varepsilon_1((A(r))^s - D_1) \otimes \Gamma) + \Upsilon_{i1}$, $d_1 > 0$ *and* $D_1 = \mathrm{diag}(2d_1, 0, \ldots, 0)$.

Theorem 1 and Corollary 1 state that a drive-response stochastic coupled networks can achieve complex projective synchronization by controlling only a fraction of the nodes, provided that its control strength is sufficiently large. It is usually much larger than the value needed. Clearly it is a natural idea to make the control strength as small as possible. Next, we will realize complex projective synchronization for relatively small control strengths by using adaptive adjustments. Let

$$\begin{cases} u_1(t) = -\varepsilon_1 d(t)\Gamma e_1(t), \\ u_i(t) = 0, \quad i = 2,\ldots,N, \\ \dot{d}(t) = \delta \sum_{i=1}^N e_i^T(t)\overline{e_i(t)}, \end{cases}$$

where $\delta > 0$ is the adaptive gain.

Theorem 2 *Suppose that Assumption* 1 *holds, P is a positive definite matrix, the complex projective synchronization in the drive-response network* (1) *and* (3) *with controllers* (12) *can be achieved if the following conditions are satisfied:*

$$\begin{aligned} &I_N \otimes \left((L+\alpha)I_m + P + \Upsilon_{i1}\right) + \varepsilon_1\left((A(r))^s - D^*\right) \otimes \Gamma < 0 \quad \textit{for } r = 1,2,\ldots,M, \\ &\frac{1}{\alpha}\varepsilon_2^2\left((B(r)) \otimes \Gamma\right)^{sT}\left((B(r)) \otimes \Gamma\right) - I_N \otimes (P - \Upsilon_{i2}) < 0 \quad \textit{for } r = 1,2,\ldots,M \end{aligned} \tag{14}$$

for a small positive constant δ *and* $D_1^* = \mathrm{diag}(2d_1^*, 0, \ldots, 0)$.

Proof Consider the Lyapunov functional candidate

$$V(t) = \sum_{i=1}^N e_i^T(t)\overline{e_i(t)} + \frac{1}{\delta}\left(d(t) - d^*\right)^2 + \sum_{i=1}^N \int_{t-\tau}^t e_i^T(s)P\overline{e_i(s)}\,ds.$$

Calculating $\mathcal{L}V(t)$ with respect to t along the solution of (4) and noticing the adaptive feedback controllers (5), for $r(t) = r$, one has

$$\begin{aligned} \mathcal{L}V(t) = &\sum_{i=1}^N \Bigg\{\Bigg[e_i^T(t)M^T(z(t)) + \varepsilon_1 \sum_{k=1}^N a_{ik}(r)e_k^T(t)\Gamma^T \\ &+ \varepsilon_2 \sum_{k=1}^N b_{ik}(r)e_k^T(t-\tau)\Gamma^T - e_i^T(t)d(t)\Bigg]\overline{e_i(t)} \\ &+ e_i^T(t)\Bigg[\overline{M(z(t))e_i(t)} + \varepsilon_1 \sum_{k=1}^N a_{ik}(r)\Gamma\overline{e_k(t)} \\ &+ \varepsilon_2 \sum_{k=1}^N b_{ik}(r)\Gamma\overline{e_k(t-\tau)} - d(t)\overline{e_i(t)}\Bigg] + e_i^T(t)P\overline{e_i(t)} - e_i^T(t-\tau)P\overline{e_i(t-\tau)}\Bigg\} \\ &+ \frac{2}{\delta}\left(d(t) - d^*\right)\delta e_1^T(t)\Gamma\overline{e_1(t)} + \frac{1}{2}\sum_{i=1}^N \mathrm{Tr}\left[\sigma_i\left(e(t), e(t-\tau)\right)^T \overline{\sigma_i\left(e(t), e(t-\tau)\right)}\right] \\ = &\sum_{i=1}^N \left[e_i^T(t)M^T(z(t))\overline{e_i(t)} + e_i^T(t)\overline{M(z(t))e_i(t)} + e_i^T(t)P\overline{e_i(t)}\right] \\ &+ \varepsilon_1 \sum_{i=1}^N \sum_{k=1}^N a_{ik}(r)\left[e_k^T(t)\Gamma^T + e_i^T(t)\Gamma\overline{e_k(t)}\right] \\ &+ \varepsilon_2 \sum_{i=1}^N \sum_{k=1}^N b_{ik}(r)\left[e_k^T(t-\tau)\Gamma^T + e_i^T(t)\Gamma\overline{e_k(t-\tau)}\right] \\ &- 2d^* e_1^T(t)\Gamma\overline{e_1(t)} - e_i^T(t-\tau)P\overline{e_i(t-\tau)} \end{aligned}$$

$$
\begin{aligned}
&+\frac{1}{2}\sum_{i=1}^{N}\mathrm{Tr}\big[\sigma_i\big(e(t),e(t-\tau)\big)^T\overline{\sigma_i\big(e(t),e(t-\tau)\big)}\big]\\
&\le\sum_{i=1}^{N}e_i^T(t)(LI_m+P)\overline{e_i(t)}+\varepsilon_1\sum_{i=1}^{N}\sum_{k=1}^{N}a_{ik}(r)\big[e_k^T(t)\Gamma^T+e_i^T(t)\Gamma\overline{e_k(t)}\big]\\
&-\sum_{i=1}^{N}e_i^T(t-\tau)P\overline{e_i(t-\tau)}-2d^*e_1^T(t)\Gamma\overline{e_1(t)}\\
&+\varepsilon_2\sum_{i=1}^{N}\sum_{k=1}^{N}b_{ik}(r)\big[e_k^T(t-\tau)\Gamma^T+e_i^T(t)\Gamma\overline{e_k(t-\tau)}\big]\\
&+\sum_{i=1}^{N}e_i^T(t)\Upsilon_{i1}\overline{e_i(t)}+\sum_{i=1}^{N}e_i^T(t-\tau)\Upsilon_{i2}\overline{e_i(t-\tau)}.
\end{aligned}
$$

Let $e(t)=(e_1^T,e_2^T,\dots,e_N^T)$, then one has

$$
\begin{aligned}
\mathcal{L}V(t)\le{}& e^T(t)\big(I_N\otimes\big((L+\alpha)I_m+P+\Upsilon_{i1}\big)+\varepsilon_1\big(\big(A(r)\big)^s-D^*\big)\otimes\Gamma\big)\overline{e(t)}\\
&+e^T(t-\tau)\Big(\frac{1}{\alpha}\varepsilon_2^2\big(B(r)\otimes\Gamma\big)^{sT}\big(B(r)\otimes\Gamma\big)-I_N\otimes(P-\Upsilon_{i2})\Big)\overline{e(t-\tau)}.
\end{aligned}
$$

In light of condition (14) of Theorem 2, we can get $\mathcal{L}V(t)<0$. In view of the LaSalle invariance principle of stochastic differential equation, which was developed in [38], we have $\lim_{t\to\infty}V(t)=0$, which in turn illustrates that $\lim_{t\to\infty}e_i(t)=0$ and, at the same time, $\lim_{t\to\infty}d(t)=d^*$. The proof is completed. □

4 Numerical simulations

In this section, we conduct some numerical simulations to illustrate the effectiveness of the theorems of the previous section.

Consider a drive-response network coupled with the following complex Lorenz systems:

$$
\begin{cases}\dot{x}=M(z)x,\\ \dot{z}=-bz+\frac{1}{2}(\bar{x}_1x_2+x_1\bar{x}_2),\end{cases}\tag{15}
$$

where

$$
M(z)=\begin{pmatrix}-\sigma & \sigma\\ r-z & -a\end{pmatrix},
$$

which exhibit chaotic behavior when $\sigma=2$, $b=0.8$, $r=60+0.02j$ and $a=1-0.06j$. Figure 1 shows a chaotic attractor of the complex Lorenz system with initial values $x_1(0)=2.040+2.020j$, $x_2(0)=5.062+4.067j$, $z=5.1$, which is the synchronization orbit in the following simulations; and the noise strength $\sigma_i(y(t),y(t-\tau))=0.001\sum_{k=1}^{N}a_{ik}(r)(y_k(t)-y_i(t))+0.001\sum_{k=1}^{N}b_{ik}(r)(y_k(t-\tau)-y_i(t-\tau))$, so we have $\Upsilon_{i1}<0.05I_{10}$, $\Upsilon_{i2}<0.02I_{10}$.

According to (9), one can easily calculate the eigenvalues of $M^s(z(t))$: $\lambda_{1,2}=-(\sigma+1)\pm\sqrt{(\sigma-1)^2+|\sigma-z+r|^2}$. From Figure 1, it is found that $20\le z\le 100$, and then one can choose $L=40$ such that Assumption 1 holds.

Firstly, consider complex projective synchronization in a drive-response network coupled with 1 + 10 identical complex Lorenz systems with switching topology via adaptive

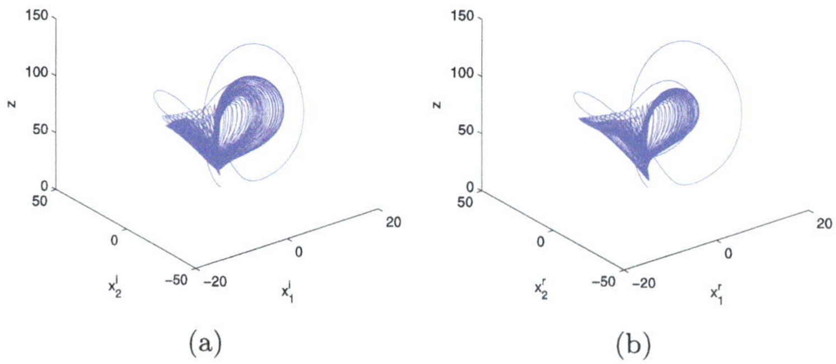

Figure 1 **A chaotic attractor of the complex Lorenz system with initial values $x_1(0) = 2.040 + 2.020j$, $x_2(0) = 5.062 + 4.067j$, $z = 5.1$.**

feedback control, where the outer coupling matrices $A(r)$ and $B(r)$ for $r = (1, 2)$ are

$$A(1) = \begin{bmatrix} -6 & 0 & 1 & 1 & 0 & 1 & 1 & 1 & 1 & 0 \\ 1 & -6 & 0 & 1 & 1 & 1 & 0 & 1 & 1 & 0 \\ 0 & 1 & -6 & 1 & 1 & 1 & 0 & 1 & 1 & 0 \\ 1 & 1 & 1 & -6 & 1 & 0 & 0 & 1 & 1 & 0 \\ 1 & 1 & 1 & 1 & -6 & 1 & 0 & 0 & 1 & 0 \\ 0 & 1 & 1 & 1 & 1 & -6 & 1 & 0 & 1 & 0 \\ 1 & 1 & 1 & 1 & 1 & 0 & -6 & 0 & 1 & 0 \\ 1 & 1 & 0 & 1 & 1 & 0 & 1 & -6 & 1 & 0 \\ 1 & 1 & 1 & 0 & 1 & 1 & 1 & 0 & -6 & 0 \\ 0 & 1 & 1 & 1 & 1 & 1 & 0 & 0 & 1 & -6 \end{bmatrix},$$

$$A(2) = \begin{bmatrix} -7 & 1 & 1 & 1 & 1 & 1 & 1 & 1 & 0 & 0 \\ 1 & -7 & 1 & 1 & 1 & 1 & 1 & 1 & 0 & 0 \\ 0 & 1 & -7 & 1 & 1 & 1 & 1 & 1 & 1 & 0 \\ 1 & 1 & 1 & -7 & 1 & 0 & 1 & 1 & 1 & 0 \\ 1 & 1 & 1 & 1 & -7 & 1 & 1 & 0 & 1 & 0 \\ 1 & 1 & 1 & 1 & 0 & -7 & 1 & 1 & 1 & 0 \\ 0 & 1 & 1 & 1 & 1 & 1 & -7 & 1 & 1 & 0 \\ 1 & 1 & 1 & 1 & 1 & 1 & 1 & -7 & 0 & 0 \\ 0 & 1 & 1 & 1 & 1 & 1 & 1 & 1 & -7 & 0 \\ 1 & 1 & 1 & 1 & 1 & 1 & 0 & 1 & 0 & -7 \end{bmatrix},$$

$$B(1) = \begin{bmatrix} -2 & 0 & 0 & 0 & 1 & 0 & 0 & 0 & 1 & 0 \\ 0 & -2 & 1 & 0 & 0 & 0 & 1 & 0 & 0 & 0 \\ 0 & 0 & -2 & 1 & 0 & 0 & 1 & 0 & 0 & 0 \\ 0 & 0 & 1 & -2 & 1 & 0 & 0 & 0 & 0 & 0 \\ 0 & 1 & 0 & 1 & -2 & 0 & 0 & 0 & 0 & 0 \\ 0 & 0 & 0 & 0 & 0 & -2 & 1 & 1 & 0 & 0 \\ 0 & 1 & 0 & 0 & 1 & 0 & -2 & 0 & 0 & 0 \\ 0 & 1 & 0 & 0 & 0 & 0 & 1 & -2 & 0 & 0 \\ 0 & 1 & 0 & 0 & 0 & 0 & 1 & 0 & -2 & 0 \\ 0 & 1 & 1 & 0 & 0 & 0 & 0 & 0 & 0 & -2 \end{bmatrix},$$

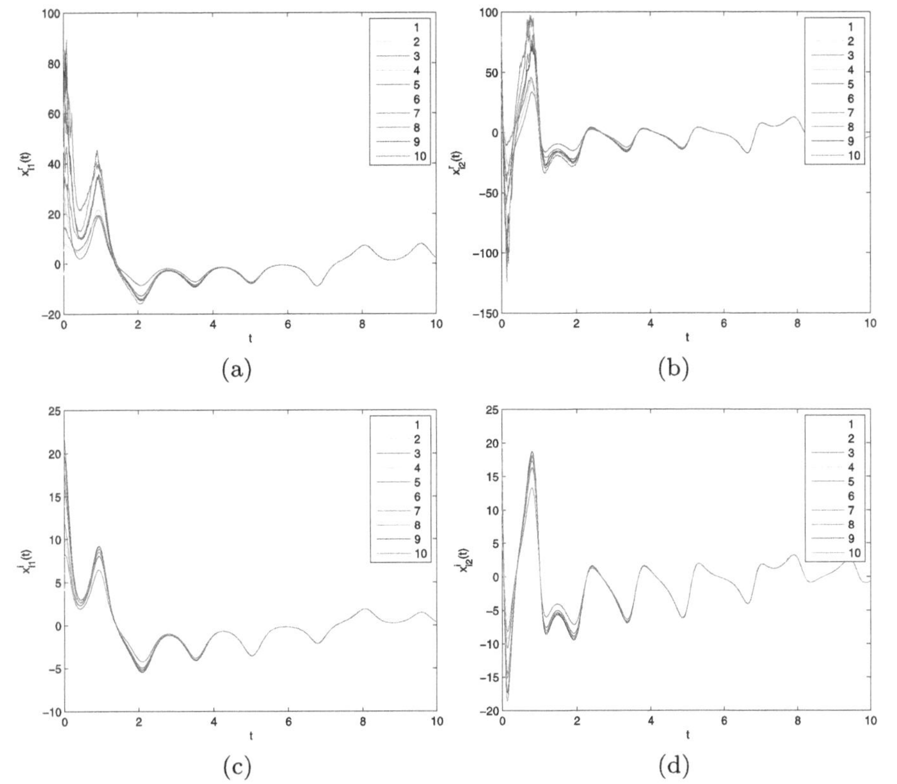

Figure 2 The evolution of the synchronization trajectory y_i ($i = 1, 2, \ldots, 10$) by pinning control.

$$B(2) = \begin{bmatrix} -3 & 0 & 1 & 1 & 0 & 1 & 0 & 0 & 0 & 0 \\ 1 & -3 & 1 & 0 & 1 & 0 & 0 & 0 & 0 & 0 \\ 0 & 0 & -3 & 1 & 0 & 0 & 1 & 1 & 0 & 0 \\ 0 & 1 & 1 & -3 & 0 & 0 & 1 & 0 & 0 & 0 \\ 0 & 1 & 0 & 0 & -3 & 0 & 0 & 1 & 1 & 0 \\ 0 & 0 & 1 & 1 & 0 & -3 & 0 & 0 & 1 & 0 \\ 1 & 1 & 1 & 0 & 0 & 0 & -3 & 0 & 0 & 0 \\ 1 & 0 & 0 & 1 & 0 & 0 & 1 & -3 & 0 & 0 \\ 1 & 0 & 0 & 1 & 0 & 1 & 0 & 0 & -3 & 0 \\ 0 & 1 & 0 & 1 & 1 & 0 & 0 & 0 & 0 & -3 \end{bmatrix},$$

respectively. Choosing $\alpha = 1 + 0.1j$, $\Gamma = \mathrm{diag}(1,1)$, $d_1 = 150$. By a simple calculation, one has that the largest eigenvalue of $A^s - D$ is -0.2605, and then one can choose $\alpha = 1$, $\varepsilon_1 = 100$ and $\varepsilon_2 = 1$ such that $a = -5.8225$ and $b = 0.0264$. Let $\tau = 0.1$, we can get $\gamma = 0.5$ such that condition (7) holds. The initial values of complex state variables $y_i(t)$ ($i = 1, 2, \ldots, 10$) are chosen as $y_{i1}(0) = 2 + 2i + j(3 + i)$ and $y_{i2}(0) = (3 + 2i) + j(4 + 2i)$. Figures 2 and 3 show the evolution of synchronization trajectory and errors by pinning control, respectively.

Finally, consider complex projective synchronization in a drive-response network coupled with $1 + 10$ identical complex Lorenz systems via a single controller and adaptive coupling strength. In numerical simulations, we choose $P = I_m \sigma = 0.001$ and the initial values of $\varepsilon(t)$ as $\varepsilon(0) = 2.3$. The other parameters are chosen the same as those in the above example such that condition (14) holds. Figures 4 and 5 show the evolution of synchronization

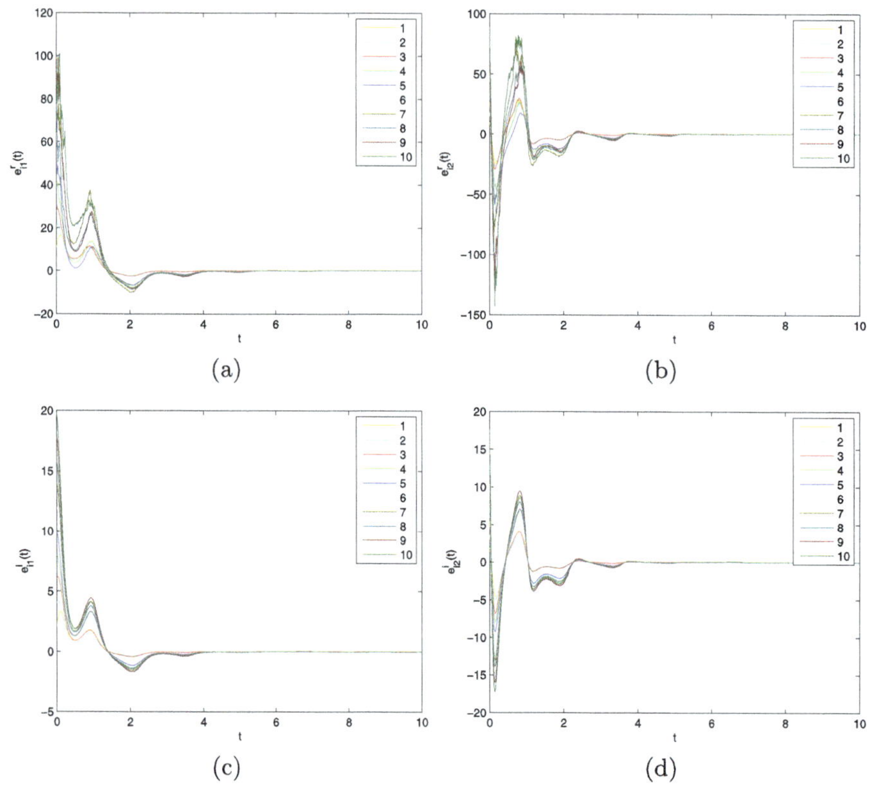

Figure 3 The evolution of the synchronization errors y_i ($i = 1, 2, \ldots, 10$) by pinning control.

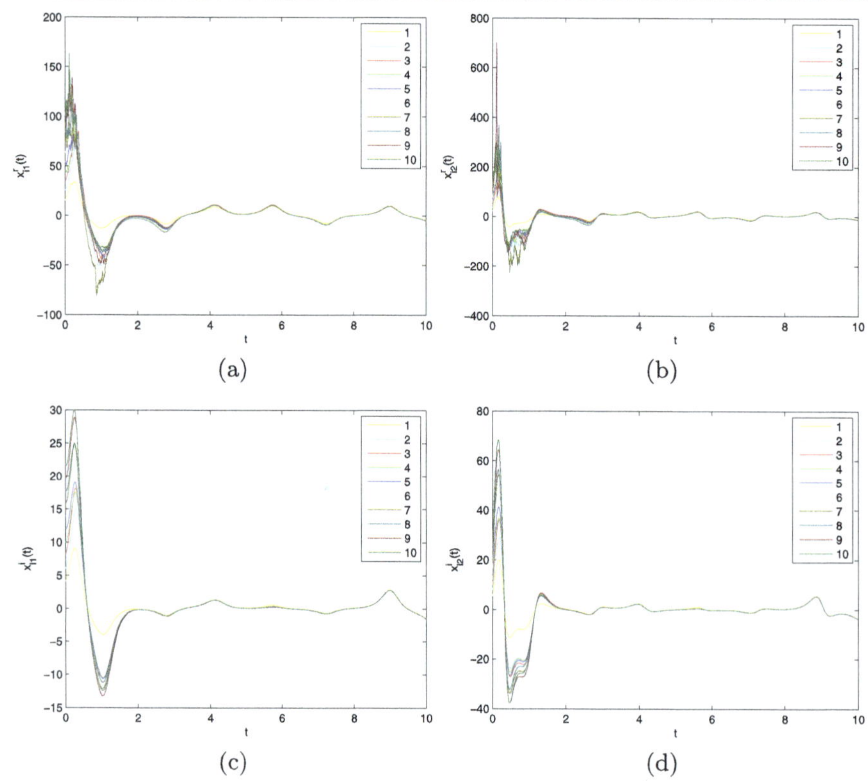

Figure 4 The evolution of the synchronization trajectory y_i ($i = 1, 2, \ldots, 10$) by adaptive control.

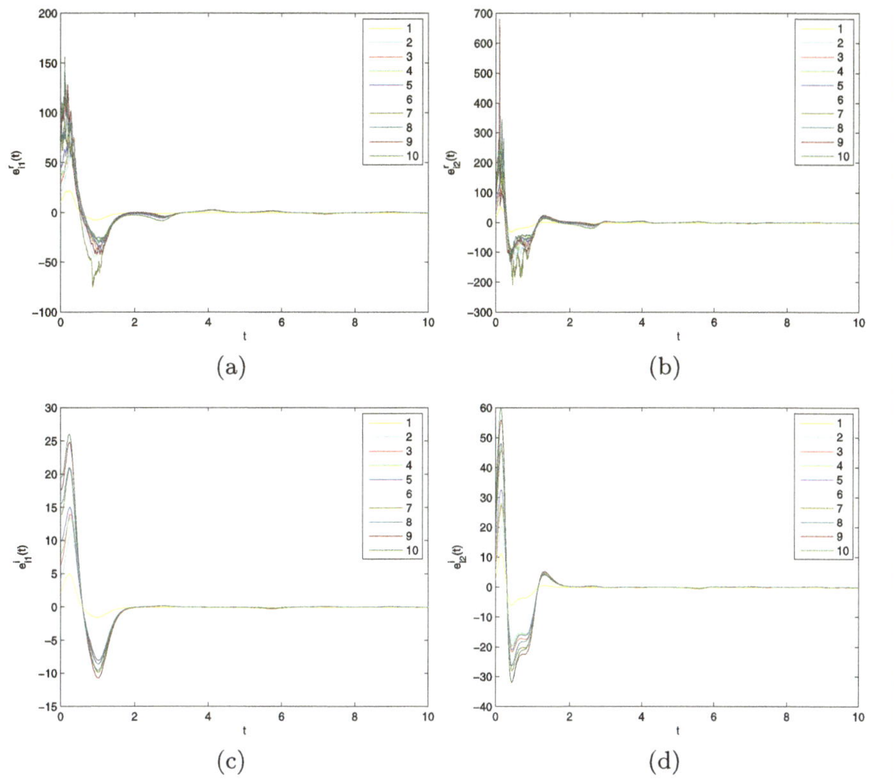

Figure 5 The evolution of the synchronization errors y_i ($i = 1, 2, \ldots, 10$) by adaptive control.

Figure 6 The evolution of adaptive control strength *d*(*t*).

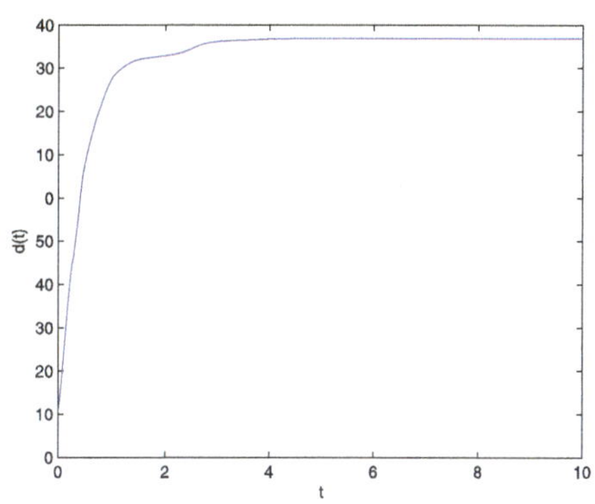

trajectory and errors by adaptive control, respectively. From Figure 6, one can see that the needed control strength value is much less than that calculated by inequality (7) in the above example.

5 Conclusion

The complex projective synchronization in drive-response stochastic coupled networks with complex-variable systems and linear coupling time delays are considered in this paper. Since drive-response systems may evolve in different directions with a constant intersection angle in many real situations, and they not simultaneously evolve along the same or inverse direction based on real number, real matrix, or even real function in a complex plane, we focus on the projective synchronization of this situation considering the time delay and stochastic disturbance for response systems through two theorems and one corollary. Eventually, several numerical simulations have verified the validity of those results.

Competing interests
The author declares that they have no competing interests.

Author's contributions
XW carried out all the work in this manuscript and read and approved the final manuscript.

Acknowledgements
The author thanks the referees and the editor for their valuable comments on this paper. This work was supported by the National Science Foundation of China under Grant Nos. 61070087 and 61273220, Shenzhen Strategic Emerging Industries Projects (ZDSY20120613125016389) and Shenzhen Basic Research Project (JCYJ20130331152625792).

References

1. Newman, MEJ: The structure and function of complex networks. SIAM Rev. **45**, 167-256 (2003)
2. Barabási, AL: Scale-free networks: a decade and beyond. Science **325**, 412-413 (2009)
3. Butts, CT: Revisiting the foundations of network analysis. Science **325**, 414-416 (2009)
4. Centola, D: The spread of behavior in an online social network experiment. Science **329**, 1194-1197 (2010)
5. Shinar, G, Feinberg, M: Structural sources of robustness in biochemical reaction networks. Science **327**, 1389-1391 (2010)
6. Wang, JW, Ma, QH, Zeng, L, Abd-Elouahab, MS: Mixed outer synchronization of coupled complex networks with time-varying coupling delay. Chaos **21**, 013121 (2011)
7. Zhu, QX, Cao, JD: Adaptive synchronization under almost every initial data for stochastic neural networks with time-varying delays and distributed delays. Commun. Nonlinear Sci. Numer. Simul. **16**, 2139-2159 (2011)
8. Gan, QT, Xu, R, Kang, XB: Synchronization of chaotic neural networks with mixed time delays. Commun. Nonlinear Sci. Numer. Simul. **16**, 966-974 (2011)
9. Kanter, L, Zigzag, M, Englert, A, Geissler, F, Kinzel, W: Synchronization of unidirectional time delay chaotic networks and the greatest common divisor. Europhys. Lett. **93**, 60003 (2011)
10. Audheer, KS, Sabir, M: Adaptive modified function projective synchronization of multiple time-delayed chaotic Rossler system. Phys. Lett. A **375**, 1176-1178 (2011)
11. Zhang, CK, He, Y, Wu, M: Exponential synchronization of neural networks with time-varying mixed delays and sampled-data. Neurocomputing **74**, 265-273 (2010)
12. Ahn, CK: Anti-synchronization of time-delayed chaotic neural networks based on adaptive control. Int. J. Theor. Phys. **48**, 3498-3509 (2009)
13. Ahn, CK: Adaptive $\mathcal{H}_\infty$ anti-synchronization for time-delayed chaotic neural networks. Prog. Theor. Phys. **122**, 1391-1403 (2009)
14. Ge, ZM, Wong, YT, Li, SY: Temporary lag and anticipated synchronization and anti-synchronization of uncoupled time-delayed chaotic systems. J. Sound Vib. **318**, 267-278 (2008)
15. Prasad, A, Kurths, J, Ramaswamy, R: The effect of time-delay on anomalous phase synchronization. Phys. Lett. A **372**, 6150-6154 (2008)
16. Senthilkumar, DV, Lakshmanan, M, Kurths, J: Transition from phase to generalized synchronization in time-delay systems. Chaos **18**, 023118 (2008)
17. Zhang, D, Xu, JA: Projective synchronization of different chaotic time-delayed neural networks based on integral sliding mode controller. Appl. Math. Comput. **217**, 164-174 (2010)
18. Ghosh, D: Generalized projective synchronization in time-delayed systems: nonlinear observer approach. Chaos **19**, 013102 (2009)
19. Ghosh, D, Banerjee, S, Chowdhury, AR: Generalized and projective synchronization in modulated time-delayed systems. Phys. Lett. A **374**, 2143-2149 (2010)
20. Ghosh, D, Saha, P, Chowdhury, AR: Linear observer based projective synchronization in delay Rössler system. Commun. Nonlinear Sci. Numer. Simul. **15**, 1640-1647 (2010)
21. Gan, QT, Xu, R, Kang, XB: Synchronization of chaotic neural networks with mixed time delays. Commun. Nonlinear Sci. Numer. Simul. **16**, 966-974 (2011)
22. Chen, JR, Jiao, LC, Wu, JS, Wang, XD: Projective synchronization with different scale factors in a driven-response complex network and its application in image encryption. Nonlinear Anal., Real World Appl. **11**, 3045-3058 (2010)
23. Wu, XJ, Lu, HT: Generalized projective synchronization between two different general complex dynamical networks with delayed coupling. Phys. Lett. A **374**, 3932-3941 (2010)
24. Wu, W, Zhou, W, Chen, TP: Cluster synchronization of linearly coupled complex networks under pinning control. IEEE Trans. Circuits Syst. I **56**, 829-839 (2009)
25. Li, GH: Generalized projective synchronization between Lorenz system and Chen's system. Chaos Solitons Fractals **32**, 1454-1458 (2007)
26. Lü, L, Li, CR, Chen, LS, Wei, LL: Lag projective synchronization of a class of complex network constituted nodes with chaotic behavior. Commun. Nonlinear Sci. Numer. Simul. **19**, 2843-2849 (2014)
27. Zhu, QY, Zhou, WN, Zhou, LW, Wu, MQ, Tong, DB: Mode-dependent projective synchronization for neutral-type neural networks with distributed time-delays. Neurocomputing **140**, 97-103 (2014)

28. Wu, ZY, Fu, XC: Complex projective synchronization in drive-response networks coupled with complex-variable chaotic systems. Nonlinear Dyn. **72**, 9-15 (2013)
29. Wu, ZY, Fu, XC: Complex projective synchronization in coupled chaotic complex dynamical systems. Nonlinear Dyn. **69**, 771-779 (2012)
30. Mahmoud, GM, Al-Kashif, MA, Farghaly, AA: Chaotic and hyperchaotic attractors of a complex nonlinear system. J. Phys. A, Math. Theor. **41**(5), 055104 (2008)
31. Yuan, C, Mao, X: Robust stability and controllability of stochastic differential delay equations with Markovian switching. Automatica **40**(3), 343-354 (2004)
32. Liu, T, Zhao, J, Hill, DJ: Exponential synchronization of complex delayed dynamical networks with switching topology. IEEE Trans. Circuits Syst. I, Regul. Pap. **57**(11), 2967-2980 (2010)
33. Wang, L, Wang, Q: Synchronization in complex networks with switching topology. Phys. Lett. A **375**(34), 3070-3074 (2011)
34. Xiao, JW, Huang, Y, Wang, YW, Yi, JO: Synchronization of complex switched networks with two types of delays. Neurocomputing **74**(17), 3151-3157 (2011)
35. Jia, Q, Tang, WKS, Halang, WA: Leader following of nonlinear agents with switching connective network and coupling delay. IEEE Trans. Circuits Syst. I, Regul. Pap. **58**(10), 2508-2519 (2011)
36. Liu, XW, Chen, TP: Exponential synchronization of nonlinear coupled dynamical networks with a delayed coupling. Physica A **381**, 82-92 (2007)
37. Perret, C: The stability of numerical simulations of complex stochastic differential equations. PhD thesis, ETH, Zurich, Number 18868 (2010)
38. Mao, X: A note on the LaSalle-type theorems for stochastic differential delay equations. J. Math. Anal. Appl. **268**(1), 125-142 (2002)

Stability analysis and observer design for discrete-time SEIR epidemic models

Asier Ibeas[1*], Manuel de la Sen[2], Santiago Alonso-Quesada[2] and Iman Zamani[3]

*Correspondence: asier.ibeas@uab.cat
[1]Departamento de Telecomunicació i Enginyeria de Sistemes, School of Engineering, Universitat Autònoma de Barcelona, Bellaterra, Barcelona, 08193, Spain
Full list of author information is available at the end of the article

Abstract

This paper applies Micken's discretization method to obtain a discrete-time SEIR epidemic model. The positivity of the model along with the existence and stability of equilibrium points is discussed for the discrete-time case. Afterwards, the design of a state observer for this discrete-time SEIR epidemic model is tackled. The analysis of the model along with the observer design is faced in an implicit way instead of obtaining first an explicit formulation of the system which is the novelty of the presented approach. Moreover, some sufficient conditions to ensure the asymptotic stability of the observer are provided in terms of a matrix inequality that can be cast in the form of a LMI. The feasibility of the matrix inequality is proved, while some simulation examples show the operation and usefulness of the observer.

Keywords: epidemics; SEIR; discrete models; stability; observer design

1 Introduction

Mathematical models have become an important tool in analyzing the causes, dynamics, and spread of epidemics [1]. Thus, its study is crucial in order to obtain valuable knowledge of the underlying aspects of the disease. Moreover, the analysis of mathematical models describing epidemics spreading allows authorities to make decisions regarding the vaccination policies, quarantine application and so on. These models can be classified into two main categories: continuous-time and discrete-time models.

Continuous-time models were the first proposed to describe epidemics spreading since modeling has traditionally been focused from a differential equations point of view. Therefore, many specific features regarding these models have been studied in many works such as the presence of bifurcations [2, 3], the construction of approximate solutions [4, 5], the fractional dynamics [6], the existence of equilibrium points and oscillatory behavior [7], and the presence of waves [8–10], to cite but a few. However, model stability has been by far the most important property to be studied [11–17].

On the other hand, discrete-time models have gained substantial importance during the last years as the increasing number of papers concerning this type of models reveals [18–22]. There are many reasons that explain this tendency. Among them we can find the simplicity of its simulation, the fact that it is easier to adjust system's parameters from statistical data in discrete-time models than in continuous-time ones and the fact that discrete-time models may exhibit a richer dynamic behavior than its continuous-time counterparts [22]. Until now, as [22] states, the study of discrete epidemic models was focused on the computation of the reproduction number, the existence and stability (both local and global) of the equilibrium points, and the extinction, permanence, and persistence of the disease. In addition, these studies have been performed basically on SI, SIR, and SIRS models.

Moreover, there are two ways to face the problem of setting up a discrete-time epidemic model. The first one consists in modeling the disease spreading directly in a discrete domain. The second one faces the problem by discretizing a previously existing continuous-time model. The latter approach will be used in this paper. In this way, several discretization methods have been used in the literature including the implicit and explicit Euler, mixed, and Runge-Kutta methods [18–22]. However, discretization procedures do not always preserve the structural properties exhibited by continuous-time models such as the conservation of the total population or the positivity of the solutions. Therefore, much research has been done in the direction of obtaining discretization methods able to preserve all these properties. One of the most important breakthroughs in this respect is the research made by Micken on the use of nonstandard finite difference (NSFD) methods [21]. These methods combine implicit and explicit Euler approximations in the incidence rate function. Therefore, this paper will apply Micken's NSFD discretization method to obtain a SEIR discrete-time epidemic model while stating its main properties such as the well-posedness of the model, the positivity of solutions and the existence and stability of equilibrium points. Since most of the previous literature is confined to SI, SIR, and SIRS models, this study will contribute to the application of NSFD approaches to a wider class of models. In addition, the properties will be proved based on an implicit representation of the equations, in comparison with previous approaches [18, 19], where an explicit description of the model is obtained prior to analysis.

Furthermore, control theory has recently emerged as one of the proposed approaches to deal with the design of vaccination campaigns [23–29]. Thus, different types of vaccination laws based on control theory have appeared in the literature during the last years, such as the state-feedback control [23, 25], robust sliding control [30], or feedback linearization [24]. Nevertheless, most of these works calculate the vaccination effort based on the values of the subpopulations, namely, susceptible, exposed, infectious and immune of the SEIR model. From a practical point of view only the total population and the number of infectious may be obtained from clinical data at health centers. Therefore, the number of susceptible, exposed, and immune cases may not be available to on-line compute the control vaccination law. Under these circumstances, a state observer is necessary to estimate the value of these subpopulations.

State observers are largely used in control applications to cope with the inability to measure some state variables. Thus, we can find many applications in chemical [31], aeronautic [32], and electrical engineering [33], to cite just a few. Also, there are few previous works where a state observer is designed for epidemic models [27, 28]. In these works, an observer is designed for a continuous-time SEIR epidemic model whose values are then used to implement a vaccination strategy. In the present work, however, a state observer is designed for a discrete-time SEIR model obtained by a NSFD discretization method, which is for the first time being considered in the literature. The design of such a state observer must face some particularities, which make it different from previous approaches to observer design:

- NSFD schemes lead to implicit systems rather than explicit. Thus, the design of the state observer must be done implicitly. This makes a big difference from traditional observer design where the problem is usually formulated in an explicit way.
- Epidemic models are singular systems so that there exists an algebraic relation between the dynamic variables of the system.
- The number of infectious can be measured, implying that there is no need to estimate this state component, which leads to a reduced-order observer.

In this way, the combination of all the above items makes the design of state observers for discrete-time SEIR epidemic models rather different from standard approaches in the literature. In addition, the asymptotic stability of the observer will be discussed, while some simulation examples will show its operation. Some preliminary results of this paper were first presented in [34] and [35].

The paper is organized as follows. Section 2 describes the model description and problem formulation. Section 3 studies the main properties of the discrete model. The design of the state observer is carried out in Section 4. Section 5 is devoted to the study of the analytic properties of the observer. Finally, Section 6 shows some numerical simulation examples while our Conclusions end the paper.

2 Model description

Consider the SEIR continuous-time epidemic model described by [23]

$$\dot{S}(t) = -\mu S(t) + \omega R(t) - \frac{\beta S(t)I(t)}{N(t)} + \nu N(t), \tag{1}$$

$$\dot{E}(t) = \frac{\beta S(t)I(t)}{N(t)} - (\mu + \sigma)E(t),$$

$$\dot{I}(t) = -(\mu + \gamma)I(t) + \sigma E(t),$$

$$\dot{R}(t) = -(\mu + \omega)R(t) + \gamma I(t), \tag{2}$$

where $S(t)$, $E(t)$, $I(t)$, and $R(t)$ denote the subpopulations of susceptible, exposed, infectious and immune, respectively, defining the state vector $x^T(t) = [S\ E\ I\ R]$. $N(t)$ denotes the total population at time t (*i.e.* $N(t) = S(t) + E(t) + I(t) + R(t)$), μ is the rate of deaths from causes unrelated to the infection, ν denotes the birth rate and ω is the rate of losing immunity. The term $\frac{\beta S(t)I(t)}{N(t)}$ is referred to as the disease incidence rate and defines the total number of new infections per unit of time at time t. σ^{-1} and γ^{-1} are, respectively, the average duration of the latent and infectious periods. All the above parameters are assumed to be positive so as to represent a real situation. This model is also referred to as the SEIRS model in the literature (see, for instance, [36]) due to the presence of the terms $\pm\omega R(t)$ in (1) and (2). The total population dynamics can be calculated by summing up all the equations (1)-(2), leading to

$$\dot{N}(t) = (\nu - \mu)N(t). \tag{3}$$

It can be deduced from (3) that the total population is constant when $\nu = \mu$, increases when $\nu > \mu$ and decreases when $\nu < \mu$. This model is discretized by using Micken's NSFD method [21], to get

$$\frac{S_{k+1}-S_k}{h} = -\mu S_{k+1} + \omega R_{k+1} - \frac{\beta S_{k+1} I_k}{N_k} + \nu N_{k+1}, \tag{4}$$

$$\frac{E_{k+1}-E_k}{h} = \frac{\beta S_{k+1} I_k}{N_k} - (\mu+\sigma)E_{k+1},$$

$$\frac{I_{k+1}-I_k}{h} = -(\mu+\gamma)I_{k+1} + \sigma E_{k+1},$$

$$\frac{R_{k+1}-R_k}{h} = -(\mu+\omega)R_{k+1} + \gamma I_{k+1}, \tag{5}$$

where $h > 0$ denotes the integration step while S_k, E_k, I_k, and R_k denote the value of the susceptible, exposed, infectious and immune at sampling time $t_k = kh$ which define the discrete-time state vector $x_k^T = [S_k\ E_k\ I_k\ R_k]$. The dynamics of the total population in discrete-time can be obtained by summing up again all the equations (4)-(5) to get

$$\frac{N_{k+1}-N_k}{h} = (\nu-\mu)N_{k+1}, \tag{6}$$

which corresponds to the NSFD discretization of the continuous-time equation (3). This fact implies that the discretization method is consistent with the total population dynamics given by the continuous-time model, property not satisfied by all discretization methods [20]. Equation (6) can be re-written as

$$N_{k+1} = \frac{1}{1+(\mu-\nu)h} N_k \tag{7}$$

which provides the explicit value of N_{k+1} in terms of the system's parameters and N_k. For the sake of simplicity, let us rearrange (4)-(5) in the following implicit matrix form:

$$\mathcal{M}(x_k)x_{k+1} = x_k + \Gamma \nu h N_{k+1} = x_k + \Gamma \frac{\nu h}{1+(\mu-\nu)h} N_k = \mathcal{D} x_k, \tag{8}$$

where

$$\Gamma^T = [1 \quad 0 \quad 0 \quad 0],$$

$$\mathcal{D} = I_{4\times 4} + \frac{\nu h}{1+(\mu-\nu)h} \begin{pmatrix} 1 & 1 & 1 & 1 \\ 0 & 0 & 0 & 0 \\ 0 & 0 & 0 & 0 \\ 0 & 0 & 0 & 0 \end{pmatrix}, \tag{9}$$

$$\mathcal{M}(x_k) = \begin{pmatrix} A & 0 & 0 & -B \\ -H & C & 0 & 0 \\ 0 & -G & D & 0 \\ 0 & 0 & -F & J \end{pmatrix} \tag{10}$$

with

$$A = 1 + \mu h + \frac{\beta h}{N_k} I_k, \qquad B = \omega h, \qquad C = 1 + (\mu+\sigma)h,$$

$$D = 1 + (\mu + \gamma)h, \qquad J = 1 + (\mu + \omega)h, \qquad F = \gamma h,$$

$$G = \sigma h, \qquad H = \frac{\beta h}{N_k} I_k. \tag{11}$$

Note that the matrix $\mathcal{M}(x_k)$ depends explicitly on x_k through entries A and H, since they depend on N_k and I_k as (11) shows. Equation (8) allows one to isolate the value of state vector x_{k+1}, when possible, as

$$x_{k+1} = \mathcal{M}(x_k)^{-1}\mathcal{D}x_k. \tag{12}$$

For the SEIR model, the closed-form explicit representation given by (12) is intricate. Therefore, all the properties of this model will be stated starting from the implicit form (8) rather than from (12) in comparison with some previous works [18, 19], where an explicit (or semi-explicit) formulation of the model is obtained prior to analysis. The following section is devoted to the study of the main properties of this scheme.

3 Main properties of the discrete-time model

Initially, the well-posedness of (12) will be proved.

Lemma 1 *$\mathcal{M}(x_k)$ is invertible, its inverse is given by* (13) *and all the entries of the matrix $\mathcal{M}(x_k)^{-1}$ are non-negative for any set of positive parameters and any $I_k \geq 0$, $N_k \geq 0$. We have*

$$\mathcal{M}(x_k)^{-1} = \frac{1}{ACDJ - HGBF} \times \begin{pmatrix} CDJ & BFG & BCF & BCD \\ DHJ & ADJ & BHF & BDH \\ GHJ & AGJ & ACJ & BGH \\ FGH & AFG & ACF & ACD \end{pmatrix}. \tag{13}$$

Proof Firstly, notice that, since all the parameters (including the time step h) are assumed to be positive and $I_k, N_k \geq 0, A, B, C, D, J, F, G > 0$ and $H \geq 0$ from (11). Now, the determinant of $\mathcal{M}(x_k)$ can be symbolically computed as

$$\det \mathcal{M}(x_k) = ACDJ - HGFB. \tag{14}$$

Hence, if $I_k, N_k \geq 0$ then $ACDJ > 0, A > H, C > G, D > F$, and $J > B$ according to (11). Therefore, $ACDJ > HGFB$, $\det \mathcal{M}(x_k) = ACDJ - HGFB > 0$, and $\mathcal{M}(x_k)$ is invertible. Moreover, its inverse can be symbolically computed as (13). Finally, it can be directly seen from (13) that all its entries are non-negative since all the quantities involved are non-negative from (11). Thus, the lemma is proved. □

Furthermore, we have the following.

Theorem 1 *S_k, E_k, I_k, and R_k remain non-negative for all discrete time $t_k = kh$, $k \in \mathbb{N}$ provided that $S_0, E_0, I_0, R_0 \geq 0$, $1 + (\mu - \nu)h > 0$, and all the system's parameters are positive.*

Proof The proof is performed by mathematical induction. Thus, let us assume that x_0 is non-negative. Now suppose that x_k is non-negative and it will be proved that so is x_{k+1}.

In this way, since x_k is non-negative so is N_k and, in particular, I_k. This fact implies that $\mathcal{M}(x_k)$ is invertible due to Lemma 1. In addition, its inverse is given by (13), whose entries are non-negative. Then the $(k+1)$th state vector, x_{k+1}, is calculated through (12) as

$$x_{k+1} = \mathcal{M}(x_k)^{-1}x_k + \mathcal{M}(x_k)^{-1}\Gamma\frac{\nu h}{1+(\mu-\nu)h}N_k. \tag{15}$$

Since $1+(\mu-\nu)h > 0$, all the elements in the right-hand side of (15) are non-negative, so is x_{k+1} and the theorem is proved. □

Remark 1 Notice that Theorem 1 states that the system is always well defined, since $I_k \geq 0$ and $N_k \geq 0$ for all $k \geq 0$ implies that $\mathcal{M}(x_k)$ is always invertible due to Lemma 1 and makes possible the calculation of x_{k+1} at each time step through (12).

Remark 2 Moreover, Theorem 1 can also be used to establish the range of values of the time step h that guarantees the well-posedness of the discrete-time model. Thus, from $1+(\mu-\nu)h > 0$ we see that h must fulfill:

$$\begin{cases} 0 < h < \frac{1}{\nu-\mu} & \text{if } \nu > \mu, \\ h \in (0,\infty) & \text{otherwise.} \end{cases} \tag{16}$$

Since h must meet this constraint, the method is a NSFD-based one, rather a pure NSFD method.

Remark 3 The non-negativity and well-posedness of the discrete-time system are crucial properties for the discrete-time model [37]. These properties are obtained easier by using the NSFD method than by using other discretization schemes. For instance, consider the forward Euler approximation of the derivative applied to the left-hand side of (1)-(2) [22], to obtain

$$\frac{S_{k+1}-S_k}{h} = -\mu S_k + \omega R_k - \frac{\beta S_k I_k}{N_k} + \nu N_k, \tag{17}$$

$$\begin{aligned} \frac{E_{k+1}-E_k}{h} &= \frac{\beta S_k I_k}{N_k} - (\mu+\sigma)E_k, \\ \frac{I_{k+1}-I_k}{h} &= -(\mu+\gamma)I_k + \sigma E_k, \\ \frac{R_{k+1}-R_k}{h} &= -(\mu+\omega)R_k + \gamma I_k. \end{aligned} \tag{18}$$

These equations can be cast in the matrix form:

$$x_{k+1} = \mathcal{M}_1 x_k + \Gamma_1 \beta h \frac{S_k I_k}{N_k}, \tag{19}$$

where

$$\mathcal{M}_1 = \begin{pmatrix} 1+(\nu-\mu)h & \nu h & \nu h & (\omega+\nu)h \\ 0 & 1-(\mu+\sigma)h & 0 & 0 \\ 0 & \sigma h & 1-(\mu+\gamma)h & 0 \\ 0 & 0 & \gamma h & 1-(\mu+\omega)h \end{pmatrix}$$

and $x_k = [S_k\ E_k\ I_k\ R_k]^T$, $\Gamma_1 = [-1\ 1\ 0\ 0]^T$. If we want to apply now the same argument as in (15) to prove the non-negativity of the system (*i.e.* the non-negativity of all the elements of the right-hand side of (15)) we find two main drawbacks: (i) the first one is that the vector Γ_1 possesses negative components, and, secondly, (ii) all the entries of the matrix $\mathcal{M}_1$ are non-negative provided that the following constraints hold:

$$1 + (\nu - \mu)h \geq 0, \tag{20}$$

$$1 - (\mu + \sigma)h \geq 0,$$

$$1 - (\mu + \gamma)h \geq 0,$$

$$1 - (\mu + \omega)h \geq 0. \tag{21}$$

Thus, the simple argument used in (15) is not directly applicable here, while it may lead to further conditions on the sampling period h arising from (20)-(21):

$$\begin{cases} 0 < h \leq \min(\frac{1}{\mu+\sigma}, \frac{1}{\mu+\gamma}, \frac{1}{\mu+\omega}, \frac{1}{\mu-\nu}) & \text{if } \nu < \mu, \\ 0 < h \leq \min(\frac{1}{\mu+\sigma}, \frac{1}{\mu+\gamma}, \frac{1}{\mu+\omega}) & \text{otherwise.} \end{cases} \tag{22}$$

Therefore, the NSFD method reveals as a very useful one to treat the discretization problem of the SEIR model since its properties are proved easily and under mild mathematical conditions.

3.1 Existence and stability of equilibrium points

This section studies the existence and stability of equilibrium points in the discrete-time SEIR model given by (4)-(5).

Lemma 2 *The discrete-time equation systems* (4)-(5) *and continuous-time equation systems* (1)-(2) *posses the same equilibrium points, given by*

(1) *If* $\mu \neq \nu$ *then the only equilibrium point is given by the trivial one* $S^* = E^* = I^* = R^* = N^* = 0$.

(2) *If* $\mu = \nu$ *the total population is constant, i.e.* $N(t) = N = N(0) = S(0) + E(0) + I(0) + R(0)$ *and there are three potential equilibrium points given by*:

  (a) *The trivial one,* $S^* = E^* = I^* = R^* = N^* = 0$.

  (b) $S^* = N$, $E^* = I^* = R^* = 0$, *i.e. the total population is susceptible. This point is usually referred to as the disease-free equilibrium point.*

  (c) *The so-called endemic equilibrium point*:

$$S^* = \frac{(\mu + \sigma)(\mu + \gamma)}{\sigma\beta} N, \tag{23}$$

$$E^* = \frac{(\mu + \omega)(\mu + \gamma)(\sigma\beta - (\mu + \gamma)(\mu + \sigma))}{\sigma\beta((\mu + \sigma + \gamma)(\mu + \omega) + \gamma\sigma)} N,$$

$$I^* = \frac{(\mu + \omega)(\sigma\beta - (\mu + \gamma)(\mu + \sigma))}{\beta((\mu + \sigma + \gamma)(\mu + \omega) + \gamma\sigma)} N,$$

$$R^* = \frac{\gamma(\sigma\beta - (\mu + \gamma)(\mu + \sigma))}{\beta((\mu + \sigma + \gamma)(\mu + \omega) + \gamma\sigma)} N. \tag{24}$$

Proof The proof of the first part is straightforward from (1)-(2) and (4)-(5) while the equilibrium points are calculated in [25]. □

Now, the stability of the equilibrium points is discussed. We will focus on the two nontrivial equilibrium points obtained when $\mu = \nu$: the disease-free and the endemic equilibrium points.[a] For this, the nonlinear system (8) is linearized around these equilibrium points and the first Lyapunov theorem (linearization theorem) is used. Thus, the linearized equation reads[b]

$$\left(\frac{\partial(\mathcal{D}^{-1}\mathcal{M}(x_k)x_{k+1})}{\partial x_{k+1}}\right)_{x^*} v_{k+1} = v_k, \tag{25}$$

where x^* denotes the corresponding equilibrium point and $v_k = x_k - x^*$ is the incremental variable. In this way, the results are obtained starting from the implicit equation (8) rather than from the explicit one (12) as some previous works do [18, 19]. The advantage of this approach is that the calculus of the Jacobian matrix $(\frac{\partial(\mathcal{D}^{-1}\mathcal{M}(x_k)x_{k+1})}{\partial x_{k+1}})_{x^*}$ is easier implicitly than explicitly. In order to perform the calculation of the Jacobian, we must firstly write $\mathcal{M}(x_k)\cdot x_{k+1}$ in terms of only x_{k+1} since, otherwise, we would be omitting some terms in the Taylor development. Therefore

$$\mathcal{M}(x_k)\cdot x_{k+1} = \begin{pmatrix} 1+\mu h+\frac{\beta h}{N}(-GE_{k+1}+DI_{k+1}) & 0 & 0 & -B \\ -\frac{\beta h}{N}(-GE_{k+1}+DI_{k+1}) & C & 0 & 0 \\ 0 & -G & D & 0 \\ 0 & 0 & -F & J \end{pmatrix}\cdot x_{k+1}. \tag{26}$$

Thus, the Jacobian is given by

$$\left(\frac{\partial(\mathcal{D}^{-1}\mathcal{M}(x_k)x_{k+1})}{\partial x_{k+1}}\right)_{x^*} = \mathcal{D}^{-1}\begin{pmatrix} 1+\mu h+\frac{\beta h}{N}(-GE^*+DI^*) & -\frac{\beta h}{N}GS^* & \frac{\beta h}{N}DS^* & -B \\ -\frac{\beta h}{N}(-GE^*+DI^*) & C+\frac{\beta h}{N}GS^* & -\frac{\beta h}{N}DS^* & 0 \\ 0 & -G & D & 0 \\ 0 & 0 & -F & J \end{pmatrix},$$

so that around the disease-free equilibrium point

$$\left(\frac{\partial(\mathcal{D}^{-1}\mathcal{M}(x_k)x_{k+1})}{\partial x_{k+1}}\right)_{x^*_{df}} = \mathcal{D}^{-1}\begin{pmatrix} 1+\mu h & -\beta hG & \beta hD & -B \\ 0 & C+\beta hG & -\beta hD & 0 \\ 0 & -G & D & 0 \\ 0 & 0 & -F & J \end{pmatrix}$$

and the linearized system is given by

$$\mathcal{D}^{-1}\begin{pmatrix} 1+\mu h & -\beta hG & \beta hD & -B \\ 0 & C+\beta hG & -\beta hD & 0 \\ 0 & -G & D & 0 \\ 0 & 0 & -F & J \end{pmatrix} v_{k+1} = v_k. \tag{27}$$

Since the incremental variables satisfy $v_{1,k} + v_{2,k} + v_{3,k} + v_{4,k} = 0$ due to the constraint $S_k + E_k + I_k + R_k = N$ we can omit one variable in our analysis, since it can be obtained from this algebraic equation. Therefore, if we omit the first one we obtain the reduced system,

$$\begin{pmatrix} C+\beta hG & -\beta hD & 0 \\ -G & D & 0 \\ 0 & -F & J \end{pmatrix} \begin{pmatrix} v_{2,k+1} \\ v_{3,k+1} \\ v_{4,k+1} \end{pmatrix} = \begin{pmatrix} v_{2,k} \\ v_{3,k} \\ v_{4,k} \end{pmatrix}. \tag{28}$$

The linearized reduced system (28) will be asymptotically stable provided that all the eigenvalues of the left-hand side matrix are strictly greater than unity in absolute value. In this way, the eigenvalues are given by

$$\begin{aligned} \lambda_1 &= J = 1 + (\mu + \omega)h, \\ \lambda_2 &= \frac{C + D + \beta hG + \sqrt{(C-D)^2 + 2\beta hG(C+D) + G^2\beta^2h^2}}{2}, \\ \lambda_3 &= \frac{C + D + \beta hG - \sqrt{(C-D)^2 + 2\beta hG(C+D) + G^2\beta^2h^2}}{2}. \end{aligned} \tag{29}$$

It can be readily seen that $\lambda_1 > 1$, while

$$\lambda_2 = 1 + \frac{(2\mu + \sigma + \gamma)h + \beta hG + \sqrt{(C-D)^2 + 2\beta hG(C+D) + G^2\beta^2h^2}}{2} > 1. \tag{30}$$

The only eigenvalue that might threaten stability is λ_3, which may be re-written as

$$\lambda_3 = \frac{C + D + \beta hG - \sqrt{(C + D + \beta hG)^2 - 4CD}}{2}. \tag{31}$$

Notice that the argument of the square root is positive from (29). Therefore, (31) implies that λ_3 must also be positive since $CD > 0$. The stability condition then reads $\lambda_3 > 1$. Some algebra on (31) shows that this condition can be met, provided that

$$\mathcal{R}_0 := \frac{\beta\sigma}{(\mu + \gamma)(\mu + \sigma)} < 1, \tag{32}$$

which corresponds to the same stability condition found in [25] for the continuous-time system. Thus, not only does the discretization method preserve positivity and total population dynamics, but also the equilibrium points and their stability. In this way, the reproduction number of the discrete-time model is given by (32). When (32) is met, the linearized system is globally stable around the disease-free equilibrium point, $v_{1,k}, v_{2,k}, v_{3,k}, v_{4,k} \to 0$ and $S_k \to N$ as $k \to \infty$, while the nonlinear system is locally stable around the equilibrium point. Thus, we have already proved the following theorem.

Theorem 2 *The discrete-time epidemic SEIR model* (4)-(5) *is locally asymptotically stable around the disease-free equilibrium point when* $\mu = \nu$ *provided that* $\mathcal{R}_0 := \frac{\beta\sigma}{(\mu+\gamma)(\mu+\sigma)} < 1$.

On the other hand, the following theorem holds when $\mathcal{R}_0 > 1$.

Theorem 3 *If* $\mu = \nu$ *and* $\mathcal{R}_0 > 1$ *then one of the following claims remains true*:

(1) *The solution of the discrete-time SEIR model* (4)-(5) *tends to the endemic equilibrium point.*
(2) *Some variables of the discrete-time model* (4)-(5) *oscillate.*

Proof Firstly note that the positivity of the model along with the constraint $S_k + E_k + I_k + R_k = N$ imply that $0 \leq S_k, E_k, I_k, R_k \leq N$. Therefore, all variables are globally bounded through time. Now, since the state variables cannot diverge as $k \to \infty$, there are only two possibilities:

(1) All the state variables converge to a constant finite value.
(2) Some of the state variables do not converge to a constant value, and therefore, oscillate.

If all the variables converge to a constant value, it must be an equilibrium point. The trivial point is not reachable due to the constraint $S_k + E_k + I_k + R_k = N$ while the disease-free equilibrium point is unstable when $\mathcal{R}_0 > 1$. In consequence, the finite constant limit must be the endemic equilibrium point. On the other hand, if any of the variables does not converge to a finite limit it must oscillate. Thus the theorem is proved. □

Theorem 3 represents an alternative-type theorem to establish the behavior of the solutions when $\mathcal{R}_0 > 1$. The particular case corresponding to one situation or another depends on the particular values of the parameters. However, the complexity of the endemic point expressions (23)-(24) makes it difficult to obtain analytical conditions based entirely on the parameters to distinguish in between. This result is also in line with those presented in [36] for the continuous-time model, where the conditions under which the endemic equilibrium point is globally stable are presented. The important feature regarding the case $\mathcal{R}_0 > 1$ is that the disease persists in the population (in a constant or oscillatory way), and this is the reason why the equilibrium point is called endemic. In this case, the application of vaccination contributes to removing the disease. Many vaccination laws make use of the values of the subpopulations to generate the vaccination effort. Nevertheless, the number of susceptible, exposed, and immune cases may be very difficult to obtain in practice since only the infectious can be registered at health centers. Therefore, a state observer is designed in the next section for the discrete-time model.

4 Design of the state observer

The purpose of the state observer is to obtain an estimation of the state, $\hat{x}_k$, from the unique available data, the infectious subpopulation, I_k. This issue is the main feature of the presented design process. In this way, the following feasible assumptions will be used in the sequel:

Assumption 1 The total population N_k is known at all $k \geq 0$.

Assumption 2 The number of infectious is available at all $k \geq 0$.

Remark 4 The total population may be obtained from (7) if birth and mortality rates are known or by direct measurement. This assumption is particularly feasible in the case of non-lethal diseases, for which the population remains almost constant during the spreading stage.

Remark 5 The number of infectious may be obtained from clinical data at health centers.

Therefore, I_k does not have to be estimated, since it is directly obtained from the actual data. This fact allows us to consider the design of a reduced-order observer. On the other hand, the constraint $N_k = S_k + E_k + I_k + R_k$ for all $k \geq 0$ also permits the algebraic direct calculation of one state variable in terms of the others. In this way, if $\hat{x}_k^T = [\hat{S}_k \; \hat{E}_k \; \hat{I}_k \; \hat{R}_k]$ denotes the observed state, there is no need to estimate $\hat{I}_k$, while $\hat{S}_k$ will be calculated from

$$\hat{S}_k = N_k - \hat{E}_k - \hat{R}_k - I_k. \tag{33}$$

Thus, we just need to observe the values of $\hat{E}_k$ and $\hat{R}_k$. The first step is to re-organize the state vector so as to put together the variables that are to be truly estimated. To this end define the re-organized state vector $z_k = [E_k \; R_k \; S_k \; I_k]^T = Tx_k$ with

$$T = \begin{pmatrix} 0 & 1 & 0 & 0 \\ 0 & 0 & 0 & 1 \\ 1 & 0 & 0 & 0 \\ 0 & 0 & 1 & 0 \end{pmatrix}, \tag{34}$$

so that only the first two components of z_k have to be observed. With this notation, the implicit epidemic model (8) can be re-written as

$$T\mathcal{D}^{-1}\mathcal{M}(x_k)T^{-1}z_{k+1} = \mathcal{K}_k z_{k+1} = z_k, \tag{35}$$

where $\mathcal{K}_k = T\mathcal{D}^{-1}\mathcal{M}(x_k)T^{-1}$. Equation (35) will be the starting point to design the reduced-order state observer. Now, (35) is expanded as

$$\begin{pmatrix} K_{11k} & K_{12k} & K_{13k} \\ K_{21k} & K_{22k} & K_{23k} \\ K_{31k} & K_{32k} & K_{33k} \end{pmatrix} \cdot \begin{pmatrix} z_{1,k+1} \\ z_{2,k+1} \\ z_{3,k+1} \end{pmatrix} = \begin{pmatrix} z_{1,k} \\ z_{2,k} \\ z_{3,k} \end{pmatrix} \tag{36}$$

with $z_{1,k}^T = [E_k \; R_k]$, $z_{2,k} = S_k$, and $z_{3,k} = I_k$. The reason to proceed with the observer design implicitly is that matrix $\mathcal{K}_k$ can be computed symbolically in an easy way. Hence, we have

$$K_{11k} = \text{diag}(C, J), \qquad K_{12k}^T = [-H \quad 0], \tag{37}$$

$$K_{13k}^T = [0 \quad -F],$$

$$K_{21k} = \frac{1}{\eta h\nu + 1} \cdot \left[\eta h\nu(G - C) \quad -(B + \eta h\nu J)\right],$$

$$K_{22k} = \frac{A + H\eta h\nu}{\eta h\nu + 1},$$

$$K_{23k} = \frac{\eta h\nu(F - D)}{\eta h\nu + 1},$$

$$K_{31k} = [-G \quad 0], \qquad K_{32k} = 0,$$

$$K_{33k} = D, \tag{38}$$

with $\eta = \frac{1}{1+(\mu-\upsilon)h}$. The last equation of (36) reads

$$K_{31k}z_{1,k+1} + K_{32k}z_{2,k+1} + K_{33k}z_{3,k+1} = z_{3,k} \tag{39}$$

and will be used to reduce the order of the observer. Since $z_{2,k+1} = S_{k+1} = N_{k+1} - E_{k+1} - R_{k+1} - I_{k+1} = N_{k+1} - [1\ 1]z_{1,k+1} - z_{3,k+1}$, (39) is transformed to

$$K_{31k}z_{1,k+1} + K_{32k}\left(N_{k+1} - [1 \quad 1]z_{1,k+1} - z_{3,k+1}\right) + K_{33k}z_{3,k+1} = z_{3,k},$$

which finally becomes

$$\begin{aligned}&\left(K_{31k} - K_{32k}[1 \quad 1]\right)z_{1,k+1}\\ &\quad = \underbrace{z_{3,k} - K_{33k}z_{3,k+1} + K_{32k}z_{3,k+1} - K_{32k}N_{k+1}}_{\delta_k}.\end{aligned} \tag{40}$$

Let us analyze the meaning of (40). The right-hand side of (40), δ_k, is known since the infectious and the total population are known by Assumptions 1 and 2. On the other hand, the left-hand side is unknown since it just involves the actual state, which is not known. This equation will be used as error signal to drive the state observer. The observer is then designed from the first equation of (36):

$$K_{11k}z_{1,k+1} + K_{12k}\left(N_{k+1} - [1 \quad 1]z_{1,k+1} - z_{3,k+1}\right) + K_{13k}z_{3,k+1} = z_{1,k}$$

by substituting the actual state by the observed one and by adding the observation error, leading to

$$\begin{aligned}&K_{11k}\hat{z}_{1,k+1} + K_{12k}\left(N_{k+1} - [1 \quad 1]\hat{z}_{1,k+1} - z_{3,k+1}\right) + K_{13k}z_{3,k+1}\\ &\quad = \hat{z}_{1,k} + L_k\left(\left(K_{31k} - K_{32k}[1 \quad 1]\right)\hat{z}_{1,k+1} - \delta_k\right),\end{aligned} \tag{41}$$

which is the reduced-order observer to estimate the only two components that remain to be observed. $L_k^T = [l_{1k}\ l_{2k}] \in \mathbb{R}^2$ is the so-called observer gain. Note that (41) is an implicit observer since the observed variable $\hat{z}_{1,k+1}$ appears at both sides of equation. Therefore, the observer is designed implicitly instead of explicitly. An explicit representation of the observer, more appropriate for implementation, can be obtained by re-arranging now (41) to

$$\begin{aligned}&Q_k\hat{z}_{1,k+1} = \hat{z}_{1,k} - L_k\delta_k - K_{12k}N_{k+1} + (K_{12k} - K_{13k})z_{3,k+1},\\ &Q_k = K_{11k} - K_{12k}[1 \quad 1] - L_kK_{31k} + L_kK_{32k}[1 \quad 1],\end{aligned} \tag{42}$$

so that

$$\hat{z}_{1,k+1} = Q_k^{-1} \cdot \left(\hat{z}_{1,k} - L_k\delta_k - K_{12k}N_{k+1} + (K_{12k} - K_{13k})z_{3,k+1}\right) \tag{43}$$

whenever Q_k^{-1} exists. It will be proved in the sequel that this inverse matrix exists if the observer is designed to be asymptotically stable. In this way, (43) and (33) provide together an estimation of all the state variables. The derivation of the observer in an implicit way allows the symbolic calculation of the matrix $\mathcal{Q}_k$ above as

$$\mathcal{Q}_k = \begin{pmatrix} C + l_{1k}G + \beta h\frac{I_k}{N_k} & \beta h\frac{I_k}{N_k} \\ Gl_{2k} & J \end{pmatrix} \tag{44}$$

with the parameters defined by (11). The explicit calculation of this matrix will allow us to analyze the stability properties of the observer.

5 Stability analysis of the observer

This section is devoted to the stability analysis of the observer. For this purpose, denote the observation error as $e_k = z_{1,k} - \hat{z}_{1,k}$. Thus, the error equation reads

$$\mathcal{Q}_k e_{k+1} = e_k. \tag{45}$$

The stability of the observer is discussed in the following theorem, which provides sufficient conditions to guarantee the observation of the states.

Theorem 4 *Consider the discrete-time model* (8) *satisfying Theorem* 1 *constraints. Then the observation error* e_k *of the observer defined by* (41) *under Assumptions* 1 *and* 2 *converges asymptotically to zero if there exist symmetric positive-definite matrices* $P_k, S_k \in \mathbb{R}^{2\times 2}$ *and vector* $L_k^T = [l_{1k}\ l_{2k}]$ *such that the matrix inequality*

$$\mathcal{Q}_k^T P_k \mathcal{Q}_k - P_{k+1} \geq S_k \tag{46}$$

holds for all $k \geq 0$.

Proof Consider the positive-definite Lyapunov candidate sequence $V_k = e_k^T P_k e_k$. Its time variation is given by

$$\begin{aligned} \Delta V_k &= V_{k+1} - V_k = e_{k+1}^T P_{k+1} e_{k+1} - e_k^T P_k e_k \\ &= e_{k+1}^T P_{k+1} e_{k+1} - e_{k+1}^T \mathcal{Q}_k^T P_k \mathcal{Q}_k e_{k+1} \\ &= e_{k+1}^T \left(P_{k+1} - \mathcal{Q}_k^T P_k \mathcal{Q}_k\right) e_{k+1} \\ &\leq -e_{k+1}^T S_k e_{k+1} \leq 0. \end{aligned} \tag{47}$$

Since V_k is positive definite and ΔV_k is negative definite, the second Lyapunov theorem guarantees the convergence to zero of the observation error and the observed variables converge to the actual ones. □

Remark 6 It will be proved below that the stability condition (46) is feasible so that it is always possible to design an asymptotically stable state observer.

Remark 7 Equation (46) can be expressed as a linear matrix inequality (LMI) by using Schur complements. Therefore, the observer design problem can be efficiently solved by available numerical suites.

Remark 8 Since $e_k \to 0$, we have $(E_k - \hat{E}_k) \to 0$ and $(R_k - \hat{R}_k) \to 0$ as $k \to \infty$. As a consequence, (33) implies $(S_k - \hat{S}_k) \to 0$ as $k \to \infty$ and all immeasurable state variables are observed.

The following corollary appears from Theorem 4.

Corollary 1 *If $\mathcal{Q}_k$ satisfies the matrix equation* (46), *then it is invertible.*

Proof Equation (46) can be re-written as

$$\mathcal{Q}_k^T P_k \mathcal{Q}_k \geq S_k + P_{k+1}. \tag{48}$$

Since S_k and P_{k+1} are symmetric positive-definite matrices, $v^T(S_k + P_{k+1})v > 0$ for any real vector $v \neq 0$ and all $k \geq 0$. Thus, from (48) we have

$$v^T \mathcal{Q}_k^T P_k \mathcal{Q}_k v \geq v^T(S_k + P_{k+1})v > 0 \tag{49}$$

for any $v \neq 0$ implying that $\mathcal{Q}_k^T P_k \mathcal{Q}_k$ is a positive-definite matrix. Therefore, $\det(\mathcal{Q}_k^T P_k \mathcal{Q}_k) > 0$. Since $\det(ABC) = \det(A)\det(B)\det(C)$ we obtain from (49) $\det(\mathcal{Q}_k)^2 \cdot \det(P_k) > 0$, implying $\det(\mathcal{Q}_k)^2 > 0$ and $\det(\mathcal{Q}_k) \neq 0$. □

In this way, if the gain vector L_k meets Theorem 4's requirements, the observer is well defined and asymptotically stable. Furthermore, we can prove now that inequality (46) is feasible.

Theorem 5 *There exists at least one gain vector $L_k^T = [l_{1k}\ l_{2k}]$ such that* (46) *holds.*

Proof It will be proved that the equality in (46) holds when $L_k = 0$ and $S_k = S$ is symmetric and positive definite. In this case, the matrix $\mathcal{Q}_k$ reads

$$\mathcal{Q}_k = \begin{pmatrix} C + \beta h \frac{I_k}{N_k} & \beta h \frac{I_k}{N_k} \\ 0 & J \end{pmatrix}, \tag{50}$$

whose eigenvalues are given by $\lambda_1 = C + \beta h \frac{I_k}{N_k} \geq C = 1 + (\mu + \sigma)h > 1$ and $\lambda_2 = J = 1 + (\mu + \omega)h > 1$ for all $k \geq 0$. Since both eigenvalues are larger than unity at all time, we see that $\mathcal{Q}_k^{-1}$ always exists and it is a convergent matrix. The proof is based on the explicit calculation of a sequence of symmetric positive-definite matrices P_k satisfying (46). The following notation will be used:

$$R\prod_{i=1}^{n} A_i = A_1 \cdot A_2 \cdot \ldots \cdot A_n,$$

$$L\prod_{i=1}^{n} A_i = A_n \cdot \ldots \cdot A_2 \cdot A_1.$$

The rationale behind the above definitions is that $R\prod$ multiplies the matrices by the right while $L\prod$ multiplies them by the left. Moreover, the following relation between the two definitions appears:

$$\left(R\prod_{i=1}^{n}A_i\right)^T = L\prod_{i=1}^{n}A_i^T. \tag{51}$$

With this notation, we can now explicitly construct a solution to (46) as

$$P_k = \sum_{q=k}^{\infty}\left(R\prod_{p=k}^{q}Q_p^{-T}\right)S\left(L\prod_{p=k}^{q}Q_p^{-1}\right). \tag{52}$$

Now we can check that (52) is a valid solution to (46):

- Convergence. Since Q_k^{-1} is a convergent matrix for all $k \geq 0$ the series (52) is convergent and well defined.
- Positive definiteness. Equation (52) can be re-written as

$$P_k = \sum_{q=k}^{\infty}\left(L\prod_{p=k}^{q}Q_p^{-1}\right)^T S\left(L\prod_{p=k}^{q}Q_p^{-1}\right) \tag{53}$$

  by using (51). Thus,

$$\begin{aligned} x^T P_k x &= x^T \sum_{q=k}^{\infty}\left(L\prod_{p=k}^{q}Q_p^{-1}\right)^T S\left(L\prod_{p=k}^{q}Q_p^{-1}\right)x \\ &= \sum_{q=k}^{\infty} x^T\left(L\prod_{p=k}^{q}Q_p^{-1}\right)^T S\left(L\prod_{p=k}^{q}Q_p^{-1}x\right) \\ &= \sum_{q=k}^{\infty}\left(L\prod_{p=k}^{q}Q_p^{-1}x\right)^T S\left(L\prod_{p=k}^{q}Q_p^{-1}x\right) \geq 0 \end{aligned}$$

  since $S = S^T > 0$.
- Solution to (46). A direct calculation yields

$$\begin{aligned} &Q_k^T P_k Q_k - P_{k+1} \\ &\quad = Q_k^T\left(Q_k^{-T}SQ_k^{-1} + Q_k^{-T}Q_{k+1}^{-T}SQ_{k+1}^{-1}Q_k^{-1} + \cdots\right)Q_k - P_{k+1} \\ &\quad = S + \underbrace{\left(Q_{k+1}^{-T}SQ_{k+1}^{-1} + \cdots\right)}_{P_{k+1}} - P_{k+1} \\ &\quad = S. \end{aligned} \tag{54}$$

As a consequence, there exists at least one solution to (46) and the observer design problem is feasible. □

Remark 9 A further interpretation of Theorem 5 is that the observer is asymptotically stable in the open-loop case since $L = 0$ implies the vanishing of the error driving term in (41).

6 Numerical examples

This section contains some simulation examples regarding the results introduced in the previous sections. The parameters of the system are assumed to be known and given by

$$\begin{aligned} &\mu^{-1} = 255 \text{ days}, \quad \sigma^{-1} = 1.2 \text{ days}, \quad \omega^{-1} = 12 \text{ days}, \\ &\gamma = \sigma, \quad \nu^{-1} = 115 \text{ days}, \quad \beta = 1.66 \text{ days}^{-1}. \end{aligned} \tag{55}$$

Model's parameters are taken from an outbreak of influenza in a British boarding school in the late 1970s, [38], where the μ and ν parameters have been modified in such a way that a short-term simulation is enough to show the dynamic behavior of the system. In this way, in a period of some days the effect of natural growth can be noticed. The initial values of the populations are $S_0 = 400$, $E_0 = 150$, $I_0 = 550$, $R_0 = 200$ for a total initial population of $N_0 = 1{,}300$. Initially, Figures 1, 2, 3, and 4 depict the dynamics of the continuous and discrete-time systems for time steps of $h = 12$ hours and $h = 24$ hours. These time steps are considered since they imply that data are gathered twice or once per day, respectively, which fits in a typical sampling procedure.

It can be seen that the transient response depends slightly on the value of the time step, while maintaining the shape of the response and tending to be the same value in the steady-state. Moreover, since $\mathcal{R}_0 > 1$ in this example, the disease is persistent and there are always a number of infectious and exposed individuals among the population.

Now, the operation of the reduced state observer will be illustrated through some numerical examples. A time step of $h = 6$ hours has been used. The observer gain has been selected as $L^T = l[1\ 1]$ for the sake of simplicity so that there is only one free parameter.

Figure 1 ***S*** **variable: susceptible population.**

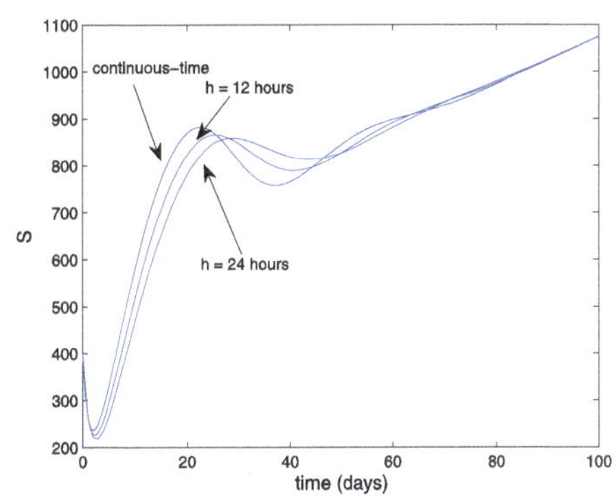

Figure 2 ***E*** **variable: exposed population.**

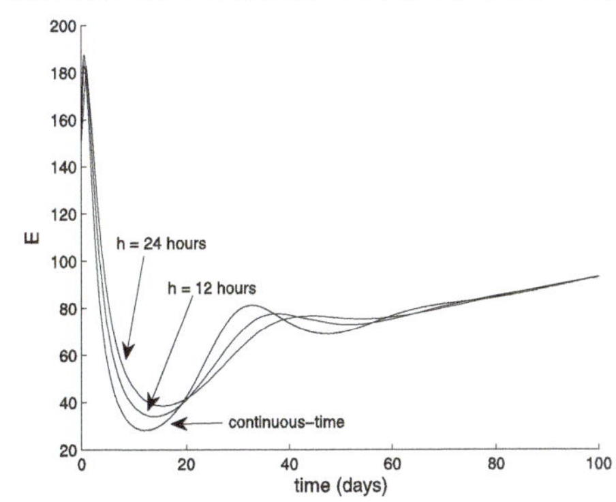

Figure 3 *I* **variable: infectious population.**

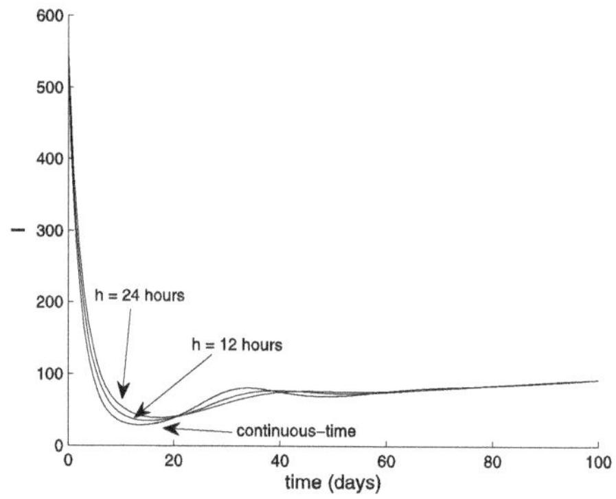

Figure 4 *R* **variable: immune population.**

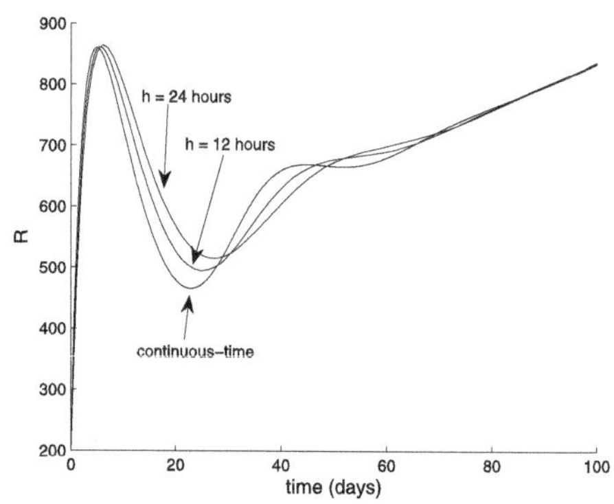

Figure 5 Evolution of the actual and observed susceptible.

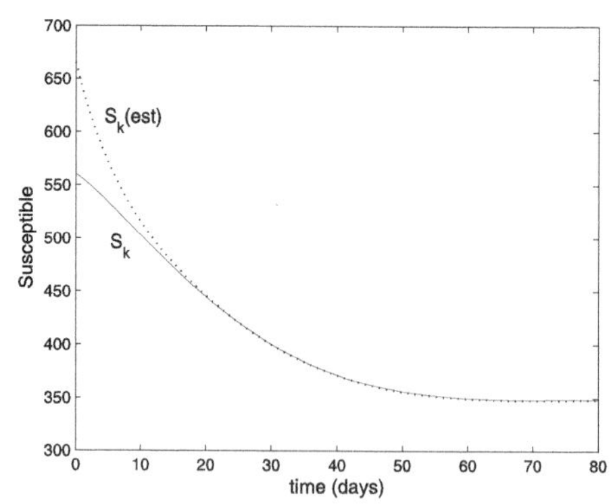

The initial values of the populations are $S_0 = 560$, $E_0 = 70$, $I_0 = 35$, $R_0 = 35$ for a total initial population of $N_0 = 700$. Since the observed populations are initially unknown we may choose the values of $\hat{S}_0$, $\hat{E}_0$, and $\hat{R}_0$ as desired while satisfying $\hat{S}_0 + \hat{E}_0 + \hat{R}_0 = N_0 - I_0 = 700 - 35 = 665$. However, as a systematic way to initialize the observer we suggest to use the initial values $\hat{S}_0 = N_0 - I_0 = 665$ and $\hat{E}_0 = \hat{R}_0 = 0$. This is always a valid selection. Figures 5-7 show the evolution of the actual and observed populations when $l = 0.5$ while Figure 8 depicts the convergence of the observation errors to zero.

It can be noticed in Figures 5-8 that the actual states are observed and the estimation errors vanish asymptotically. Now, we may analyze the effect of varying the observer gain l into the dynamics of the observation. Thus, l will be changed from $l = -0.2$ to $l = 1$. Figure 9 depicts the behavior of the observer for the susceptible when the gain is varied

Figure 6 Evolution of the actual and observed exposed cases.

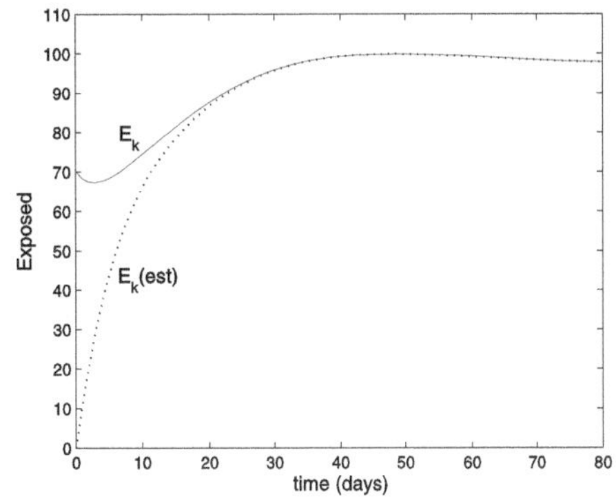

Figure 7 Evolution of the actual and observed immune cases.

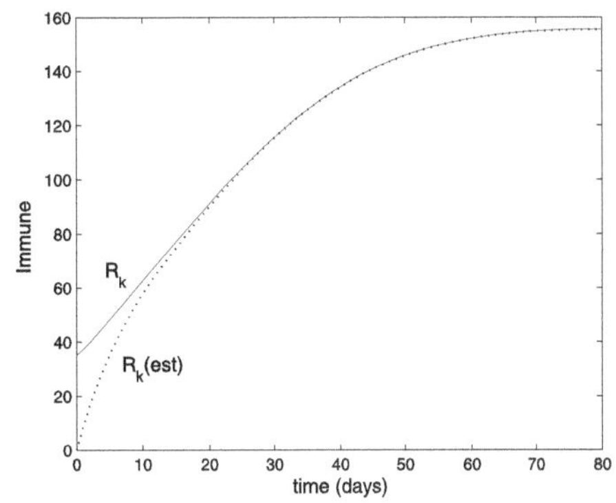

Figure 8 Dynamics of the observation errors.

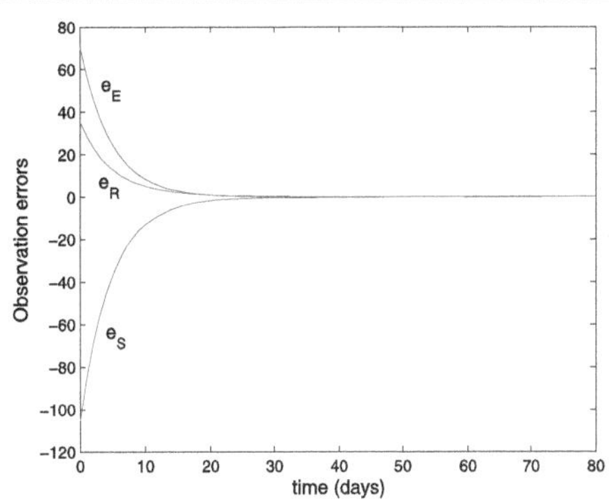

from $l = -0.2$ to $l = 1$, while Figure 10 displays the observation error. Also, Figures 11 and 12 show the observation error for the remaining variables. As can be deduced, the larger the observer gain is the faster the estimation of the state variables is. Hence, the most advisable tuning for the observer gain is the largest constant l satisfying the stability constraint (46).

7 Conclusions

In this paper a discrete-time SEIR model has been obtained from a continuous-time one by using Micken's discretization method. Afterwards, the well-posedness of the model along with the existence and stability of the equilibrium points have been studied. Moreover, an implicit method has been used to analyze these properties instead of the explicit one used in previous works. It has been shown that this method preserves all the structural prop-

Figure 9 Behavior of the observer in *S*.

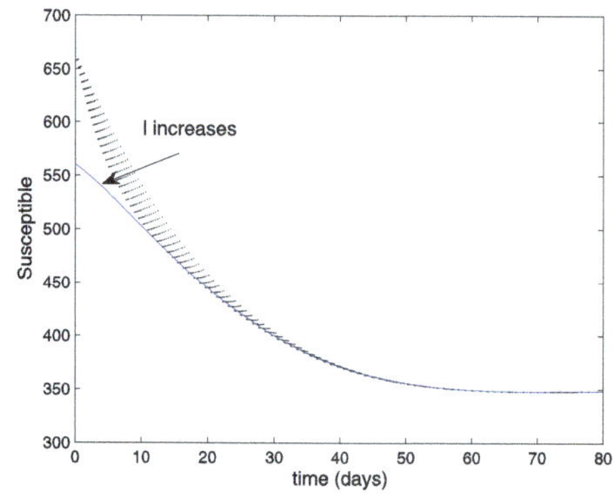

Figure 10 Observation error in *S*.

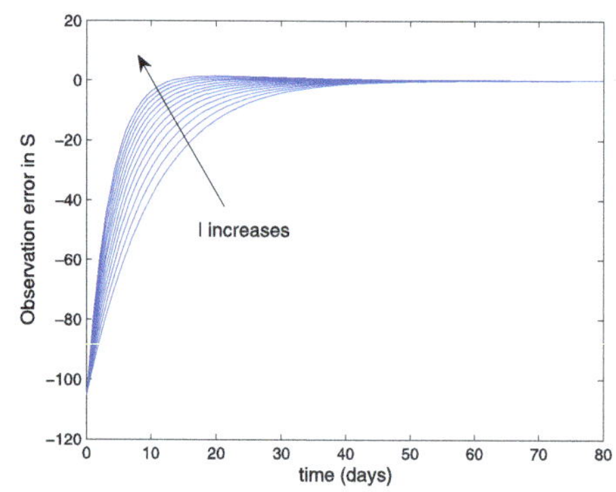

Figure 11 Observation error in *E*.

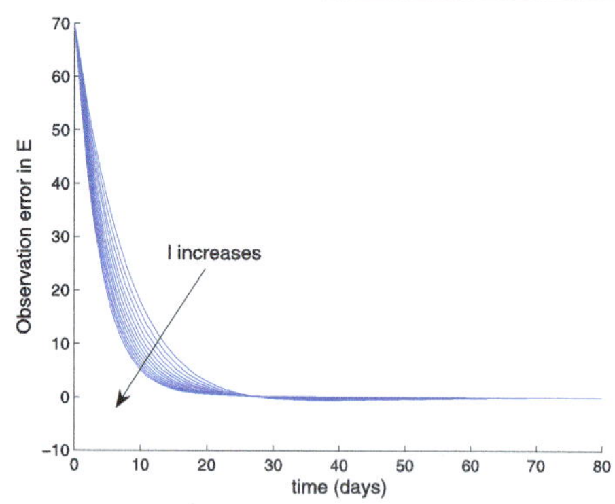

erties exhibited by the continuous-time model such as the expression of the reproduction number and the stability properties of the equilibrium points. In addition, the design of a state observer has been introduced for a discrete-time SEIR epidemic model. The observer was designed in an implicit way since the derivation of its stability properties as well as the design process itself are easier than when proceeding with an explicit model of the system. Moreover, some sufficient conditions to ensure the asymptotic stability of the observer have been provided in terms of a matrix inequality that can be cast in the form of a LMI. The feasibility of these conditions has been proved while some simulation examples showed the operation and usefulness of the observer. The observer is shown to be capable of estimating the actual state variables, while all the observation errors are converging to zero.

Figure 12 Observation error in *R*.

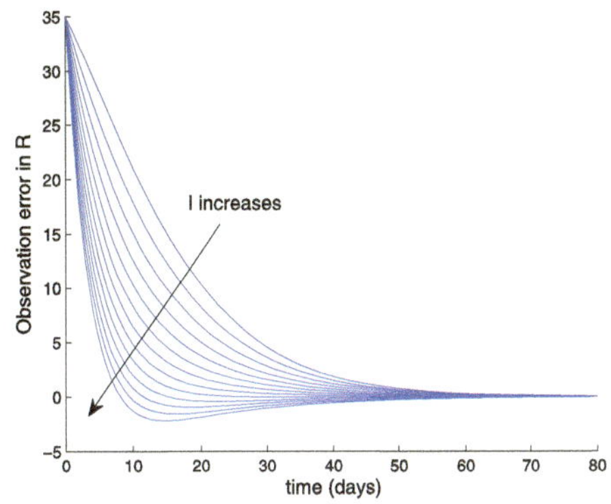

Competing interests
The authors declare that they have no competing interests.

Authors' contributions
All authors contributed equally to the writing of this paper. All authors read and approved the final manuscript.

Author details
[1]Departamento de Telecomunicació i Enginyeria de Sistemes, School of Engineering, Universitat Autònoma de Barcelona, Bellaterra, Barcelona, 08193, Spain. [2]Department of Electricity and Electronics, Faculty of Science and Technology, University of the Basque Country, P.O. Box 644, Bilbao, 48080, Spain. [3]Young Researchers and Elite Club, Yasooj Branch, Islamic Azad University, Yasooj, Iran.

Acknowledgements
This work was partially supported by the Spanish Ministry of Economy and Competitiveness through grants DPI2012-30651 and DPI2013-47825-C3-1-R, by the Basque Government (Gobierno Vasco) through grant IE378-10 and by the University of the Basque Country through grant UFI 11/07.

Endnotes
[a] Note that this implies that the total population is constant, $N_k = N = N_0$.

[b] The inverse of matrix $\mathcal{D}$ exists due to (9) and the constraint $1 + (\mu - \nu)h > 0$.

References

1. Keeling, MJ, Rohani, P: Modeling Infectious Diseases in Humans and Animals. Princeton University Press, Princeton (2008)
2. Zhou, X, Cui, J: Analysis of stability and bifurcation for a SEIR epidemic model with saturated recovery rate. Commun. Nonlinear Sci. Numer. Simul. **16**(11), 4438-4450 (2011)
3. Zhou, X, Cui, J: Stability and Hopf bifurcation analysis of an eco-epidemiological model with delay. J. Franklin Inst. **347**(9), 1654-1680 (2010)
4. Atangana, A, Bildik, N: Approximate solution of tuberculosis disease population dynamics model. Abstr. Appl. Anal. **2013**, Article ID 759801 (2013)
5. Khan, Y, Vazquez-Leal, H, Wu, Q: An efficient iterated method for mathematical biology model. Neural Comput. Appl. **23**(3-4), 677-682 (2013)
6. Rihan, FA, Baleanu, D, Lakshmanan, S, Rakkiyappan, R: On fractional SIRC model with salmonella bacterial infection. Abstr. Appl. Anal. **2014**, Article ID 136263 (2014)
7. de la Sen, M, Agarwal, RP, Ibeas, A, Alonso-Quesada, S: On the existence of equilibrium points, boundedness, oscillating behavior and positivity of a SVEIRS epidemic model under constant and impulsive vaccination. Adv. Differ. Equ. **2011**, Article ID 748608 (2011). doi:10.1155/2011/74860
8. Mukhopadhyay, B, Bhattacharyya, R: Existence of epidemic waves in a disease transmission model with two-habitat population. Int. J. Syst. Sci. **38**(9), 699-707 (2007)
9. Zhou, K, Wang, Q-R: Existence of traveling waves for a delayed SIRS epidemic diffusion model with saturation incidence rate. Abstr. Appl. Anal. **2014**, Article ID 369072 (2014)
10. Gan, Q, Xu, R, Li, Y, Hu, R: Travelling waves in an infectious disease model with a fixed latent period and a spatio-temporal delay. Math. Comput. Model. **53**(5-6), 814-823 (2011)
11. Li, J, Xiao, Y, Zhang, F, Yang, Y: An algebraic approach to proving the global stability of a class of epidemic models. Nonlinear Anal., Real World Appl. **13**(5), 2006-2016 (2012)
12. Bowong, S, Tewa, JJ: Global analysis of a dynamical model for transmission of tuberculosis with a general contact rate. Commun. Nonlinear Sci. Numer. Simul. **15**(11), 3621-3631 (2010)
13. Fan, X: Global stability of multigroup SIRS epidemic model with varying population sizes and stochastic perturbation around equilibrium. Abstr. Appl. Anal. **2014**, Article ID 154725 (2014)
14. Chen, H, Sun, J: Global stability of delay multigroup epidemic models with group mixing and nonlinear incidence rates. Appl. Math. Comput. **218**(8), 4391-4400 (2011)

15. Tian, JP, Wang, J: Global stability for cholera epidemic models. Math. Biosci. **232**(1), 31-41 (2011)
16. Shu, H, Fan, D, Wei, J: Global stability of multi-group SEIR epidemic models with distributed delays and nonlinear transmission. Nonlinear Anal., Real World Appl. **13**(4), 1581-1592 (2012)
17. Canto, B, Coll, C, Sanchez, E: A study on vaccination models for a seasonal epidemic process. Appl. Math. Comput. **243**, 152-160 (2014)
18. Arenas, AJ, Gonzalez-Parra, G, Chen-Charpentier, BM: A nonstandard numerical scheme of predictor-corrector type for epidemic models. Comput. Math. Appl. **59**, 3740-3749 (2010)
19. Jodar, L, Villanueva, RJ, Arenas, AJ, Gonzalez, GC: Nonstandard numerical methods for a mathematical model for influenza disease. Math. Comput. Simul. **79**, 622-633 (2008)
20. Izzo, G, Muroya, Y, Vecchio, A: A general discrete-time model of population dynamics in the presence of an infection. Discrete Dyn. Nat. Soc. **2009**, Article ID 143019 (2009)
21. Micken, RE: Numerical integration of population models satisfying conservation laws: NSFD methods. Journal of Biological Dynamics **1**(4), 427-436 (2007)
22. Hu, Z, Teng, Z, Jiang, H: Stability analysis in a class of discrete SIRS epidemic models. Nonlinear Anal., Real World Appl. **13**(5), 2017-2033 (2012)
23. de la Sen, M, Ibeas, A, Alonso-Quesada, S: On vaccination controls for the SEIR epidemic model. Commun. Nonlinear Sci. Numer. Simul. **17**(6), 2637-2658 (2012)
24. de la Sen, M, Ibeas, A, Alonso-Quesada, S: Feedback linearization based vaccination control strategies for true-mass action type SEIR epidemic models. Nonlinear Anal., Model. Control **16**(3), 283-314 (2011)
25. de la Sen, M, Alonso-Quesada, S: Vaccination strategies based on feedback control techniques for a general SEIR-epidemic model. Appl. Math. Comput. **218**, 3888-3904 (2011)
26. Yi, N, Zhang, Q, Mao, K, Yang, D, Li, Q: Analysis and control of an SEIR epidemic system with nonlinear transmission rate. Math. Comput. Model. **50**(9-10), 1498-1513 (2009)
27. de la Sen, M, Ibeas, A, Alonso-Quesada, S: Observer-based vaccination strategy for a true mass action SEIR epidemic model with potential estimation of all the populations. Discrete Dyn. Nat. Soc. **2011**, Article ID 743067 (2011)
28. Alonso-Quesada, S, de la Sen, M, Agarwal, RP, Ibeas, A: An observer-based vaccination control law for an SEIR epidemic model based on feedback linearization techniques for nonlinear systems. Adv. Differ. Equ. **2012**, Article ID 161 (2012)
29. Huang, S-Z: A new SEIR epidemic model with applications to the theory of eradication and control of diseases, and to the calculation of R_0. Math. Biosci. **215**, 84-104 (2008)
30. Ibeas, A, de la Sen, M, Alonso-Quesada, S: Robust sliding control of SEIR epidemic models. Math. Probl. Eng. **2014**, Article ID 104764 (2014)
31. Banerjee, S, Jana, AK: High gain observer based extended generic model control with application to a reactive distillation column. J. Process Control **24**(4), 235-248 (2014)
32. Zong, Q, Wang, J, Tian, B, Tao, Y: Quasi-continuous high-order sliding mode controller and observer design for flexible hypersonic vehicle. Aerosp. Sci. Technol. **27**(1), 127-137 (2013)
33. Mahmud, MA, Hossain, MJ, Pota, HR: Transient stability enhancement of multimachine power systems using nonlinear observer-based excitation controller. Int. J. Electr. Power Energy Syst. **58**, 57-63 (2014)
34. Ibeas, A, de la Sen, M, Alonso-Quesada, S: Adaptive control of discrete-time SEIR epidemic models. In: ICNPAA Congress: Mathematical Problems in Engineering, Aerospace and Sciences, Narvik, Norway, 15-18 July (2014)
35. Ibeas, A, de la Sen, M, Alonso-Quesada, S, Zamani, I, Safiee, M: Observer design for SEIR discrete-time epidemic models. In: The 13th International Conference on Control, Automation, Robotics and Vision (ICARCV 2014), Singapore, 10-12 December, pp. 10-12 (2014)
36. Cheng, Y, Yang, X: On the global stability of SEIRS models in epidemiology. Can. Appl. Math. Q. **20**, 115-133 (2012)
37. Canto, B, Coll, C, Sanchez, E: Positive solutions of a discrete-time descriptor system. Int. J. Syst. Sci. **39**(1), 81-88 (2008)
38. de la Sen, M, Alonso-Quesada, S: A simple vaccination control strategy for the SEIR epidemic model. In: Proceedings of the 5th IEEE International Conference on Management of Innovation and Technology, pp. 1037-1044 (2010)

Periodic solutions for Liénard type equation with time-variable coefficient

Fabao Gao[1,2*], Shiping Lu[3] and Minghui Yao[4]

*Correspondence:
gaofabao@sina.com;
gaofabao@vt.edu
[1]School of Mathematical Science, Yangzhou University, Yangzhou, 225002, China
[2]Department of Engineering Science and Mechanics, Virginia Tech, Blacksburg, VA 24061, USA
Full list of author information is available at the end of the article

Abstract

By means of topological theory, a class of Liénard type equations with a deviating argument of the form $(\varphi_p(x'(t)))' + f(x(t-\tau))x'(t-\tau) + \beta(t)g(x(t-\tau)) = e(t)$ is studied. It is notable that the coefficient $\beta(t)$ in front of $g(x(t-\tau))$ is allowed to change sign in this paper. Moreover, a numerical simulation is carried out to verify the validity of the obtained results. In addition, the generalized form of the above equation with time-varying delays is also discussed briefly.

Keywords: periodic solution; p-Laplacian; topological theory; time-varying coefficient

1 Introduction

Consider the Liénard type p-Laplacian differential equation

$$(\varphi_p(x'(t)))' + f(x(t-\tau))x'(t-\tau) + \beta(t)g(x(t-\tau)) = e(t), \tag{1.1}$$

where $p > 1$ and $\varphi_p : R \mapsto R$ is given by $\varphi_p(s) = |s|^{p-2}s$ for $s \neq 0$ and $\varphi_p(0) = 0$, $\int_0^T \beta(s)\,ds \neq 0$, $f, g, e \in C(R, R)$, and $e(t)$ is T-periodic, $\int_0^T e(s)\,ds = 0$, $T, \tau > 0$ are given constants. The equation can be applied to the dynamics of fluids. An example is Euler's equation, governing the flow of an ideal fluid in a conservative force field. While there are plenty of results on the existence of periodic solutions for the p-Laplacian equation (see [1–10] and references therein), studying delay Liénard equations with a variable coefficient in front of the nonlinear term is relatively uncommon. The main difficulty lies in finding *a priori bounds* for (1.1) requiring the coefficient of the nonlinear term to keep a fixed sign.

In the aforementioned literature, based on the Leray-Schauder degree theory, Amster *et al.* [1] considered the existence of at least one periodic solution for the p-Laplacian-like system with a fixed delay. Gao and Zhang [2] discussed an n-dimensional p-Laplacian-like neutral functional differential equation, and they established some criteria to guarantee the existence of periodic solutions for the equation by using Mawhin's continuation theorem. Manásevich and Sędziwy [3] were concerned with the existence and uniqueness of a limit circle for a generalized Liénard type equation, which involves the one-dimensional p-Laplacian operator and a positive small parameter. Manásevich and Mawhin [4] studied the existence of periodic solutions to some system cases involving the fairly general vector-valued p-Laplacian operator. Aizicovici *et al.* [5] investigated a nonlinear periodic problem driven by the scalar p-Laplacian with a nonsmooth potential via the degree map; they proved the existence of at least three distinct nontrivial solutions, two of which have a constant sign. In Gaines and Mawhin's monograph [6], coincidence degree theory was described and used to study alternative problems. Aizicovici *et al.* [7] conducted periodic problems driven by the scalar p-Laplacian with a multivalued right-hand side nonlinearity. Gao *et al.* [8, 9] considered the existence of periodic solutions for two kinds of Rayleigh type p-Laplacian equations by using the continuation theorem. Cheung [10] proposed the existence of periodic solutions of a p-Laplacian Rayleigh equation with two deviating arguments.

Based on topological theory and some analysis techniques, the existence of periodic solutions for (1.1) is investigated in the present paper. It is significant that the coefficient $\beta(t)$ of the nonlinear term can change sign, which cannot be achieved in most of the previous papers. Furthermore, a numerical simulation is performed to validate the feasibility of the obtained results. In addition, the fixed delays in (1.1) are also extended to the time-varying delays and we briefly discuss them. Moreover, the approaches used to estimate *a priori bounds* of periodic solutions are different from the corresponding ones in the literature.

2 Preliminaries

Let us consider the problem: find $u \in C_T^1$ such that

$$(\varphi_p(u'(t)))' = \widetilde{f}(t, u(t-\tau), u'(t-\tau)). \tag{2.1}$$

Lemma 2.1 (see Amster *et al.* [1]) *Let $\Omega \in C_T^1$ be an open set. Assume that:*

(A$_1$) *For $\lambda \in (0,1]$ the problem*

$$(\varphi_p(u'(t)))' = \lambda\widetilde{f}(t, u(t-\tau), u'(t-\tau))$$

has no solution on $\partial\Omega$.

(A$_2$) *The equation*

$$\widetilde{F}(a) \triangleq \frac{1}{T}\int_0^T \widetilde{f}(t, a, 0) = 0$$

has no solution on $\partial\Omega \cap R^n$.

(A$_3$) *The Brouwer degree*

$$\deg(\widetilde{F}, \Omega \cap R^n, 0) \neq 0.$$

Then the problem (2.1) *has at least one solution in C_T^1.*

3 Main results

For the sake of convenience, we only study the periodic solutions of (1.1) in the case $\int_0^T \beta(t)\,dt > 0$ ($\int_0^T \beta(t)\,dt < 0$ can be discussed in the same way).

Theorem 3.1 *Assume that the following conditions hold:*

(H_1) *There are positive constants* m_1, m_2, *and* d *such that*

$$m_1|u|^{p-1} \leq |g(u)| \leq m_2|u|^{p-1}, \quad \forall |u| > d.$$

(H_2) *There is a constant* $r \geq 0$ *such that* $\lim_{|u|\to\infty} \frac{|F(u)|}{|u|^{p-1}} \leq r$, *where* $F(x) = \int_0^x f(s)\,ds$.

(H_3)

$$A := \begin{cases} [\frac{(\beta_\infty+\varepsilon_1)m_2 T}{m_1\int_0^T(\beta(t)+\beta_\infty+\varepsilon_1)\,dt}]^{\frac{1}{p-1}} 2^{\frac{2-p}{p-1}} < 1, & 1 < p \leq 2, \\ [\frac{(\beta_\infty+\varepsilon_1)m_2 T}{m_1\int_0^T(\beta(t)+\beta_\infty+\varepsilon_1)\,dt}]^{\frac{1}{p-1}} < 1, & p > 2, \end{cases}$$

where $\varepsilon_1 > 0$, $\beta_\infty = \max_{t\in[0,T]} |\beta(t)|$.

(H_4) $ug(u) > 0$, $\forall |u| > d$.

Then (1.1) *has at least one* T*-periodic solution, if* $C_p r[\frac{T}{2(1-A)}]^{p-1} + \frac{\beta_\infty m_2 T}{2(1-A)^p} < 1$, *where*

$$C_p = \begin{cases} 1, & 1 < p \leq 2, \\ 2^{p-2}, & p > 2. \end{cases}$$

Proof Consider the homotopic equation of (1.1) as follows:

$$(\varphi_p(x'(t)))' + \lambda f(x(t-\tau))x'(t-\tau) + \lambda\beta(t)g(x(t-\tau)) = \lambda e(t), \quad \lambda \in (0,1]. \tag{3.1}$$

For $\forall \varepsilon_1 > 0$, $\beta_\infty = \max_{t\in[0,T]} |\beta(t)|$, (3.1) can be written in the following form:

$$\begin{aligned}(\varphi_p(x'(t)))' &+ \lambda f(x(t-\tau))x'(t-\tau) + \lambda(\beta(t) + \beta_\infty + \varepsilon_1)g(x(t-\tau)) \\ &= \lambda(\beta_\infty + \varepsilon_1)g(x(t-\tau)) + \lambda e(t), \quad \lambda \in (0,1].\end{aligned} \tag{3.2}$$

Integrating both sides of (3.2) from 0 to T and using the integral mean value theorem, there exists a constant $\xi \in (0,T)$ such that

$$g(x(\xi-\tau)) \int_0^T (\beta(t) + \beta_\infty + \varepsilon_1)\,dt = \int_0^T (\beta_\infty + \varepsilon_1)g(x(t-\tau))\,dt. \tag{3.3}$$

Now, we claim that

$$|x(\xi-\tau)| \leq A|x|_\infty + B, \tag{3.4}$$

where

$$A = \begin{cases} [\frac{(\beta_\infty+\varepsilon_1)m_2 T}{m_1\int_0^T(\beta(t)+\beta_\infty+\varepsilon_1)\,dt}]^{\frac{1}{p-1}} 2^{\frac{2-p}{p-1}}, & 1 < p \leq 2, \\ [\frac{(\beta_\infty+\varepsilon_1)m_2 T}{m_1\int_0^T(\beta(t)+\beta_\infty+\varepsilon_1)\,dt}]^{\frac{1}{p-1}}, & p > 2, \end{cases}$$

$$B = \begin{cases} [\frac{(\beta_\infty+\varepsilon_1)M_g T}{m_1\int_0^T(\beta(t)+\beta_\infty+\varepsilon_1)\,dt}]^{\frac{1}{p-1}} 2^{\frac{2-p}{p-1}} + d, & 1 < p \leq 2, \\ [\frac{(\beta_\infty+\varepsilon_1)M_g T}{m_1\int_0^T(\beta(t)+\beta_\infty+\varepsilon_1)\,dt}]^{\frac{1}{p-1}} + d, & p > 2, \end{cases}$$

$$M_g = \max_{|u|\leq d}|g(u)|.$$

Case 1: If $|x(\xi - \tau)| \le d$, then (3.4) holds clearly.
Case 2: If $|x(\xi - \tau)| > d$. Denote

$$E_1 = \{t : t \in [0, T], |x(t-\tau)| \le d\}, \qquad E_2 = \{t : t \in [0, T], |x(t-\tau)| > d\}.$$

From (3.3), we have

$$|g(x(\xi-\tau))| \int_0^T (\beta(t) + \beta_\infty + \varepsilon_1)\, dt \le (\beta_\infty + \varepsilon_1)\left(\int_{E_1} + \int_{E_2}\right)|g(x(t-\tau))|\, dt,$$

which, together with assumption (H_1), leads to

$$\begin{aligned} |x(\xi-\tau)|^{p-1} m_1 \int_0^T (\beta(t) + \beta_\infty + \varepsilon_1)\, dt &\le (\beta_\infty + \varepsilon_1)\left(\int_{E_2} m_2 |x(t-\tau)|^{p-1}\, dt + M_g T\right) \\ &\le (\beta_\infty + \varepsilon_1) m_2 T |x|_\infty^{p-1} + (\beta_\infty + \varepsilon_1) M_g T, \end{aligned}$$

i.e.,

$$\begin{aligned} |x(\xi-\tau)| &\le \left[\frac{(\beta_\infty + \varepsilon_1) m_2 T}{m_1 \int_0^T (\beta(t) + \beta_\infty + \varepsilon_1)\, dt}\right]^{\frac{1}{p-1}} 2^{\frac{2-p}{p-1}} |x|_\infty \\ &\quad + \left[\frac{(\beta_\infty + \varepsilon_1) M_g T}{m_1 \int_0^T (\beta(t) + \beta_\infty + \varepsilon_1)\, dt}\right]^{\frac{1}{p-1}} 2^{\frac{2-p}{p-1}}, \quad 1 < p \le 2, \\ |x(\xi-\tau)| &\le \left[\frac{(\beta_\infty + \varepsilon_1) m_2 T}{m_1 \int_0^T (\beta(t) + \beta_\infty + \varepsilon_1)\, dt}\right]^{\frac{1}{p-1}} |x|_\infty \\ &\quad + \left[\frac{(\beta_\infty + \varepsilon_1) M_g T}{m_1 \int_0^T (\beta(t) + \beta_\infty + \varepsilon_1)\, dt}\right]^{\frac{1}{p-1}}, \quad p > 2. \end{aligned}$$

Thus, from Case 1 and Case 2, we see that (3.4) holds.
Let $\xi - \tau = kT + \bar{\xi}$, where k is an integer and $\bar{\xi} \in [0, T]$; noticing (3.4), we get

$$|x(t)| \le A|x|_\infty + B + \int_{\bar{\xi}}^t |x'(s)|\, ds, \quad t \in [\bar{\xi}, \bar{\xi} + T],$$

and

$$|x(t-T)| \le A|x|_\infty + B + \int_{t-T}^{\bar{\xi}} |x'(s)|\, ds, \quad t \in [\bar{\xi}, \bar{\xi} + T].$$

Combining the above two inequalities, we obtain

$$\begin{aligned} |x|_\infty &= \max_{t \in [\bar{\xi}, \bar{\xi}+T]} |x(t)| \\ &\le \max_{t \in [\bar{\xi}, \bar{\xi}+T]} \left\{A|x|_\infty + B + \frac{1}{2}\left(\int_{\bar{\xi}}^t |x'(s)|\, ds + \int_{t-T}^{\bar{\xi}} |x'(s)|\, ds\right)\right\} \\ &\le A|x|_\infty + B + \frac{1}{2}\int_0^T |x'(s)|\, ds. \end{aligned}$$

In view of (H_3), we have

$$|x|_\infty \leq \frac{B}{1-A} + \frac{\int_0^T |x'(s)|\,ds}{2(1-A)}. \tag{3.5}$$

On the other hand, noticing $C_p r[\frac{T}{2(1-A)}]^{p-1} + \frac{\beta_\infty m_2 T}{2(1-A)^p} < 1$, we easily see that there is a sufficiently small constant $\varepsilon_2 > 0$ such that

$$C_p(r+\varepsilon_2)\left[\frac{T}{2(1-A)}\right]^{p-1} + \frac{\beta_\infty m_2 T}{2(1-A)^p} < 1. \tag{3.6}$$

By assumption (H_2) and for such ε_2, we know there exists a constant $\rho > d$ (independent of λ) such that

$$|F(u)| \leq (r+\varepsilon_2)|u|^{p-1}, \quad |u| > \rho. \tag{3.7}$$

Let

$$\Delta_1 = \{t : t \in [0,T], |x(t-\tau)| \leq \rho\}, \qquad \Delta_2 = \{t : t \in [0,T], |x(t-\tau)| > \rho\}.$$

Multiplying both sides of (3.1) by $x(t)$ and integrating them with $[0,T]$, noticing (3.7), we get

$$\begin{aligned}
\int_0^T |x'(t)|^p\,dt &= -\int_0^T (\varphi_p(x'(t)))' x(t)\,dt \\
&= \lambda \int_0^T f(x(t-\tau))x'(t-\tau)x(t)\,dt \\
&\quad + \lambda \int_0^T \beta(t) g(t, x(t-\tau))x(t)\,dt - \lambda \int_0^T e(t)x(t)\,dt \\
&\leq \left(\int_{\Delta_1} + \int_{\Delta_2}\right) |F(x(t-\tau))x'(t)|\,dt \\
&\quad + \beta_\infty |x|_\infty \left(\int_{E_1} + \int_{E_2}\right) |g(x(t-\tau))|\,dt + e_\infty |x|_\infty T \\
&\leq [M_\rho + (r+\varepsilon_2)|x|_\infty^{p-1}] \int_0^T |x'(t)|\,dt + \beta_\infty m_2 T |x|_\infty^p \\
&\quad + \beta_\infty M_g T |x|_\infty + e_\infty T |x|_\infty,
\end{aligned} \tag{3.8}$$

where $M_\rho = \max_{|u|\leq\rho} |F(u)|$, $e_\infty = \max_{t\in[0,T]} |e(t)|$.

Substituting (3.5) into (3.8) yields

$$\begin{aligned}
\int_0^T |x'(t)|^p\,dt &\leq (r+\varepsilon_2)\left[\frac{B}{1-A} + \frac{\int_0^T |x'(t)|\,dt}{2(1-A)}\right]^{p-1} \int_0^T |x'(t)|\,dt \\
&\quad + \beta_\infty m_2 T \left[\frac{B}{1-A} + \frac{\int_0^T |x'(t)|\,dt}{2(1-A)}\right]^p \\
&\quad + \left[M_\rho + \frac{\beta_\infty M_g T}{2(1-A)} + \frac{e_\infty T}{2(1-A)}\right] \int_0^T |x'(t)|^p\,dt
\end{aligned}$$

$$
\begin{aligned}
&+ (\beta_\infty M_g + e_\infty)\frac{BT}{1-A} \\
&\le C_p(r+\varepsilon_2)\left[\left(\frac{B}{1-A}\right)^{p-1} + \left(\frac{\int_0^T |x'(t)|\,dt}{2(1-A)}\right)^{p-1}\right]\int_0^T |x'(t)|\,dt \\
&+ \frac{\beta_\infty m_2 T}{2(1-A)^p}\left(\int_0^T |x'(t)|\,dt\right)^p \\
&+ \left[M_\rho + \frac{\beta_\infty M_g T}{2(1-A)} + \frac{e_\infty T}{2(1-A)}\right]T^{\frac{1}{q}}\left(\int_0^T |x'(t)|^p\,dt\right)^{\frac{1}{p}} \\
&+ (\beta_\infty M_g + e_\infty)\frac{BT}{1-A} + \frac{2^{p-1}\beta_\infty m_2 B^p T}{(1-A)^p} \\
&\le \left[C_p(r+\varepsilon_2)\left[\frac{T}{2(1-A)}\right]^{p-1} + \frac{\beta_\infty m_2 T}{2(1-A)^p}\right]\int_0^T |x'(t)|^p\,dt \\
&+ \theta_1\left(\int_0^T |x'(t)|^p\,dt\right)^{\frac{1}{p}} + \theta_2,
\end{aligned}
\tag{3.9}
$$

where

$$
C_p = \begin{cases} 1, & 1 < p \le 2, \\ 2^{p-2}, & p > 2, \end{cases}
$$

$$
\theta_1 = C_p(r+\varepsilon_2)\left(\frac{B}{1-A}\right)^{p-1} T^{\frac{1}{q}} + \left[M_\rho + \frac{\beta_\infty M_g T}{2(1-A)} + \frac{e_\infty T}{2(1-A)}\right]T^{\frac{1}{q}},
$$

$$
\theta_2 = (\beta_\infty M_g + e_\infty)\frac{BT}{1-A} + \frac{2^{p-1}\beta_\infty m_2 B^p T}{(1-A)^p},
$$

$$
e_\infty = \max_{t\in[0,T]} |e(t)|.
$$

In view of (3.6) and $\frac{1}{p} < 1$, it follows from (3.9) that there is a constant $M > 0$ such that

$$
\int_0^T |x'(t)|^p\,dt \le M, \tag{3.10}
$$

which implies that there exists a constant $M_1 > 0$ such that

$$
|x'|_\infty \le M_1. \tag{3.11}
$$

From (3.5) and (3.10), we can see that there exists a constant M_2 such that

$$
|x|_\infty \le \frac{B}{1-A} + \frac{T^{\frac{1}{q}} M^{\frac{1}{p}}}{2(1-A)} \triangleq M_2. \tag{3.12}
$$

According to (3.11) and (3.12), set $\Omega = \{x : |x'|_\infty < M_1 + 1, |x|_\infty < M_2 + 1\}$, then we see that (1.1) has no solution on $\partial\Omega$ for $\lambda \in (0,1]$, and when $x(t) = M_2 + 1$ or $-M_2 - 1$, from (H_4), we can get

$$
-\frac{1}{T}\int_0^T \beta(t) g(M_2+1)\,dt < 0,
$$

$$-\frac{1}{T}\int_0^T \beta(t)g(-M_2-1)\,dt < 0.$$

So, condition (A_2) of Lemma 2.1 is also satisfied.

Denote

$$H(x,\mu) = \mu x + \frac{1-\mu}{T}\int_0^T \beta(t)g(x)\,dt,$$

and when $x \in \partial\Omega \cap R$, $\mu \in [0,1]$, in view of $\int_0^T \beta(s)\,ds > 0$, we have

$$xH(x,\mu) = \mu x^2 + \frac{(1-\mu)x}{T}\int_0^T \beta(t)g(x)\,dt > 0.$$

Thus, $H(x,\mu)$ is a homotopic transformation and

$$\begin{aligned}\deg\{\widetilde{F},\Omega\cap R,0\} &= \deg\left\{-\frac{1}{T}\int_0^T \beta(t)g(x)\,dt,\Omega\cap R,0\right\}\\ &= \deg\{x,\Omega\cap R,0\} \neq 0.\end{aligned}$$

Therefore, condition (A_3) of Lemma 2.1 is also satisfied. By using Lemma 2.1, we conclude that (1.1) has at least one T-periodic solution $x(t)$ on $\bar{\Omega}$ with $|x|_\infty \le M_2$. This completes the proof of Theorem 3.1. □

4 Generalization

As a matter of fact, (1.1) can also be extended to the time-varying case, which admits the following form:

$$(\varphi_p(x'(t)))' + f(x(t-\tau))x'(t-\tau) + \beta(t)g(x(t-\delta(t))) = e(t), \tag{4.1}$$

where $\delta(t+T) \equiv \delta(t)$, $\delta'(t) < 1$, and $\int_0^T \beta(s)\,ds \neq 0$ (the case of $\int_0^T \beta(s)\,ds > 0$ will be considered in this section, and the other case, $\int_0^T \beta(s)\,ds < 0$, can be studied in the same way). Then, according to Mawhin's continuation theorem, the following result is obtained with a similar process to Theorem 3.1.

Theorem 4.1 *Assume that the following conditions hold:*

(C_1) *There exist constants $r_1 > 0$, $r_2 > 0$, and $d_1 \ge 0$ such that*

$$r_1|u|^{p-1} \le |g(u)| \le r_2|u|^{p-1} \quad and \quad ug(u) > 0, \quad \forall |u| > d_1.$$

(C_2) *There are constant $r_3 > 0$ and $d_2 \ge 0$ such that $|f(u)| \le r_3|u|^{p-1}$, $\forall |u| > d_2$.*

(C_3) *There is a constant $\varepsilon_2 > 0$ such that*

$$M := \begin{cases} \left[\frac{(|\beta^-|_\infty+\varepsilon_2)r_2T}{r_1\int_0^T(\beta^++\varepsilon_2)dt}\right]^{\frac{1}{p-1}} 2^{\frac{2-p}{p-1}} < 1, & 1 < p \le 2,\\ \left[\frac{(|\beta^-|_\infty+\varepsilon_2)r_2T}{r_1\int_0^T(\beta^++\varepsilon_2)dt}\right]^{\frac{1}{p-1}} < 1, & p > 2,\end{cases}$$

where $\beta^+ = \max_{t\in[0,T]}\{\beta(t),0\}$, $\beta^- = \max_{t\in[0,T]}\{-\beta(t),0\}$.

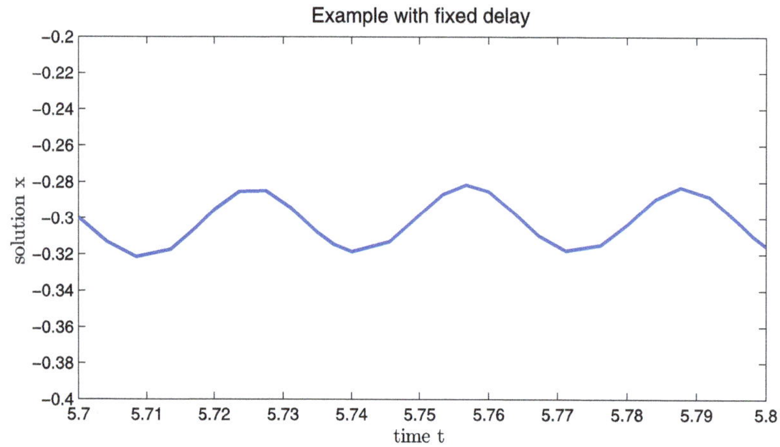

Figure 1 Phase diagram of the system for $\tau = 0.02$.

Then (4.1) *has at least one* T*-periodic solution if*

$$C_p r_3 T^{p-1} + 2^{p-1} r_2 \theta |\beta|_\infty T^p < (1-M)^{p-1},$$

where $\theta = \max_{t\in[0,T]} \frac{1}{1-\delta'(\gamma(t))}$ *and* $\gamma(t)$ *denotes the inverse of function* $t - \delta(t)$.

5 Example

For convenience, as an application of Theorem 3.1, we consider the following example:

$$\left(\varphi_4\left(x'(t)\right)\right)' + f\left(x(t-\tau)\right)x'(t-\tau) + \left(\sin t + \frac{1}{2}\right)g\left(x(t-\tau)\right) = \cos 200t, \tag{5.1}$$

where $p = 4$, $f(t) = \frac{t^2 \sin t}{500{,}000}$, $\beta(t) = \sin t + \frac{1}{2}$, $g(u) = \frac{u^3}{10{,}000} + 1$, $e(t) = \cos 200t$, then $F(u) = \int_0^u f(s)\,ds = \frac{1}{500{,}000}(-u^2\cos u + 2u\sin u + 2\cos u - 2)$, and $\beta_\infty = \frac{3}{2}$. So we can choose $m_1 = \frac{1}{10{,}000}$, $m_2 = \frac{1}{5{,}000}$, $d = 10\sqrt[3]{10}$, $r = \frac{1}{500{,}000}$, $\varepsilon_1 = \frac{1}{2}$ so that the conditions of Theorem 3.1 hold. Therefore, by Theorem 3.1 we find that (5.1) has at least one $\frac{\pi}{100}$-periodic solution, which can also be illustrated by numerical simulation.

By using MATLAB© 7.12.0 (R2011a) toolkit: *dde*23, which can be used to solve delay differential equations with constant delays, (5.1) is simulated on tspan = [5.7, 5.8] with history = [0.5, −0.3] for $t \le 0$ (see Figure 1). It can be found from Figure 1 that the equation admits one periodic solution with periodicity 0.0314, which is around $\frac{\pi}{100}$. Therefore, the results achieved in this paper are significant.

Competing interests
The authors declare that they have no competing interests.

Authors' contributions
All authors contributed equally to this work. All authors read and approved the final version of the manuscript.

Author details
[1]School of Mathematical Science, Yangzhou University, Yangzhou, 225002, China. [2]Department of Engineering Science and Mechanics, Virginia Tech, Blacksburg, VA 24061, USA. [3]College of Mathematics & Statistics, Nanjing University of Information Science & Technology, Nanjing, 210044, China. [4]College of Mechanical Engineering, Beijing University of Technology, Beijing, 100124, China.

Acknowledgements
This work is supported by the National Natural Science Foundation of China (NSFC) through grant Nos. 11302187, 11271197, 11172009, 11372015 and 61473340. The authors are extremely grateful to the referees for their careful review of the manuscript and insightful comments, which helped to improve the paper greatly. The authors are also very grateful to Ms. Xiaojing Li for her original insightful ideas.

References

1. Amster, P, De Nápoli, PL, Mariani, MC: Periodic solutions for p-Laplacian-like systems with delay. Dyn. Contin. Discrete Impuls. Syst., Ser. A Math. Anal. **13**, 311-319 (2006)
2. Gao, FB, Zhang, W: Periodic solutions for a p-Laplacian-like NFDE system. J. Franklin Inst. **348**, 1020-1034 (2011)
3. Manásevich, R, Sędziwy, S: p-Laplacian and Liénard type equation. Rocky Mt. J. Math. **27**, 611-617 (1997)
4. Manásevich, R, Mawhin, J: Periodic solutions for nonlinear systems with p-Laplacian-like operators. J. Differ. Equ. **45**, 367-393 (1998)
5. Aizicovici, S, Papageorgiou, NS, Staicu, V: Multiple nontrivial solutions for nonlinear periodic problems with the p-Laplacian. J. Differ. Equ. **243**, 504-535 (2007)
6. Gaines, RE, Mawhin, J: Coincidence Degree and Nonlinear Differential Equations. Springer, Berlin (1977)
7. Aizicovici, S, Papageorgiou, NS, Staicu, V: Periodic solutions for second order differential inclusions with the scalar p-Laplacian. J. Math. Anal. Appl. **322**, 913-929 (2006)
8. Gao, FB, Lu, SP, Zhang, W: Periodic solutions for a Rayleigh type p-Laplacian equation with sign-variable coefficient of nonlinear term. Appl. Math. Comput. **216**, 2010-2015 (2010)
9. Gao, FB, Lu, SP: Periodic solutions for a Rayleigh type equation with a variable coefficient ahead of the nonlinear term. Nonlinear Anal., Real World Appl. **10**, 254-258 (2009)
10. Cheung, HM: On p-Laplacian equations with deviating arguments. Master thesis, The University of Hong Kong (2009)

Permissions

All chapters in this book were first published in ADE, by Springer; hereby published with permission under the Creative Commons Attribution License or equivalent. Every chapter published in this book has been scrutinized by our experts. Their significance has been extensively debated. The topics covered herein carry significant findings which will fuel the growth of the discipline. They may even be implemented as practical applications or may be referred to as a beginning point for another development.

The contributors of this book come from diverse backgrounds, making this book a truly international effort. This book will bring forth new frontiers with its revolutionizing research information and detailed analysis of the nascent developments around the world.

We would like to thank all the contributing authors for lending their expertise to make the book truly unique. They have played a crucial role in the development of this book. Without their invaluable contributions this book wouldn't have been possible. They have made vital efforts to compile up to date information on the varied aspects of this subject to make this book a valuable addition to the collection of many professionals and students.

This book was conceptualized with the vision of imparting up-to-date information and advanced data in this field. To ensure the same, a matchless editorial board was set up. Every individual on the board went through rigorous rounds of assessment to prove their worth. After which they invested a large part of their time researching and compiling the most relevant data for our readers.

The editorial board has been involved in producing this book since its inception. They have spent rigorous hours researching and exploring the diverse topics which have resulted in the successful publishing of this book. They have passed on their knowledge of decades through this book. To expedite this challenging task, the publisher supported the team at every step. A small team of assistant editors was also appointed to further simplify the editing procedure and attain best results for the readers.

Apart from the editorial board, the designing team has also invested a significant amount of their time in understanding the subject and creating the most relevant covers. They scrutinized every image to scout for the most suitable representation of the subject and create an appropriate cover for the book.

The publishing team has been an ardent support to the editorial, designing and production team. Their endless efforts to recruit the best for this project, has resulted in the accomplishment of this book. They are a veteran in the field of academics and their pool of knowledge is as vast as their experience in printing. Their expertise and guidance has proved useful at every step. Their uncompromising quality standards have made this book an exceptional effort. Their encouragement from time to time has been an inspiration for everyone.

The publisher and the editorial board hope that this book will prove to be a valuable piece of knowledge for researchers, students, practitioners and scholars across the globe.

List of Contributors

Xin Zhao
College of Information Technology, Jilin Agricultural University, Changchun, 130118, P.R. China

Yin Li

School of Mathematics and Statistics, Shaoguan University, Shaoguan, 512005, P.R. China
Department of Mathematics, Sun Yat-sen (Zhongshan) University, Guangzhou, 510275, P.R. China

Na Li
Department of Mathematics, QiLu Normal University, Jinan, 250013, China
School of Mathematics, Shandong University, Jinan, 250100, China

Zhiyong Yu
School of Mathematics, Shandong University, Jinan, 250100, China

Zuomao Yan and Xiumei Jia
Department of Mathematics, Hexi University, Zhangye, Gansu 734000, P.R. China

Guoying Miao and Tao Li
School of Information and Control, Nanjing University of Information Science and Technology, Nanjing 210044, P.R. China

Rui Wang and Xiaoyue Li
School of Mathematics and Statistics, Northeast Normal University, Changchun, 130024, P.R. China

Wenxue Li
Department of Mathematics, Harbin Institute of Technology, Weihai, 264209, P.R. China

Irada Dzhalladova
Kiev National Economics Vadym Hetman University, Kyiv, Ukraine

Miroslava Růžičkova
University of Žilina, Žilina, Slovakia

Viera Štoudkova Růžičkova
Brno University of Technology, Brno, Czech Republic

Hui Cao and Hongwu Tan
Department of Mathematics, Shaanxi University of Science & Technology, Xi'an, 710021, China

Qamar Din
Department of Mathematics, Faculty of Basic and Applied Sciences, University of Poonch Rawalakot, Rawalakot, Pakistan

Bo Du
Department of Mathematics, Huaiyin Normal University, Huaian, Jiangsu 223300, China
Department of Mathematics, Yangzhou University, Yangzhou, Jiangsu 225002, China

YaminWang
Department of Basis Course, Lianyungang Technical College, Lianyungang, Jiangsu 222006, China

Xiuguo Lian
Department of Mathematics, Huaiyin Normal University, Huaian, Jiangsu 223300, China

S Narayanamoorthy and S Sowmiya
Department of Applied Mathematics, Bharathiar University, Coimbatore, 641 046, India

Tianlan Chen and Ruyun Ma
Department of Mathematics, Northwest Normal University, Lanzhou, 730070, P.R. China

Rongzhong Yu, Baidi Chen and Dingqin Shi
College of Sciences, Jiujiang University, Qianjin East Road, Jiujiang, 332005, China

Kezan Li
School of Mathematics and Computing Science, Guilin University of Electronic Technology, Guilin, 541004, China

Xuefei Wu
School of Computer and Software Engineering, Shenzhen Polytechnic, Shenzhen, 518055, P.R. China

Asier Ibeas
Departamento de Telecomunicació i Enginyeria de Sistemes, School of Engineering, Universitat Autònoma de Barcelona, Bellaterra, Barcelona, 08193, Spain

Manuel de la Sen and Santiago Alonso-Quesada
Department of Electricity and Electronics, Faculty of Science and Technology, University of the Basque Country, P.O. Box 644, Bilbao, 48080, Spain

Iman Zamani
Young Researchers and Elite Club, Yasooj Branch, Islamic Azad University, Yasooj, Iran

Fabao Gao
School of Mathematical Science, Yangzhou University, Yangzhou, 225002, China

Department of Engineering Science and Mechanics, Virginia Tech, Blacksburg, VA 24061, USA

Shiping Lu
College of Mathematics & Statistics, Nanjing University of Information Science & Technology, Nanjing, 210044, China

Minghui Yao
College of Mechanical Engineering, Beijing University of Technology, Beijing, 100124, China